Pharmaceutical Capsules

Free Pharmaceutical Press

e-alerts

Our latest product news *straight to your inbox*
register@ **www.pharmpress.com/alerts**

Pharmaceutical Press is the publishing division of the Royal Pharmaceutical Society

Pharmaceutical Capsules

SECOND EDITION

Edited by

Fridrun Podczeck

PhD, DSc
Professor of Pharmaceutics
Sunderland University
UK

Brian E Jones

BPharm, MPharm, FRPharmS
Consultant
Shionogi Qualicaps SA
Spain

Pharmaceutical Press

Published by the Pharmaceutical Press
Publications division of the Royal Pharmaceutical Society of Great Britain

1 Lambeth High Street, London SE1 7JN, UK

© Pharmaceutical Press 2004

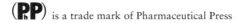 is a trade mark of Pharmaceutical Press

First edition published as *Hard Capsules*, 1987
Second edition published 2004. Reprinted 2007, 2014

Text design by Barker/Hilsdon, Lyme Regis, Dorset
Typeset by Type Study, Scarborough, North Yorkshire
Printed in Great Britain by TJ International, Padstow, Cornwall

ISBN 978 0 85711 165 4

A catalogue record for this book is available from the British Library

To our friend and mentor Professor J. Michael Newton for his pioneering work and continuing research into the development and manufacture of hard-shell capsule products

Contents

Preface

IN THE 1980s, a team of scientists set out to write the book *Hard Capsules – Development and Technology*, which was published by The Pharmaceutical Press in 1987. At that time, there was little technical information available on the manufacture, filling and formulation of hard capsules because publications in the area were few. The book was intended to fill this gap and to report on all aspects of hard capsule technology. It covered everything from the history of capsule development through the making of gelatin, and empty capsule properties to the formulation of powders to be filled. It described the machinery available and looked at all aspects of drug release from capsules, i.e. highlighting the advantages of this dosage form in terms of drug bioavailability. It also provided an extensive bibliography on these topics.

Meanwhile capsules, hard and soft, filled with either dry solids or liquids have increased in their variety and usage. Consequently the technologies for manufacturing capsules and knowledge about the development of shells and fills have similarly changed significantly. The second edition of this book sets out to readdress this by reporting on the advances made in hard capsule technology made since 1987 and it has been extended to include the technology, formulation and development of soft capsules. The contents of the book were completely rethought in order to keep the book to a reasonable size. Authors who contributed chapters to the first edition were asked to focus on updating the technology and knowledge, rather than simply expanding their previous texts. A number of new authors were recruited to provide chapters with extra information. The editors decided not to update the bibliography, which was a significant feature of the first edition, because in this age of information technology and on-line literature search engines it was thought not to be such a necessity. Instead, each chapter provides full details of the relevant current literature in the field.

We are very grateful to Mrs Michelle Wake (Library of the School of Pharmacy, University of London, UK) for her persistence in helping us with the literature research for the book. We also wish to acknowledge the assistance of many manufacturers in allowing the publication of illustrations and details of their machinery and capsules. We are also grateful to the editorial team of the Pharmaceutical Press, who helped us at all stages, from planning to the publication of this book.

Fridrun Podczeck and
Brian E Jones

About the editors

Fridrun Podczeck PhD, DSc is Professor of Pharmaceutics at Sunderland University. Her main areas of research are powder technology, oral solid dosage forms and the role of interparticulate forces in the performance and manufacture of solid oral dosage forms. She has written numerous publications in peer-reviewed journals of materials science, applied physics and pharmaceutics. She acts as peer-reviewer for many journals and as a consultant to pharmaceutical companies and machine manufacturers.

Brian E Jones BPharm, MPharm, FRPharmS works as a consultant on hard capsules and is a scientific advisor for Shionogi Qualicaps SA. He has 29 years of experience with Eli Lilly & Co. Ltd., where he was involved with all aspects of capsules, from shell manufacture through to their filling and formulation. He has published a large number of papers on various aspects of capsule usage. He is an honorary lecturer in the Welsh School of Pharmacy, Cardiff University.

Contributors

Norman Anthony Armstrong
Harpenden, UK; formerly Senior Lecturer, The
Welsh School of Pharmacy, Cardiff University,
UK

Brian E Jones
Consultant, Shionogi Qualicaps SA, Alcobendas,
Madrid, Spain

Roger T Jones
Formerly Director of Research and Development,
Croda Colloids Ltd, Widnes, Cheshire, UK

J Michael Newton
Emeritus Professor, The School of Pharmacy,
University of London, UK

Walther Pietsch
Engineer, Robert Bosch GmbH, Waiblingen,
Germany

Fridrun Podczeck
Professor of Pharmaceutics, The Sunderland
Pharmacy School, Sunderland University, UK

Gabriele Reich
Senior Lecturer, Institute of Pharmacy and
Molecular Biotechnology, University of
Heidelberg, Germany

Geoff Rowley
Emeritus Professor of Pharmaceutics, Sunderland University, UK

Kristina Schlauch
Engineer, Robert Bosch GmbH, Waiblingen,
Germany

Ralf Schmied
Engineer, Robert Bosch GmbH, Waiblingen,
Germany

Katja Vollmer
Engineer, Robert Bosch GmbH, Waiblingen,
Germany

1

The history of the medicinal capsule

Brian E Jones

Introduction

The word 'capsule' in the English language is derived from the Latin word *'capsula'*, which means a small box or container. The word occurs in many scientific disciplines, ranging from anatomy, as an enclosing membrane, and botany, as a descriptive word for fruit, to astrophysics, as a space vehicle. In pharmacy, capsule has been used to describe a glass ampoule (e.g. amyl nitrite capsules) and also as a name for a protective cap over the stopper of a bottle of medicine. In more recent times, capsule has been used primarily to describe a solid oral dosage form, which consists of a container, usually made of gelatin, filled with a medicinal substance. The word can be used to refer either to the container itself or to the whole object, container plus drug. In this book it will be used chiefly to mean a single-dose medicine container.

There are many forms of capsule and they can be divided into two main categories, which in current English usage are described by the adjectives 'hard' and 'soft'. The 'hard capsule' consists of two separate parts, each a semi-closed cylinder in shape. One part, the 'cap', has a slightly larger diameter than the other, which is called the 'body' and is longer. The cap fits closely over the body to form a sealed unit. The 'soft capsule' is a one-piece container, which has a variable shape and owing to its method of manufacture, may be either seamed, along its axis, or seamless. The adjectives hard and soft are sometimes applied to soft gelatin capsules, and in this context the terms refer to whether the capsule wall contains glycerol or another plasticizer which makes it soft and elastic, if included, or hard and rigid, if omitted. In current manufacturing practice all one-piece capsules are soft and elastic.

Other languages did not have this problem because they use different nouns for each type. For example the French have used the word 'capsule' from the beginning for the one-piece capsule, the adjective 'molle' (soft) was used only much later after glycerin had been added to the shell formulation to make the wall pliable. This was used to distinguish it from the original one-piece capsule, which was 'dure' (hard) because its walls were made from a mixture of gelatin, sugar and acacia. This was reflected in common English usage at the turn of the 19th century when there were three kinds of capsule described; hard capsules, soft capsules, which were both one-piece, and empty capsules, which were two-piece (Alpers, 1896a). The French developed a special word, for the hard two-piece capsule 'gélule'. However, the spread of 'Franglais' has lead to the use of the phrase 'gélule dure', a mistranslation taken from the English. A similar thing has happened in German. The original word, 'Kapseln', was used for the one-piece capsule; however, German has succumbed to English usage and now 'Hartgelatinekapseln' and 'Weichgelatine-kapseln' are used for the two forms. A variety of commercial names have been used to try and emphasise certain aspects of capsules, e.g. Pulvules for the powder-filled two-piece capsules of Eli Lilly and Company, and Softgels for filled one-piece capsules produced by members of the Soft Gelatin Capsule Manufacturers Association. In this book the currently accepted English usage will be used, i.e. the hard capsule referring to the capsule made in two separate parts and the soft capsule referring to the capsule made and filled in a single operation.

Invention of the gelatin capsule

The medicinal capsule is an established medicinal dosage form. It has the image of being relatively modern but like most such pharmaceutical ideas, when the records are examined, the famous quotation 'there is nothing new under the sun' is shown to be true. It has been in use for over 160 years. The gelatin capsule was invented early in the 19th century as a result of the need to mask the obnoxious taste of many of the medicinal substances, which were in vogue at that time. One important such drug was the oleoresin of copaiba, which is extremely nauseating when taken by mouth. It was used in the treatment of venereal disease, the incidence of which appears to have been high as a result of the Napoleonic wars and the associated social unrest in Europe. Many unsuccessful attempts were made to overcome this taste problem. Mixtures with the standard pharmaceutical vehicles, aromatic waters, essential oils, honey and syrups were made, but to no avail. All the culinary arts were employed, even going so far as the making of a 'copaiba custard' (Acton, 1845–6). The first successful solution to the problem was the invention of the capsule, which covered the individual dose in a bland tasteless film of gelatin, enabling it to be swallowed easily and without interfering with its activity.

The first recorded patent was granted on 25th March 1834 for an instrument to make gelatin capsules and for the capsules themselves to Messieurs Dublanc and Mothes. The idea was quickly acclaimed and its use spread rapidly both inside and outside of France. This rapid spread might have been caused either by simultaneous discovery by other workers, or, what is more probable, by the fact that it filled an undoubted need and was a good commercial proposition. In 1835, the year following the first patent, capsules were being manufactured in places as far apart as Berlin and New York (Anon., 1896b; Schlenz-Cassel, 1897).

The actual inventor of the gelatin capsule appears to have been Mr F. A. B. Mothes. In a further patent of Mothes granted in December 1834, it refers to the patent of Mothes and Dublanc and states that Mothes was a pharmacy student and so was under the legal age to obtain his pharmacy diploma; in order to present the patent he became associated with Dublanc, who was already an established pharmacist (Mothes, 1834). This patent is further described as being made by Mothes, 'in his personal name and in his own private interest'. In 1837 Mothes applied for an extension to the period of validity of the original patent. The French Academy in reviewing his case noted that he had broken off his association with Dublanc (Planche and Guéneau de Mussy, 1837), and granted to him a ten-year extension as the sole proprietor of the invention. Dublanc's name was soon omitted in the many references to capsules made in the contemporary literature.

No copy of the original patent is available as it was removed illegally from the Patent Office in Paris sometime during the 1980s (Cole, 1992). However, Mothes's patent described the method of manufacture in full, said that, 'It was not without its own inconveniences,' and went on to describe improvements (Mothes, 1834). The capsules were formed on moulds, which were small round pouches made of soft leather tied to a small long-necked metal funnel by a waxed string. The moulds were filled with mercury to make them firm, then dipped into a solution of gelatin, removed and placed in a heated box at 40°C to dry. To remove the capsules from the moulds the mercury was emptied out and the gelatin shape was then carefully taken off the mould. To prevent the gelatin adhering to the moulds they were covered in 'baudruche', goldbeater's skin. As a result the capsules had walls consisting of two layers. Mothes considered that this method was expensive and not capable of producing capsules of a large enough size. The improved method of his patent was to use solid moulds, in the shape of an elongated sphere, made of burnished brass to prevent oxidation. They were dipped into concentrated solutions of flavoured and sweetened gelatin, then placed upright on a special board and allowed to cool. The capsules were removed before they were completely dry and finished off by being placed on a sieve in a slightly heated room. They were carefully filled with the balsam and then sealed with a drop of gelatin solution.

In France, capsules acquired an immediate popularity. An idea of the extent of this can be gained from the quantities of materials which Mothes was reported as having used in his process: in 1835, 3500 kg of gelatin and in 1836, 1500 kg of copaiba balsam (equivalent to 750 000 doses) (Herpin, 1837). To purchase this large quantity of copaiba he had to travel to London because Paris was unable to supply his needs (Neufeld, 1913). His business was extended by increasing the number of types of oil filled into capsules and also by supplying empty capsules to pharmacists.

Capsules rapidly achieved official recognition. When Mothes presented his invention to the Académie Royale de Médecine they paid him a high compliment, saying that he had rendered an immense service to science and humanity (Anon, 1889). In 1837 Dr Ratier made an entry in the *Dictionnaire de Médecine et de Chirurgie pratiques* to the effect that capsules were a means of directly administering copaiba without in any way altering its virtues (Ratier, 1837).

In 1838 Mothes ceased to supply empty capsules to the pharmaceutical trade and, as a result, several attempts were made to overcome his patent, which he was rigorously enforcing. In the previous year, when the Economic Committee of the French Academy investigated his application for a patent extension, they found he was employing 20 workmen in his business (Herpin, 1837). They surmised that if the patent were not renewed there would be a rush of other companies into the field. They held the capsule in such high regard that it was proposed that it should be brought to the attention of other industries. Herpin suggested to the Academy that because of the importance of capsules, they should be filled with native resinous medicines that were similar to copaiba, such as turpentine oil. This would have a favourable economic effect and would also be of strategic importance. He obviously suggested this because of the very successful English naval blockade during the Napoleonic wars, which was then still fresh in people's memories. In 1838 a method was published for masking the flavour of cubeb, copaiba and other nauseous materials using gelatin, but not as a capsule as defined in the patent (Garot, 1838). The drugs were first formed into standard pill masses, each of which was mounted on a pin and then dipped into a gelatin solution. After drying, the pins were removed and the resulting

holes sealed with a drop of gelatin. Garot stated that he had devised this method only because of Mothes's refusal to sell empty capsules.

For some reason, the method mainly quoted in the literature as being used by Mothes was the cumbersome one described in the first patent (Dublanc and Mothes, 1834). This may have been deliberately brought about for commercial reasons, or may have been due to the fact that French patents at that time were hand-written documents and it would have been necessary to attend the patent office in person to read them. The standard source for information was a report on the original patent by Cottereau published in the journal *Traite de Pharmacologie* (Anon, 1835–6). As a result, several workers published 'new methods' for making capsules using metal moulds.

In 1846 in the *Journal de Pharmacie et de Chimie*, Giraud claimed a method using iron moulds (Giraud, 1846). The editor in a footnote to the article pointed out that this infringed Mothes's patent. In 1850, in the same journal, there was a report on Mothes's improved method of making capsules using metal moulds which, rather strangely, described the method used by him in his patent published 15 years previously (Anon, 1850). Alpers, in a history of capsules published in 1896, upbraided Mothes for taking Giraud's idea and claiming it as his own (Alpers, 1896a).

In the United States of America the first mention of capsules in the pharmaceutical literature was in 1835 in the *American Journal of Pharmacy*, which published an abstract of the paper by Cottereau describing their method of manufacture (Anon, 1835–6). In the same journal two years later, Alfred Guillou of Philadelphia reported the results of a series of experiments on capsules (Guillou, 1837). These enabled him to make capsules that looked identical to those made in France, so much so that they could not be distinguished from them. In fact he sold them as being of French origin. He used metallic ovoid moulds mounted on tin dishes, a method exactly as described in Mothes's addition to his patent. Alpers credited Guillou with being the first to suggest this method of manufacture (Alpers, 1896a). No practical use appears to have been made of this invention and there is no further record of Guillou's association with capsules in the pharmaceutical literature.

The first American capsule business was started in New York in 1836 by H. Planten, a Dutchman who had emigrated there in 1835 (Alpers, 1896a; Anon, 1896b). He was described as using French manufacturing methods, selling his capsules as 'Mothes's capsules' and packing them with directions for use in both English and French. The medical profession in America had very rapidly accepted that capsules were an excellent method of administration for copaiba and that France was the source of high-quality material. It was many years before capsules of American origin were sold openly and made with American-manufactured gelatin.

In the German-speaking states the invention appears to have been first reported by Buchner in 1837 (Schlenz-Cassel, 1897). He described Mothes's capsules as copaiba balsam covered in a bubble of glue and said that the method of making them was unknown. Later the same year, a Munich physician, Dr Feder, disclosed the secret of their manufacture to those interested (Schlenz-Cassel, 1897). He described a method similar to that of Mothes, using solid moulds of hard wood or, later, iron, which were lubricated by dipping in soap spirit. Feldhaus (1912), commenting on the spread of capsule usage in Germany, referred to the fact that there was no uniform patent law at that time and it was only in Prussia, where patents were examined carefully, that the patent of Mothes and Dublanc was upheld. The first large-scale manufacture in Germany is credited to a Berlin pharmacist, J. E. Simon, also in 1837. Schlenz-Cassel remarked in 1897 that the rapid spread of capsules in Germany was due to the high profit margin in their manufacture; in that year there were 13 capsule factories listed in the *Pharmaceutical Calendar*.

In Britain, although capsules appear to have been widely used from an early date, there are few references to their manufacture. The first was an abstract in the *Lancet* in 1840 of a paper by Desfontenelles from the *Journal de Chimie* (Anon, 1839). This described a method of making 'gelatinous capsules' by an adaptation of the original Mothes and Dublanc patent but using a rather strange mould, the swimming bladder of a tench or other fish which are from 5 to 7 inches long. This was tied by a ligature to a copper tube,

inflated, greased with lard, and then dipped in a hot solution of gelatin made to the formula of Garot. When set, the capsule was removed by deflating the bladder.

An evening meeting at the Pharmaceutical Society in 1842 appears to have been the first public meeting at which capsules were discussed (Anon, 1842). Gelatine capsules manufactured by a Mr Chaston, from Norfolk, were demonstrated that were designed to administer fluid medicines to horses and dogs. They were 'open at one extremity, which required a covering of skin, after the introduction of the fluid'. They were recommended for the administration of substances that did not dissolve gelatin and were described as being about the size of an ordinary horse-ball. A comment in the 'Art of Dispensing' states that there was a demand for capsules from the very beginning and in 1843 Cooley described them as the 'common gelatine capsules' (MacEwan, 1915). London agents were appointed for both Mothes and for Denoual and a Mr Bateman and a Mr Turner started local manufacture at this time. The second article in the *Pharmaceutical Journal* came in 1843 with the translation of an article by Adolph Steege, Court Apothecary at Bucharest, called 'On the formation of the gelatinous capsules of Balsam of Copaiba' (Steege, 1842). The moulds were made of iron, mounted on a wooden plate and, after lubrication with almond oil, were dipped in a gelatin solution (see Figure 1.1). When the gelatin had set, the capsules were trimmed to the correct length, carefully pulled off the moulds and dried on a loose hair sieve. The balsam was filled into them using a glass dropper tube and they were sealed by carefully dipping them into gelatin solution. Steege claimed that capsules could be made elastic by adding sugar to the gelatin solution, and made transparent by using isinglass instead of gelatin.

In France, the commercial success of the gelatin capsule stimulated many workers to try to devise ways of overcoming Mothes's patent, which he was enforcing with litigation. Mothes's precise methods were apparently known only by a few people, which led to his improved process, using metallic moulds, being reinvented several times. Sinninoir from Nancy in 1841 suggested the use of wax moulds mounted on metallic pins

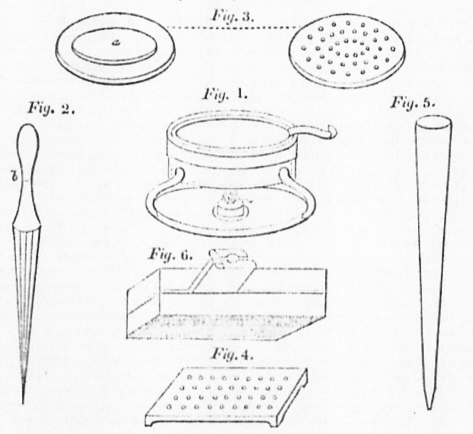

ON THE FORMATION OF THE GELATINOUS CAPSULES OF
BALSAM OF COPAIBA.

BY ADOLPH STEEGE,
Court Apothecary at Bukarest.

Fig. 3.

Fig. 1.

Fig. 2.

Fig. 5.

Fig. 6.

Fig. 4.

Figure 1.1 Apparatus of Adolph Steege (1842) for making one-piece gelatin capsules. Key to figures: 1, heated water bath; 2, metal capsule mould; 3, wooden disc to carry moulds for dipping; 4, tin stand to hold capsules for filling; 5, glass dropping tube for filling capsules; 6, tin press for holding filled capsules for dipping into gelatin solution to seal the open ends.

(Schlenz-Cassel, 1897). The moulds were re-moved from the capsule by heating them and melting the wax, a tedious process, which did not prove viable. Moulds for dipping were best made from metal and there was no success in the search for alternative materials.

Improved methods of manufacture of one-piece gelatin capsules

Many inventors applied themselves to devising improved mechanical systems. In 1859, Viel obtained a French patent (Viel, 1859) for a method of making and filling the capsule on the same piece of apparatus. This was further modified by an addition made by Viel (1861). The use of the apparatus was described by the inventor in the *Journal de Pharmacie et de Chimie* (Viel, 1864). Sheets of gelatin, prepared to the formulation used to coat pills, were cast in strips 2 m long, 7 cm wide and 0.5 mm thick. This was partly dried so that it remained elastic and did not stick to metal surfaces. The strips were brought into contact with each other and sealed by pressure on three sides to form a long tube opened at one end. This was filled with liquid from a hopper and individual capsules were formed from this tube by passing it through a pair of steel mould plates to form a lenticular-shaped liquid-filled capsule. Viel called these 'globules', later referred to as a type of pressure capsule. Viel's process was described in a German journal by Schlenz-Cassel who likened the equipment for cutting the filled tubes to pincers developed from the castration scissors used by farmers (Schlenz-Cassel, 1897). Mothes and his new partner Lamouroux claimed infringement, and a panel of three experts considered the case (Thuret, 1943–5). Two of the experts considered that Viel's process was an infringement of Mothes's patent, but the third, Professor Chevallier of the School of Pharmacy, Paris, thought Viel's method was significantly different from Mothes's because of the use of a single mould for both making and filling the capsules. He complained also of Mothes's restrictive practices in refusing to sell empty capsules and trying to obtain a monopoly in the supply of copaiba capsules. The Chambre du Conseil accepted the minority opinion and gave judgement in favour of Viel. Mothes and Lamouroux appealed against the decision, but at the same time purchased Viel's process from him for 37 000 francs. Their appeal was subsequently rejected.

In Italy, also in 1844, Pegna made capsules by casting in moulds instead of by dipping. He used moulds that consisted of a closely fitting pair of brass plates with the forms of the capsules on their inside surfaces. A gelatin solution was poured into the moulds to form the capsule. A similar process was patented in France in 1846 by Lavalle, a doctor, and Thévenot, a pharmacist, from Dijon (Lavelle and Thévenot, 1846; Guéneau

de Mussy and Guibourt, 1848). Their capsule moulds were a pair of hexagonal copper or iron plates in which suitable cavities had been formed and a plate that acted as a guide to locate the moulds accurately together. Their capsule mass was a solution of gum, sugar, gelatin, and a little honey, which they claimed was more soluble than the mixture used by Mothes. It was cast into thin sheets and a sheet was laid on top of one of the plates, so that it took up the shape of the cavities. The medicament was placed in the hollows. A second sheet was placed on the first and the second plate was put into position by means of the guide plate. The whole assembly was placed in a press and the capsules stamped out. These capsules were much more regular in shape than Mothes's and they later became known as 'perles'. The inventors made capsules containing liquids, such as copaiba, cod-liver oil and turpentine oil, and powders such as quinine sulfate and rhubarb. In an addition to this patent in 1850 in the names of Clertan and Lavalle (1850), the capsules were coated with syrup and presented as a bonbon.

Mothes, Lamouroux & Company (1846), using some of Viel's ideas, obtained a patent in December 1846 for a 'machine to prepare and manufacture a large number of Mothes capsules'. Their machine was a rotary die type apparatus, a forerunner to the main process in use today. Gelatin ribbons were passed between two sets of rollers. The first set positioned the ribbons. The second set had capsule-shaped cavities in their surfaces so that as the rollers came together capsules were cut out of the ribbons. They called this machine a 'capsulateur mecanique', but do not appear to have used it because Mothes obtained another patent for making capsules by dipping in 1850 (Anon, 1850). Their problem was probably the difficulty in filling the capsules during manufacture. However, Viel, who had obviously not relinquished his interest in capsules, obtained a patent in 1859 for a 'capsulateur Viel'. He used large ribbons of gelatin that were passed between rollers turned by hand in a device similar to a clothes mangle. The liquid to be filled was held in a pear-shaped flask and was delivered into the nip of the rollers just where the capsules were formed. Berthiot (1860), a Parisian pharmacist, patented another method in 1860. A gelatin solution was forced through an annular orifice to form a tube that was continuously filled with liquid from a concentric tube in the centre. The gelatin tube was chilled to set it and then formed into capsules by passing it between two engraved rollers. The process was the forerunner of that used to make seamless soft gelatin capsules today. Many other patents were granted about this time for improvements to the systems for making capsules, which is an indication of the commercial success of the capsule as a dosage form (Thuret, 1943–5).

Other workers explored the use of alternative materials from which to prepare capsules to defeat Mothes's limitations on trade. Joseau in 1848 suggested the use of casein for capsule manufacture because it could mask strong odours better than gelatin (Anon, 1848a). Other substances were chosen on the basis of alleged improvements in performance *in vivo* over gelatin, perhaps anticipating the present day interest in drug release. In 1843, Savaresse, father and son, and Douillet obtained a French patent for the manufacture of organic capsules called 'Savaresse' (Thuret, 1943–5). The elder Savaresse was a manufacturer of musical strings and they used the catgut process to prepare thin membranes from the small intestines of sheep. These were cut into small lengths, dipped into sulfurous acid solution as a preservative and then placed over capsule-shaped moulds. When dry they were removed and filled, the open end being closed with a knot of fine thread and sealed with a drop of gelatin. The inventors argued that a gelatin capsule dissolved too quickly, perhaps even in the mouth, and as a result patients could experience 'nauseous eructations'. They claimed that the membrane passed through the stomach unaffected and disintegrated only in the intestines. This description of an enteric product was made 40 years before Unna, the generally acknowledged inventor of this idea, published it (Dumez, 1921). It also described the use of sulphur dioxide as a preservative anticipating a use in capsule production that occurred during the 1960s and 1970s. Cooley in 1844 suggested modifying the shell of capsules containing copaiba. He proposed a mixture of three parts of rhatany root extract, 1 part of syrup and 1 part of acacia. This could either be used to coat gelatin capsules or be used to form a shell and he claimed

that 'these capsules sit easily on the stomach, the tone of which they continue to improve' (Cooley, 1884).

In 1845 a patent for an identical membrane capsule was granted in Britain to Evans and Lescher (Anon, 1845–6). Alpers refers to them as French pharmacists (Alpers, 1896a). The same journal also published a letter from a Dr Garrod reporting that a trial of membrane capsules showed that they protected his patients from the taste of copaiba without detracting from its efficacy. A few months later, in the same journal, Dr William Acton wrote a paper in which he stated that capsules were the best modern method of giving nauseous liquids and that membrane capsules were an improvement over gelatin ones because the latter were often too thin and burst on swallowing (Acton, 1845–6). A contemporary advertisement for gelatin capsules made the counter-claim, 'They are very easily dissolved in the most delicate stomachs without any effort (the envelope being gum, sugar, and gelatin); they are superior to those made of gluten or membrane, which frequently pass through the stomach without dissolving or are thrown up after uselessly fatiguing the stomach' (Anon, 1868). The relationship between Savaresse, Evans and Lescher is unknown, but in a letter to the *American Journal of Pharmacy* in 1896, the firm of Evans and Sons of Boston are reported to have been manufacturers of Savaresse capsules (Alpers, 1896b).

Mothes appears to have been unsuccessful in maintaining the exclusive use of his patent for making soft gelatin capsules, despite resorting to litigation. In countries outside France, capsules were sold openly under the guise of being made in France, or of French origin, even when manufactured locally. His efforts are illustrated by an amusing story from the *Pharmaceutical Journal* (Anon, 1857–8). A London pharmacist was taken to court by Mothes's agents for selling counterfeit capsules. These had been purchased from a recognised wholesaler and there was an out-of-court settlement of £19 3s 4d. It was found that a London company had been making and selling capsules as 'genuine Mothes' for at least 10 years previously, packing them as Mothes's with a green label, seal and signature. A solicitor's letter warned all pharmacists that they should not be fooled by such crude imitations and that they would be prosecuted if they sold counterfeit capsules. The editor noted that during his inquiries he had purchased two genuine boxes of Mothes's capsules that were dissimilar and that the London 'imitation' capsules were closer in appearance to one of the genuine specimens than the genuine ones were to each other.

Developments in the use of one-piece gelatin capsules

Following on from the numerous patents granted in the 1850s and 1860s the one-piece gelatin capsule became a well-established pharmaceutical item. The extent of the spread of usage can be gauged from the numerous advertisements that appeared in the pharmaceutical press, and from the increasing mention of capsules in books of reference. The first pharmacopoeia to include a monograph on capsules was the French Codex of 1884, which was closely followed in Europe by the Dutch Pharmacopoeia of 1889 and the German Pharmacopoeia of 1896 (Jones and Törnblom, 1975). In North America the first entry was in the Mexican Pharmacopoeia (Anon, 1885). Alpers remarked in 1896 that, 'Our pharmacopœia (USA) ignores them completely'. Unlike other Pharmacopoeiae the United States has never included a separate monograph on capsules and has only included them more recently in the general notices.

At the end of the 19th century the only other significant development was the production of an elastic soft gelatin capsule, which is the form that we know today. This was achieved by the addition of glycerol to the mass. Several authors have credited Detenhoff with being the originator of this idea (Alpers, 1896a; Wilkie, 1913). His work was published in the *Pharmazeutische Zeitung für Russland* in 1878 (Anon, 1878; Cotzhausen, 1878). He suggested using a capsule mix consisting of 1 part of gelatin, 2 parts of water and 2 parts of glycerol, and prepared capsules from this by the standard technique. However, in 1875, Taetz had been granted a French patent for 'an elastic capsule to make it easier to swallow medicines'. His formula was 1 part of gelatin, 1½ parts of glycerol and 2½ parts of water. The glycerol in these

formulations acts as a plasticizer. Since this time other substances have been used to obtain the same effect.

In France, capsules were well established by the end of the 19th century. In Britain, the spread was similar but slower. In the pharmaceutical press several attempts were made to popularise their use, particularly for extemporaneous dispensing. In 1888 an article on capsule manufacture was published in the *Chemist and Druggist* (Anon, 1888). It explained that if it were known that a dozen capsules could be made in the same time as a dozen suppositories then every pharmacist would adopt this art. It gave a detailed description of how to make moulds and contained recipes for making both hard and soft capsules. It referred to the perle or globule, which was only popular in France, and included a diagram of how it was made. In 1889 there was an article in the same journal on the Paris Exhibition that described the part concerning capsules (Anon, 1889). It posed the question, 'Why should English pharmacists have lived for so long with the idea that the industry is a secret one?' It described the small-scale equipment available. Amongst the exhibitors was a company, Capgrand-Mothes and Company, Paris, which was apparently the successor to Mothes's previous business ventures. They were offering for sale an 'encapsuleur' for filling powders into capsules and also pepsinised capsules, with pepsin in their walls. This article was followed in 1890 by another (Forret, 1890), on the extemporaneous manufacture of flexible gelatin capsules. Ten years later, in the same journal, there was another article on flexible gelatin capsules by a manufacturer who commented that, 'It is no unusual thing nowadays to see a prescription for gelatin capsules in place of pills' (Anon, 1900).

The rate of increase of capsule usage in Britain can be clearly seen from an examination of the 'Extra Pharmacopœia' started by Martindale and Westcott. The first edition was published in 1883 and listed only one capsule product, oil of santal. The fifth edition of 1888 listed nine products and also included the first references to a hard two-piece gelatin capsule. This rate of increase gradually accelerated in succeeding years: in the seventh edition of 1892, there were 14 products listed; in the ninth edition of 1898 there were 24; in the tenth edition of 1901 there were 31 and 95 in the 11th edition of 1904.

There was a gradual decline in the extemporaneous manufacture of capsules as more and more manufacturers offered for sale an extensive list of capsule products, covering nearly all the possible requirements. In 1906, A. M. Hance presented a paper to the American Pharmaceutical Association which was reported upon in the *Pharmaceutical Journal* in the following year (Hance, 1907). He said that the improvements in manufacturing technique had almost rendered the hand-made capsule obsolete. He listed nine points of superiority of the machine-made variety that included uniformity, appearance, solubility, stability and profit.

An account of how the industrial manufacturers of soft gelatin capsules had revolutionised their output was given several years earlier (Anon, 1890). It reported that an American firm of capsule makers had got a new machine for making these useful aids to medication. It had four iron plates, each of which, bearing from 200 to 250 olives, was dipped automatically into gelatin and sent rolling down a track to the cutter which automatically cut off the waste. Without manual help, these plates rolled on to an elevator and were hoisted on to another track, which rolled them back to the place they started off from, after the capsules had been cleaned. There they were pulled off and tipped into bins. The whole process required only a short time, the greater part of which was the drying step. The machine did the work of about 175 people.

The invention that led to the present large-scale industrial manufacture of capsules was the perfection of the rotary die technique in 1933 by R. P. Scherer, an American citizen of German birth. This process produced capsules of high quality and significantly reduced manufacturing costs. Scherer established a large business, both in making generic preparations and in filling other companies' specialities. After 1945 he established companies in Europe to extend the availability of his service. Since then other companies in many countries have entered the field (Ebert, 1977).

Invention of the hard gelatin capsule

One successful way of overcoming Mothes's patent resulted in the production of a new type of capsule, the hard two-piece, which was the forerunner to the modern hard two-piece capsule. This was invented by another Parisian pharmacist, J. C. Lehuby, who was granted a patent on 20th October 1846, for '*Mes enveloppes médicamenteuses*'. The name 'medicinal covering' was probably chosen to avoid any patent problems with the capsules of Mothes. Lehuby described them as cylindrical in shape, like silk-worm cocoons, composed of two compartments or cups, which fitted one inside the other to form a box. The moulds used were silver-plated metal cylinders about 4 to 5 cm long, varying in diameter from several millimetres to several centimetres. The diameter of the two moulds was different, so that the cast halves fitted together snugly. Capsules were made by dipping the moulds into a decoction of starch or tapioca the consistency of a bouillie, which was sweetened with a little sugar and coloured with 'fish silver'. The patent was quickly followed by three additions. The first was granted on 20th January 1847, extending the range of materials to be used for making the envelopes and referring to them for the first time as capsules. The prime material recommended was carragheen, to which could be added, if necessary, various gelatins, gums and glues. On the same day he was granted a completely separate patent for a new gelatin derived from carragheen moss, *Fucus crispus*. He described this as having superior properties to animal gelatins, which were too brittle. His second addition was made in November 1847 and here he called his envelopes '*capsules-boîtes*', capsule boxes. He described improvements to his equipment for making capsules, using several moulds mounted on a disc and an improved method for making his new gelatin from carragheen (see Figure 1.2). The third addition, granted in 1850, was described as being the collective property of Lehuby, Silbermann and Garnier. This commented that earlier capsules were either brittle or had an unpleasant smell and a further-improved process was given for extracting gelatin from

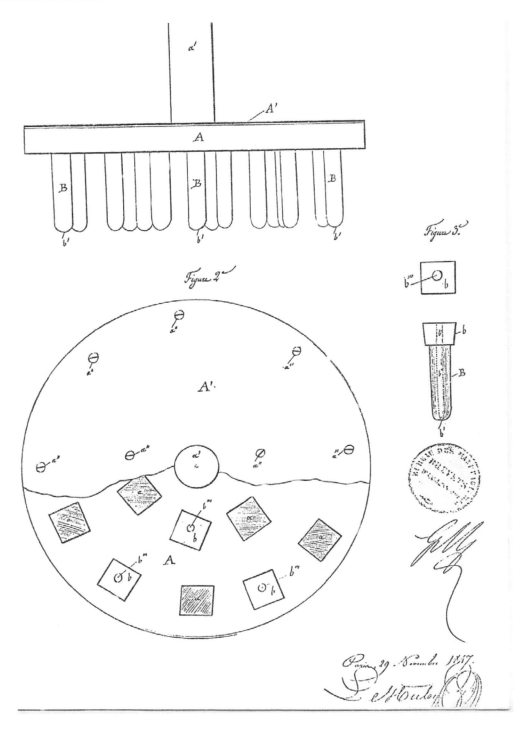

Figure 1.2   Moulds for making two-piece capsules made from polished tin with central air hole to allow easy removal of dried capsule part. J. C. Lehuby, drawing from 2nd patent addition (1848).

carragheen moss. This was still the preferred material for the manufacture of capsules which they called 'capsules en lichen'. The formulation for the solution for making capsules was: 10 L of water, 1.8 kg of dried carragheen extract, 0.9 kg of animal gelatin to reduce the hygroscopicity of the capsules and 2 or 3 L of a decoction of marshmallow root to overcome the brittleness caused by the gelatin.

Although Lehuby was the undoubted inventor of the two-piece gelatin capsule, many authors have credited it to a Mr. J. Murdoch (1847), who was granted British Patent 11 937 in London on 2nd May 1847 for 'an improved capsule or small case for protecting matters enclosed therein from the action of air; and an improved material to be used in their manufacture' (Anon, 1848b, 1848c). The patent described, in the same wording as used in Lehuby's patent, two-piece capsules, a method for their manufacture and how to make a vegetable gelatin from carragheen moss. It was also the first patent that admitted the sole use of animal gelatin for making capsules.

Murdoch was a patent agent by profession, a fact stated in the patent. Thuret (1943–5) was the first to point out the similarities between the patents of Lehuby and Murdoch, particularly in the use of vegetable gelatin. Murdoch's patent preamble states that, 'It was communicated to me from abroad,' and all the evidence points to Lehuby being the original inventor. The probable reason for Murdoch being credited with the invention was the commercial and communication problems of that time. The use of Mothes's capsules spread quickly into many countries, which resulted in many references to them in the literature and caused them to be sold under a French guise, acknowledging their origin. Lehuby's capsules, on the other hand, were not so quickly taken up by the pharmaceutical trade. There was a 30-year gap between the patents and the proper commercial use of hard gelatin capsules and very few references to them appeared in the literature. In fact two authoritative papers on the history of capsules, by Alpers (1896a) and Wilkie (1913), failed to mention either Lehuby's or Murdoch's part in the development of capsules.

In both Britain and Germany at the beginning of the 20th century, hard gelatin capsules were referred to as being of American origin. The first published reference to them outside Britain and France was in 1848 in *Dingler's Polytechnisches Journal* which had an abstract taken from the *London Journal of the Arts* recording Murdoch's patent (Anon, 1848d). Feldhaus (1912) in Germany, in a history of capsules, dismissed the claims of a Viennese pharmacist, Gross von Figely, to have invented them and recorded Murdoch as the inventor. He used *Dingler's Polytechnisches Journal* as his information source. Two years later, Urdang (1914) in the USA also gave credit to this belief. These works have become the standard historical reference sources for others and as a result the original error has been perpetuated (Griffenhagen, 1956).

Development of hard gelatin capsules

After the first patents, the spread in the use of hard gelatin capsules appears to have been delayed for two reasons. First, the most popular medicinal substances in use were galenicals, which were frequently liquid or semi-solid and thus were more suited to being filled into soft gelatin capsules. Secondly, to make a two-piece capsule that fitted together properly required large numbers of accurately made moulds, which were an expensive item.

The first recorded large-scale manufacture of hard two-piece capsules was by an American company, H. Planten of New York (Anon, 1896b). They were an old-established capsule house, having started making soft gelatin capsules as early as 1836. In the early 1860s they produced two-piece capsules made from jujube paste, a material used in the confectionery trade. This was similar to the material described by Lehuby in the first patent. Later references complained that these capsules 'fell apart' and their use in the early 1860s was short lived.

The first successful manufacturer of hard gelatin capsules on a commercial scale was a Detroit pharmacist, F. A. Hubel (Wilkie, 1913). He overcame the problem of making low-cost but accurate moulds by using pieces of gauged iron rod or wire set in wooden blocks. It is recorded by Wilkie that Hubel became interested in such capsules after being told of them by one of his friends who had watched them being made in Italy. Crude lead moulds had been used and the two parts of the capsules, which had the same diameter, had to be dampened after filling to allow them to be forced together. In 1874, Hubel started manufacturing capsules. For each size he used two sets of mould pins, one for the cap and one for the body, with different diameters, so that a good telescopic fit of the two halves was obtained. The mould pins were set in wooden blocks. They were dipped into gelatin solution and withdrawn to produce a film that, whilst still wet, was cut to the required length using a penknife. The coated moulds were left to stand for the films to air dry; when ready, the capsule pieces were removed from the moulds with brass tongs and joined together.

Hubel was a community pharmacist and must have realised that low-cost gelatin capsules would be useful in the dispensary to replace paper wrapped powders, sachets and other extemporaneously dispensed medicines. The capsule quickly achieved popularity, particularly after 1875, when the whole of Hubel's output was sold for him by another Detroit company, Parke, Davis & Company, who applied aggressive sales and promotional methods (Stadler, 1959). Hubel continually improved his methods and in February 1877 was granted the first of several patents for his equipment. In the period 1877 to 1883 a large number of patents were granted for capsule-making machines (Alpers, 1896a). Disputes arose over priority, but in the lawsuits that followed, Hubel was successful in establishing his claims. When he retired in 1900 his plant had a production capacity of one million capsules per day.

The commercial success of hard gelatin capsules soon brought other companies into the field. Several of these were in Detroit and one of them was the Merz Capsule Company. This company, which was founded in 1887, started manufacture in a 1200-square foot loft and in 1894 was one of several companies that successfully opposed the 'Capsule Trust', a group of companies that tried to force up the price of capsules (Anon, 1896c). The Trust was defeated after extensive litigation, which had an unsettling effect on the market. By 1896, however, Merz's output had expanded and the plant occupied a three-storey factory with a floor space of 25 200 square feet. Its capacity was then 750 000 capsules per day, and it was exporting to Canada, the West Indies, Central and South America and the Hawaiian Islands.

The two major pioneers in the development of hard capsules during the 20th century were Parke, Davis & Company and Eli Lilly & Company. Between them they developed the manufacturing process and spread the manufacture of empty capsules outside of the USA, first to Europe and then to other parts of the world. Their history exemplifies the development of the hard capsule.

Parke, Davis & Company started in Detroit, USA, in the 1860s (Taylor, 1915). It made a range of standard pharmaceutical products. The company became involved with capsules when they started selling the output of Mr. Hubel's company. In 1881 they introduced into their price list 'soluble elastic filled capsules' (Anon, 1881). Their advertisements for these informed the public that these capsules were superior to any which had been sold previously because they used a new method of manufacture and imported the machines and the workmen from Germany. They offered capsules in sizes from 10 minims (c. 0.6 mL), apparently the largest size available previously, up to half an ounce (240 minims, or c. 15 mL). The larger ones they claimed were swallowed as easily 'as an oyster'. The marketing efforts that they put into their sales effort, helped them spread knowledge of the uses to which 'gelatin pharmaceuticals' could be put. An advertisement of theirs in 1882 extolled the advantages of gelatin in helping to disguise nauseous tastes (Anon, 1882). Their price list showed coated pills, hard and soft gelatin capsules, and the hard capsules were supplied either empty or filled.

In 1902 the company bought the US Capsule Company and its subsidiary the M. L. Capsule Company and for the first time controlled the manufacture of two-piece gelatin capsules (Taylor, 1915). These capsules became an important part of their business. In 1915 their advertisements claimed that 'Empty capsules are built "like a watch"' (Anon, 1915). The foreman of the capsule works at this time was Warren Wilkie, who was a pioneer in capsule production (Taylor,

1915). He had apparently first entered the business because he was a salesman for the gauged iron rod used for making capsule moulds. He introduced an improved material phosphor bronze that overcame the problem of capsules becoming discoloured owing to contact with the iron moulds. He joined the company in 1901 when they purchased the US Capsule Company. He developed improved machines for their manufacture.

In 1924, Parke, Davis & Company commissioned the Arthur Colton Company of Detroit to design an improved capsule manufacturing machine (Stadler, 1959). This machine was known as the 'C1' and was the first to make the caps and bodies on separate parts of the same machine, which enabled two-colour capsules to be made (see Figure 1.3). In 1959 they opened their first plant outside the USA in the UK at Hounslow near London (Norris, 1961). This capsule-manufacturing company evolved into the company Capsugel that is now part of the Pfizer Corporation.

The Eli Lilly Company of Indianapolis, USA, was founded by Colonel Eli Lilly in 1868 (Kahn,

1975). They manufactured a range of standardised galenical preparations for supply to the pharmaceutical trade. In 1894 they decided to build a capsule-manufacturing plant. This was started in 1895 and was finished in 1897. They employed an engineer to oversee its construction, a Mr Riley P. Hobbs, who had the distinction of patenting the first 'self-locking' capsule (Hobbs, 1894) (see Figure 1.4). It was in advance of its time because capsules were then being sold in boxes of 100 or 1000 for filling by hand in the dispensary. The need for this type of capsule only arose when capsules were being filled and packaged on an industrial scale. Eli Lilly (1963) was the first company successfully to manufacture and sell this type of capsule, the LOK-CAPS. Since then capsules with different locking features have been made by other companies. This type of capsule is now the industry standard because it can be handled on automatic filling and packaging machines with a minimal risk of separation and product loss.

An idea of the growth of the use of the two-piece hard capsule usage in the United States can be obtained from the increase in output recorded

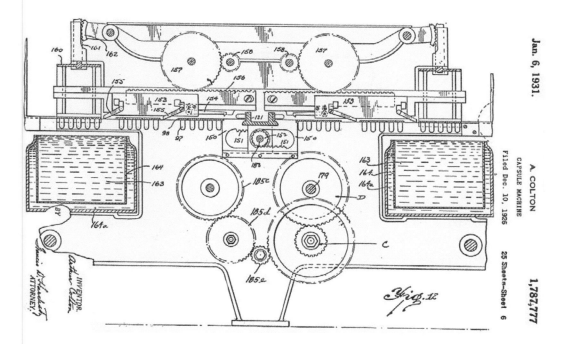

Figure 1.3 C1 patent drawing showing machine with two dipping pans containing gelatin solution (A. Colton, 1931).

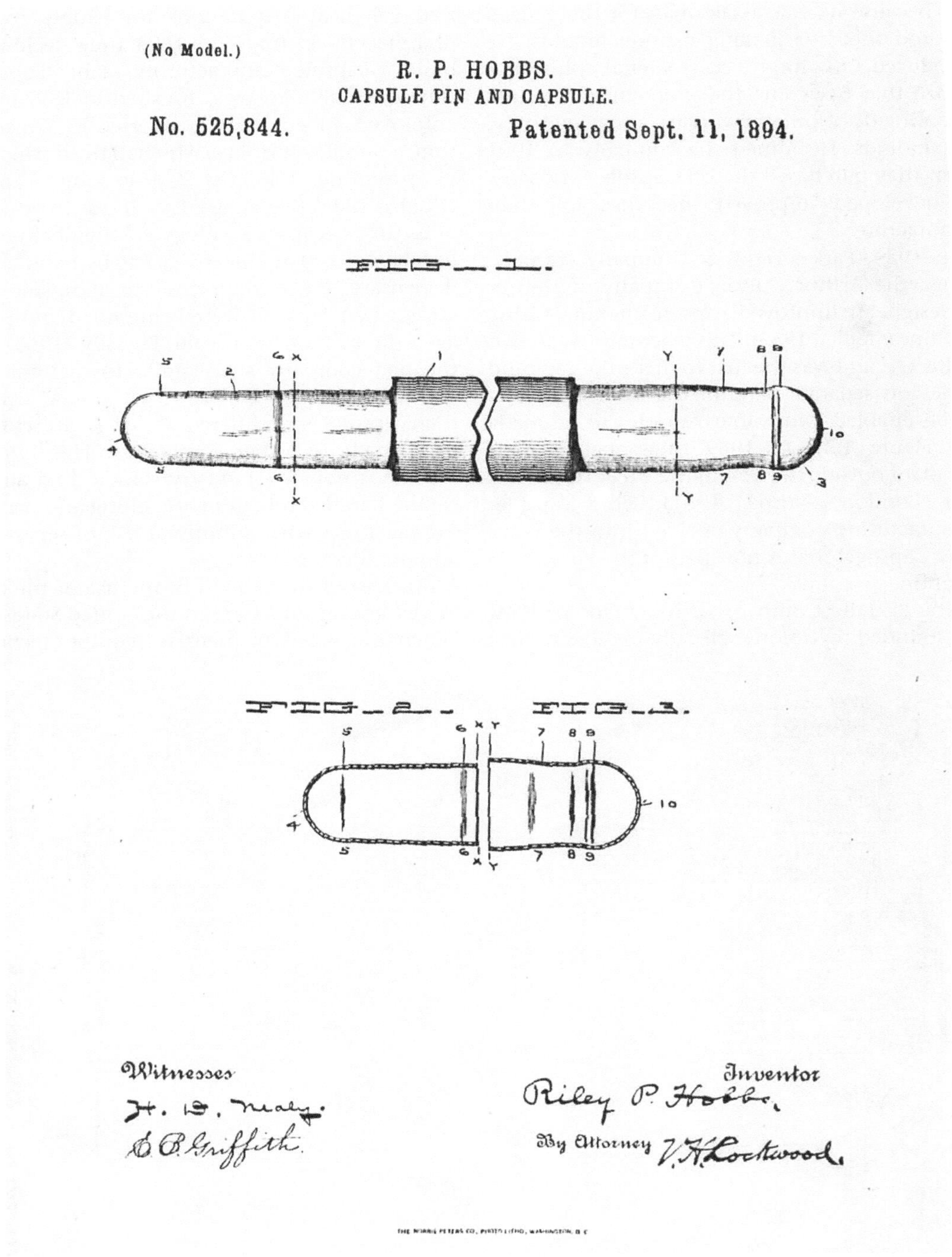

Figure 1.4   First self-locking hard capsule designed for the cap to grip the body due to the constriction '7–9' in the cap shape, with provision for the air to escape during closing (Hobbs, 1894).

by Eli Lilly & Company (Davidson, 1964). Their annual outputs were: 1910, 112 million; 1911, 241 million; 1912, 319 million; 1919, 650 million. In 1920 the record for one day's production (7.30 am to 5.00 pm) was 3 091 000 capsules. This rapid increase in output was aided by developments in the machinery for their manufacture.

Their first manufacturing machines were semiautomatic and required that the bars carrying the mould pins be transferred by hand from the section in which the moulds were dipped to the drying kilns and then to the section for removing, cutting and assembly (see Figure 1.5). This type of machine was in general use at the start of the 20th century (Anon, 1896b; Anon, 1896c).

Initially capsules were cut to the correct length immediately after dipping, but eventually they were cut at a later stage, when the film had dried, in order to improve the quality of the cut edge. This facet of capsule quality was frequently mentioned in sales literature at the turn of the 19th century (Eli Lilly & Company, 1899). Each capsule machine operator could produce between 10 000 to 12 000 capsules per day.

Eli Lilly worked closely on machine development with the Colton Company of Detroit, which was lead by the doyen of pharmaceutical engineering, Arthur Colton. In 1909 they purchased their patents for capsule-making machines from Colton. In 1913 they commissioned them to build an improved machine that was designed for

Figure 1.5 The capsule plant of Eli Lilly & Co., Indianapolis 1900. In the right fore-ground the dipping operation, in the right rear the drying kilns and in the left centre rear the machine for stripping the capsule shells from the moulds (reprinted with permission from Eli Lilly & Company).

Figure 1.6   The capsule plant of Eli Lilly & Co., Indianapolis 1931, showing the Colton Stacker machines that were the first fully automatic hard capsule machines (reprinted with permission from Eli Lilly & Company).

them by Burton W. Scott (1913). It was completely automatic in operation and was called the 'stacker' machine (see Figure 1.6). This is because the bars with the dried capsules were stacked in a row, alternatively caps and bodies, so that they could be fed simultaneously through a device that stripped them from the mould pins, cut them to the correct length and joined the two parts together. The capsule moulds were mounted 30 to each bar and there were 175 pairs of bars on the machine. The whole cycle was now automatic and the output was about 8000 capsules per hour. Originally there was only one gelatin container and as a result only single-coloured capsules could be produced.

Many years later in the 1950s, the machine was modified to enable two-coloured capsules to be made. A second gelatin container was added and a device incorporated to separate the body and cap mould bars for the dipping operation. Another feature of the machine was that the pin bars were held stationary in an inverted position and the gelatin solution container was raised up to them. Another major industry first was the installation in 1913 of a type of air-conditioning plant in their factory. They had the assistance of the Carrier Corporation of Buffalo to build an air washing machine, which could cool and warm air. During that summer they claimed that this was the first plant in the world that worked without an interruption caused by the weather. Previously in the summer when the temperatures rose to 30 to 40°C (80 to 100°F) the gelatin would not set on the mould pins and production ceased. Eli Lilly was the first American company to build a factory outside the USA, opening a plant in Basingstoke in Great Britain in 1952 (Norris, 1959). This was built to manufacture methyl cellulose capsules, which the company hoped would be a substitute for gelatin. However, their *in vivo* performance was not as good as gelatin and after a couple of years the plant changed over to the production of gelatin capsules. In 1992 Eli Lilly & Company Limited sold their capsule business, Elanco Qualicaps, to their long-time Japanese partner Shionogi & Company Limited.

The manufacture of capsules was largely, but not wholly, restricted to the USA during the latter part of the 19th century. In 1910 the estimated world production was over one thousand million capsules per year, of which 90% were made in the USA. The only other country in which there appears to have been production activity was Germany. In the 1890s, 13 companies were listed in the German pharmaceutical calendar as manufacturers of capsules, the bulk of their output being the various forms of soft gelatin capsule, but at least one manufacturer, G. G. Pohl of Schönbaum near Danzig, made hard gelatin capsules (Anon, 1892). These were recommended for oils, though surprisingly not powders, and appear to have been widely used. A stand for holding capsules whilst filling them with oil, devised by Tschanter, was shown in several journals (Anon, 1896d).

The design on which nearly all modern large-scale machines are based was patented by Arthur Colton in 1931 and assigned to Parke, Davis & Company (Colton, 1931). This differed from previous machines in that the machine was divided into two parts through which the cap and body pins passed separately but synchronously. The machine was much larger, holding over 700 pairs of pin bars and had an output of 30 000 capsules per hour. It was able to produce two-coloured capsules because there were separate gelatin containers. The capsules produced were of a more uniform thickness, owing to the procedure of lowering the moulds into the gelatin

solution and then removing them to spread the gelatin film more evenly over the pins. Modern machines, as well as running faster, have been improved so that they can run on a 24-hour day, all-year-round basis with only short periodic stops for maintenance and holidays (Martyn, 1974). The machines are constructed either by the manufacturers themselves or by specialist machinery manufacturers such as Technopharm of Canada.

Use of capsules as a dosage form

Capsules, because of their elongated shape, are easy to swallow, which is one reason for the number of capsule-shaped tablets manufactured today. Patients are sometimes timid about their ability to swallow a capsule because they look at its largest dimension, its length, and imagine that it will never go down their throats. They do not realise that the tongue reflexively lines up a capsule end-on for swallowing. Advice on how to take capsules has been given since earliest times. Dr William Acton (1845–6) said, 'Recommend such sceptics (people unable to swallow) to take about a dessert spoonful of water in their mouth, and then place the capsule on the tongue, when the whole will be swallowed without difficulty, whereas, if the capsule be placed on the tongue and water be drunk, the patient will often swallow the water, but the capsule will remain and produce convulsive action of the pharynx'.

The consumers of hard gelatin capsules are still not too familiar with them despite their long history. Many patients are under the misapprehension that they are made of plastic. This, in the loosest chemical terms, could be said to be true, in that gelatin is a polymer prepared from an organic starting material. The apocryphal story about the use of capsules is of the patient who thinks the shells are inedible and empties out the contents to take them. Like all tales of this nature its origins go back many years. Taylor (1915) quoted an item from the *Louisville Medical News* of 1877, 'Physician states that an intelligent gentleman has been given capsules to take without specific directions and later informed the physician that he did not like them at all, his idea of taking them being to peel off the hulls and put the stuff in water'.

Soft gelatin capsules come in a wide range of shapes and sizes, governed solely by the dose required and the method of manufacture. However, because hard gelatin capsules are manufactured empty by one supplier and filled by another there has been the need to produce standard sizes, particularly when machine filling is involved. Hubel in 1875 was already making capsules in three standard sizes known as numbers 1, 2 and 3 (Stadler, 1959). No reference appears in the literature as to how the standard sizes 000, 00, 0, 1, 2, 3, 4 and 5 were arrived at. It has been suggested that they were simply based on standard engineering dimensions for iron rods (Jones and Turner, 1974). Wilkie, one of the early capsule experts, is reported to have entered capsule manufacturing because he had been a salesman supplying capsule makers with accurately gauged iron for their moulds. Lehuby described capsules as being preferably cylindrical in shape with hemispherical ends. Over the years several variations on that theme have been patented. In 1865 an American, D. Dick, was granted a patent for a cone-shaped capsule (Alpers, 1896a). Parke, Davis & Company in 1896 offered a rectal capsule for sale in Britain, which differed from the normal one in that it had a pointed cap-half to facilitate insertion (Anon, 1896a). In 1914, R. J. Estes of Oklahoma was granted a patent for a capsule that had its body-end edge cut away at an oblique angle to facilitate closing the two halves together (Anon, 1915). Non-standard shapes of capsules have been produced by several large pharmaceutical companies in the USA in order to prevent counterfeiting of their products. For example, Smith Kline & French used capsules with flat ends and bevelled shoulders and Eli Lilly & Company used capsules with paraboloidal ends to the bodies (Physicians' Desk Reference, 1985).

Probably the only significant change in basic design since the time of Lehuby has been the development of the self-locking capsule. One of the drawbacks in the industrial use of hard gelatin capsules was that when filled they could be vibrated apart, with consequent powder spillage, especially when handled on automatic packaging machines. To overcome this, two methods of

sealing capsules were used, banding and dot-sealing, but both had the disadvantages of being time-consuming, costly and spoiling the appearance of the capsule (Jones, 1969). It was pointed out in the *Pharmaceutical Journal* that capsules which were trimmed before they were dry had a rough body edge, which acted as a sort of spring lock to hold the capsules together (Anon, 1912). The best solution to the problem was the development of the self-locking capsule, which enabled the two halves to stay closed during rough handling without the need for a post-filling operation.

The first recorded patents, as mentioned earlier, for a self-locking capsule were granted in 1894, to R. P. Hobbs of Indianapolis (Hobbs, 1894). His capsules had a circular constriction ring around the cap and a slightly bell-shaped body. The patent stated, 'This tight fit or lock fit will prevent separation during joining or handling'. In 1914, M. Pollock of New York obtained a US patent for a cylindrical flat-ended capsule which had engaging notches in the cap and body to prevent the capsule coming apart (Anon, 1914). However, either these ideas presented production problems or they were not required by the users, because they were not commercially exploited. The first modern patent for such a system was granted to Eli Lilly & Company in 1963, for capsules called 'Lok-Caps'. A series of indentations inside the capsule cap forms a friction seal when the capsule is closed after filling. Since then other companies, Parke, Davis & Company and R. P. Scherer have patented similar systems to perform this function. Practically all industrially manufactured hard capsules produced currently have a locking feature.

Enteric capsules

Capsules are usually required to dissolve in the stomach as rapidly as possible, releasing their contents, but for certain purposes they are designed to pass through the stomach and into the intestine before dissolving. Such products are described by a variety of terms, including gastric-resistant, entero-soluble and enteric. Their production was first suggested in the 1840s as a means of administering medicines that were very irritant to the gastric mucosa. The first true enteric coating of products is credited to a German, Dr Unna (Dumez, 1921). Prior to this time people had used fats and waxes to coat the products or used less-digestible materials to try to prevent disintegration in the stomach. Unna suggested the use of a keratin coating on pills so that they were insoluble in the acid secretion of the stomach but soluble in the alkaline secretions of the intestine. Unna's pills were first sold in Germany in 1884. Subsequently the firm of Pohl, in Schönbaum, offered for sale keratinised capsules 'kapsulae keratinosae', prepared not from gelatin but from a mixture of keratin, shellac, borax and colophony (Anon, 1892). They did not perform well *in vivo* because the contents tended to absorb water and swell, bursting the capsule shell.

In 1895 a Swiss pharmacist, Dr Weyland, offered a different solution to the problem by suggesting the use of formaldehyde to 'harden' the capsules so that the consequent increase in their solution time would carry them through the stomach before they dissolved (Dumez, 1921). The capsules were treated by soaking them in an aqueous solution of formaldehyde for a short period then drying them in a warm place. This process was patented in Germany by Hausmann, Weyland's employer, in 1895. Hausmann obtained a second patent in the same year for a similar process which used another aldehyde, acrolein. In 1904, Rumpel suggested a means whereby empty capsules could be treated with formaldehyde. However, his method lacked some logic because he recommended that after filling, the capsule should be sealed by painting the joint with collodion or gelatin solution and re-treating; the latter process, it would seem, obviated the need for the initial treatment. Capsules produced by Rumpel's method were called 'kapsulae geloderatae' and were sold in Germany by the firm of Pohl of Schönbaum. They were also sold in America where they were known as 'Pohl's enteric capsules'. In 1907 the firm of Evans Sons Lescher and Webb obtained a British patent for 'improvements relating to gelatine capsules containing medicines for internal use'. They described the manufacture of a two-layered capsule from prepared animal intestinal membranes, which were rendered enteric by coating them with cocoa butter, paraffin wax, or other suitable vegetable, animal or mineral fat.

Enteric products fell into disrepute for two reasons, both of which led to poor *in vivo* results. The first was that the formaldehyde treatment, despite being a simple process, had its limitations. The reaction between gelatin and formaldehyde proceeds by a cross-linking mechanism that, once started, continues even when the capsules are removed from the aldehyde. In 'The Extra Pharmacopoeia' in 1910, there is an entry for formagules, formolised gelatin capsules, which states that 'they cannot be relied upon if stored for too long' (Martindale and Westcott, 1910). Dumez, in 1921, in his history of the development of enteric-coated capsules, pointed out that despite what certain workers had said there was evidence pointing to the instability of formaldehyde-treated capsules and concluded that 'only time would tell'. Gorley and Lee (1938) carried out one of the first *in vivo* disintegration studies using an X-ray machine. Their study compared the methods to produce enteric-coated products and commented that 'capsules treated with formaldehyde become insoluble upon aging and are, therefore, unreliable for enteric uses'. However, for all of its drawbacks the method must have been popular because of its simplicity. This is shown by the number of references to the process in pharmacopoeias right up until the 1970s (Jones and Törnblom, 1975).

The second reason was that the coating materials, whose function relied upon the changes in solubility with pH, were chosen on the basis of an incorrect physiological assumption. They were intended to be insoluble at the acid pH of the stomach but to be soluble at an alkaline pH, which was assumed to be the condition in the duodenum. It has more recently been shown that the pH of the duodenum is seldom, if ever, alkaline. During the first half of the 20th century many workers searched for suitable materials and amongst others, shellac, tolu, mastic, resin, carnauba wax and salol were tried. It was not until the development of modified cellulose derivatives such as cellulose acetate phthalate (Malm *et al.*, 1951), which had the correct solubility properties, i.e. insoluble in strongly acid solution but soluble at slightly acid pH values, that satisfactory products were produced (Jones, 1970).

References

Acton, W. (1845–6). On the best means of disguising the taste of nauseous medicines. *Pharmaceutical Journal*, 5, 502–505.

Alpers, W. C. (1896a). Gelatine capsules. *Am. J. Pharm.*, 68, 481–494.

Alpers, W. C. (1896b). Letter to the editor. *Am. J. Pharm.*, 68, 575.

Anon (1835–36). Gelatine capsules, report on article by Cottereau in *Traite de Pharmacologie*. *Am. J. Pharm.*, 7, 351–352.

Anon (1840). Method for making gelatinous capsules, report on method of M. Desfontenelles in *Journal de Chimie*. *Lancet*, 1, 901.

Anon (1842). Evening meeting, 12th October. *Pharmaceutical Journal*, 2, 343.

Anon (1845–6). The mode of preparing the patent organic capsules, patented by Evans and Lescher. *Pharmaceutical Journal*, 5, 361–363.

Anon (1848a). Capsules médicinales au caséum de M. Joseau en *Journal de Chimie et de Médicine*. *Journal de Pharmacie et de Chimie*, 14, 42–43.

Anon (1848b). To James Murdoch, of Staple Inn, in the county of Middlesex, for an improved capsule or small case, for protecting matters from the air, and an improved material to be used in the manufacture of said capsules, – being a communication. [sealed 2nd November 1847]. *London Journal for the Arts*, 2(5), 42–44.

Anon (1848c). Über die Gelatine-Kapseln für den Arzneigebrauch und zu anderen Zwecken. *Dingler's Polytech. J.*, 65, 238–239.

Anon (1848d). Murdoch's Verfahren Gelatinekapseln zum Einhüllen von Arzneistoffen u. zu verfertgen. *Dingler's Polytech. J.*, 109, 397.

Anon (1850). Perfectionnement apporté par M. Mothes à son procédé de fabrication de capsules gélatineuse. *Journal de Pharmacie et de Chimie*, 17–18, 204–205.

Anon (1857–8). Mothes' capsules – extraordinary proceedings. *Pharmaceutical Journal*, 17, 43–44.

Anon (1868). Denoual's superior medicinal capsules, advertisement. *Chemist Drugg.*, 9, 835.

Anon (1878). Préparation des capsules pharmaceutiques, report on Detenhoff's capsules in *Pharmazeutische Zeitung für Russland* 1878. *Journal de Pharmacie et de Chimie*, 27–28, 73–74.

Anon (1881). Soluble elastic filled capsules. *Am. J. Pharm.*, 53(9), advertising sheet 1.

Anon (1882). Fine pharmaceutical products from the laboratories of Parke, Davis & Company. *Am. J. Pharm.*, 54(12), advertising sheet 4.

Anon (1885). Pharmaceutical preparations in the Mexican Pharmacopoeia. *Am. J. Pharm.*, 57, 373.

Anon (1888). Capsules and capsule-making. *Chemist Drugg.*, 32, 496.

Anon (1889). Gelatine capsules. *Chemist Drugg.*, 35, 214.

Anon (1890). Reports. *Chemist Drugg.*, 36, 64.

Anon (1892). Verschiedene Mittheilungen, Über Gelatinekapseln. *Pharmazeutische Centralhalle für Deutschland*, 33, 512–514.

Anon (1896a). Pharmaceutical novelties, some Detroit specialities. *Chemist Drugg.*, 48, 584–585.

Anon (1896b). Platen's American medical dispensary. *Pharmaceutical Era*, 29, 992–993.

Anon (1896c). The Merz capsule company. *Pharmaceutical Era*, 29, 994–995.

Anon (1896d). Pharmaceutical apparatus, stand for filling capsules, report of apparatus of Tschanter of Oppeln, from *Pharmazeutische Zeitung*. *Pharmaceutical Journal*, 62, 307.

Anon (1900). Flexible gelatin capsules, notes on their manufacture by a manufacturer. *Chemist Drugg.*, 56, 131–133.

Anon (1912). Chapter in practical pharmacy (26). Gelatin capsules – continued – Hard gelatin capsules. *Pharmaceutical Journal*, 88, 778–779.

Anon (1914). Soluble capsules for medicines, Montague Pollock, US Patent No. 1 079 438. *Pharmaceutical Era*, 47, 43.

Anon (1915). Capsule of Robert J. Estes, US Patent No. 1 122 089. *Pharmaceutical Era*, 48, 91.

Berthiot, C. (1860). French Patent No. 43 963.

Clertan and Lavelle (1850). French Patent No. 3 960.

Cole, E. T. (1992). Personal communication.

Colton, A. (1931), US Patent No. 1 787 777.

Cooley, A. J. (1844). Capsules of copaiba and rhatany. *Am. J. Pharm.*, 9, 238.

Cotzhausen, von L. (1878). Gleanings from German journals, elastic gelatin capsules of Detenhoff reported in *Pharmazeutische Zeitung für Russland* 1876. *Am. J. Pharm.*, 50, 295.

Davidson, H. L. (1964). Capsule production a chronological development. In: *Archives report, 5/11/64*, Eli Lilly & Company, Indianapolis, USA.

Dublanc, I. G. A. and Mothes, F. A. B. (1834). French Patent No. 5 648.

Dumez, A. G. (1921). A contribution to the history of the development of enteric capsules. *J. Am. Pharm. Assoc*, 10, 372–376.

Ebert, W. E. (1977). Soft elastic gelatin capsules: a unique dosage form. *Pharm. Technol.*, 1, 44–50.

Eli Lilly & Company (1899). Lilly's empty gelatin capsules are peerless. *Sales Bulletin*, May, page 14.

Eli Lilly & Company (1963). French Patent No. 1 343 698, and 1964, British Patent No. 970 761, and 1966 US Patent No. 3 285 408.

Feldhaus, F. M. (1912). Zur Geschichte der Arneikapseln. *Chemikerzeitung*, 35, 697, and through Jottings from German journals, The inventor of gelatin capsules. *Pharmaceutical Era*, 45, 524.

Forret, J. A. (1890). Flexible gelatin capsules. *Chemist Drugg.*, 37, 291–292.

Garot, M. (1838). Art. XXXVII. – New method of covering pills with a coating of gelatin. *Journal de Pharmacie et de Chimie*, through *Am. J. Pharm.*, 4, 229–231.

Giraud, A. J. (1846). Sur la fabrication des capsules pour renfermer les substances médicamenteuses. *Journal de Pharmacie et de Chimie*, Paris, 9–10, 354–355.

Gorley, J. T. and Lee, C. O. (1938). A study of enteric coatings. *J. Am. Pharm. Assoc.*, 27, 379–384.

Griffenhagen, G. (1956). Tools of the apothecary. 10. Lozenges, tablets and capsules. *J. Am. Pharm. Assoc., Pract. Pharm. Edn*, 17, 810–813.

Guéneau de Mussy and Guibourt (1848). Sur les capsules médicamenteuses de MM. Lavelle et Thévenot. *Journal de Pharmacie et de Chimie, Paris,* 14, 350–352.

Guillou, A. (1837). Art. III – Capsules of gelatin (Capsules gelatineuses: Dublanc et Mothes, à Paris). *Am. J. Pharm.*, 9, 20–22.

Hance, A. M. (1907). Machine-made v hand-made capsules. *Pharmaceutical Journal*, 78, 205–206.

Herpin (1837). Sur les capsules gélatineuses présentées par M. Mothes, pharmacien, rue Sainte-Anne, no. 20, à Paris. *Bulletin de la Société d'Encouragement pour l'Industrie Nationale*, 36, 219–221.

Hobbs, R. P. (1894). US Patent No. 525 844 and 525 845.

Jones, B. E. (1969). Hard gelatin capsules, a literature review. *Manuf. Chem.*, 40(2), 25–28.

Jones, B. E. (1970). Production of enteric coated capsules. *Manuf. Chem.*, 41(5), 53–54, 57.

Jones, B. E., and Törnblom, J.-F. (1975). Gelatin capsules in the pharmacopoeiae. *Pharm. Acta Helv.*, 50, 33–45.

Jones, B. E. and Turner, T. D. (1974). A century of commercial hard gelatin capsules. *Pharmaceutical Journal*, 213, 614–617.

Kahn E. J. (1975). *All in a century, the first 100 years of Eli Lilly and Company*. Eli Lilly & Company, Indianapolis, USA, pp 270.

Lavalle and Thévenot (1846). French Patent No. 3 906.

Lehuby, J. C. (1846). French Patent No. 4435, 1st addition, 20th January 1847, 2nd addition, 12th February 1848, 3rd addition, 10th May 1850 (collective property of Lehuby, Silbermann & Garnier).

MacEwan, P. (1915). In: *The Art Of Dispensing*, 9th edition. London: Chemist & Druggist, 156–164.

Malm, C. J., Emerson, J. and Hiatt, G. D. (1951). Cellulose acetate phthalate as an enteric coating material. *J. Am. Pharm. Assoc., Scient. Edn*, 40, 520–525.

Martindale, W. H. and Westcott, W. W. (1910). In Lewis, H. K. (ed.) *The Extra Pharmacopoeia*, 14th edition. London: Pharmaceutical Press, 542–545.

Martyn, G. W. (1974). The computer interface in a capsule molding operation. *Drug Dev. Comm.*, 1, 39–49.

Mothes, F. A. B. (1834). French Patent No. 9 690.

Mothes, Lamouroux & Company (1846). French Patent No. 4 780.

Murdoch, J. (1847). British Patent No. 11 937.

Neufeld, M. W. (1913). Mothes, Erfinder der Gelatinekapseln. *Apotheker-Zeitung*, 37, 917.

Norris, W. G. (1959). Hard gelatin capsules – how Eli Lilly make 500 million a year. *Manuf. Chem.*, 30, 233–236.

Norris, W. G. (1961). P. D.'s new capsule plant. *Manuf. Chem.*, 32, 249–252.

Physicians Desk Reference (1985). 39th edition, Oradell, New Jersey, Medical Economics Company Inc.

Planche and Guenéau de Mussy (1837). Capsules gélatineuses de M. Mothes, pharmacien, à Paris. *Bulletin de l'Académie Royale de Médecine*, 442–443.

Ratier, M. F. (1837). Thérebinthine de copahu. Diction-naire de Médecine et de Chirurgie pratiques, Vol. XV, pp. 285–288, through Alpers, W. C., *Am. J. Pharm.*, 68, 481–494.

Schlenz-Cassel (1897). Zur Geschichte der Gelatine-kapseln. *Apotheker-Zeitung, Berlin*, 34, 275–276.

Scott, B. W. (1913). US Patent No. 1 076 459.

Stadler, L. B. (1959). The gelatin capsule. *J. Am. Pharm. Assoc., Pract. Pharm. Edn*, 20, 723–724.

Steege, A. (1842–3). On the formation of the gelatinous capsules of balsam of copaiba. *Repertorium für die Pharmazie*, through *Pharmaceutical Journal*, 2, 769–770.

Taetz, R. (1875). French Patent No. 106 325.

Taylor, F. O. (1915). Forty-five years of manufacturing pharmacy. *J. Am. Pharm. Assoc.*, 4, 468–481.

Thuret, A. (1943–5). Les capsules medicamenteuses. *Travaux de Laboratoires de Matière Medicale et de Pharmacie Galenique de la Faculté de Pharmacie de Paris*, 32(76), Pt 1, 7–257.

Urdang, G. (1914). The invention of the gelatin cap-sules. *Pharm. Archs*, 14, 58–59.

Viel, J. (1859). French Patent No. 43 022.

Viel, J. (1861). Addition to French Patent No. 43 022, 29th November.

Viel, J. (1864). Description et usage d'un appariel propre à capsuler les liquides. *Journal de Pharmacie et de Chimie*, Paris, 30, 490–491.

Wilkie, W. (1913). The manufacture of gelatin capsules. *Bull. Pharm., Detroit*, 27, 382–384.

2

Gelatin: manufacture and physico-chemical properties

Roger T Jones

Gelatin is the commercial protein derived from the native protein collagen which is present in animal skin and bone and the term 'gelatin' originates from the Latin *'gelatus'*, meaning stiff or frozen. Gelatin has all the properties required to meet the technical needs of the pharmaceutical capsule industry. These include solubility, solution viscosity and thermally reversible gelation properties in aqueous solution. It produces strong, clear, flexible, high-gloss films, which dissolve readily under the conditions existing in the stomach. Until recently, these properties, unmatchable by other film-forming hydrocolloids, have meant that gelatin has been the raw material of choice for manufacture of both hard and soft capsules.

The origins of gelatin manufacture can be traced back as far as 4000 BC. Crude gelatin was extracted by cooking hide pieces in water and its interest lay in the adhesive properties of this animal glue. The history of gelatin manufacture has been traced by Bogue (1922), by Smith (1929) and by Koepff (1985). Today, gelatin finds applications in many industries as well as pharmaceuticals. There are no official published figures for the total worldwide consumption of gelatin, however, the Gelatin Manufacturers of Europe (GME) reliably report a total worldwide gelatin production for 2001 as 269 000 tonnes. The production was divided up by geographical regions: western Europe (43.5%), eastern Europe (1.8%), North America (22.5%), South America (14.5%), Asia–Oceanic (16.9%) and 'others' (0.8%). The proportion of gelatin produced by raw material type was: pigskin (41%), hide (28.6%), bone (30%) and 'others' (0.4 %) (GME, 2002). The food industry uses about 70% of the gelatin produced and pharmaceutical industry about 15%. Within the pharmaceutical industry, hard and soft capsules constitute the major use, about 80%, for gelatin and the remainder being used for microencapsulation, tablet binding, tablet coating, emulsion stabilisation and the manufacture of pastes and suppositories. The pharmaceutical and medical applications of gelatin have been reviewed (Jones, 1971).

Gelatin manufacture, despite being a traditional industry, has been subject to extreme pressures and changes in recent years, since the first case of bovine spongiform encephalopathy (BSE) was identified in the UK in 1986. This has had a profound influence and effect on many aspects of gelatin manufacture and its markets, including the acceptability of different types of bovine raw materials and their countries of origin, standardisation by GME manufacturers on the conditions of certain key (with respect to their ability to remove potential BSE infectivity) processing stages and registration and inspection of gelatin plants by national authorities. Although current scientific evidence indicates that gelatin is a safe raw material, the BSE 'scare' has presented an opportunity for other materials from non-animal sources to be considered for capsule manufacture. It has also created an interest in the use of fish gelatin for these applications, although the limited production of this type of gelatin (less than 1% of world production in 2001) is likely to restrict its use to 'speciality products'.

Structure of collagen

Gelatin is not a naturally occurring protein but is derived from the fibrous protein collagen, which is the principal constituent of animal skin, bone, sinew and connective tissue. Any discussion of the structure of gelatin requires, therefore, an understanding of the nature and structure of collagen and of its conversion to gelatin.

The primary structure of collagen arises from the linkage of α-amino and imino acids by peptide bonds to form a polymer. The amino acid composition of a variety of mammalian collagens has been reviewed (Jones, 1989a) and, although some differences are apparent between collagens from different sources, there are certain features that are common to, and uniquely characteristic of, collagen. Collagen is the only mammalian protein containing large amounts of hydroxyproline and hydroxylysine and the total imino acid (proline and hydroxyproline) content is high; approximately one-third of the residues consist of glycine and methionine is the only sulphur-containing amino acid present; cysteine and cystine are absent, as is tryptophan. Collagens present in fish skins show a wider range of amino acid composition than for mammalian collagens (Pikkarainen, 1968). In particular, the imino acid content is generally lower than for mammalian collagens, and is lower for cold-water fish, such as cod, whiting and halibut, than for warm-water fish, such as tuna and tilapia.

In considering commercial gelatin, 'Type 1' collagen constitutes practically all the collagen present in bone and tendon and predominates in hide (about 85%). Wide angle X-ray diffraction studies show that collagen has a pattern of long intertwined filaments (Jones, 1989a) and narrow angle X-ray diffraction shows that there is a characteristic spacing of about 64 nm for dry collagen. This pattern can also be seen under the electron microscope, indicating that the microscopic fibrils are composed of much thinner filaments, or protofibrils, which possess specifically ordered chemical and structural variations along their main axes (Jones, 1989a).

The structure of collagen has been studied intensively and several useful reviews have been published (Jones, 1989a). There is now a generally accepted model of the collagen molecular structure, although certain details remain vague. The collagen unit, or monomer (tropocollagen), consists of a triple helix of three polypeptide chains, each of which has a helically coiled configuration. This is a threefold left-handed screw, in which each amino acid extends 120° around the coil axis and 0.312 nm along the direction of the axis. Thus each chain is built up of triplets of amino acids, with every third residue being in a similar environment and glycine appearing as every third residue. The three coiled chains are coiled around each other to give a right-handed coil of about 1.4 nm diameter, forming a tropocollagen molecule, which behaves as a rigid rod with a length of 300 nm and molecular weight of about 300 000 Da. The triple helical structure arises from two factors. First, the glycine residues permit triple chain packing by allowing the chains to come close enough for hydrogen bonding to occur and, second, the high content of imino acids directs the individual chains into a poly-L-proline II-type helix through steric hindrance to rotation (Cowan and McGavin, 1955). The tropocollagen molecules are joined together in an end-to-end fashion to form protofibrils and these, in turn, form fibrils by aggregation. Freely dangling, single peptide chains (telopeptides) at one or both ends of the tropocollagen molecule are believed to be responsible for 'cementing' together the basic units. A periodicity of 64 nm in the axial direction of the collagen fibril is apparent when the fibril has been negatively stained with phosphotungstic acid and observed under the electron microscope. Hodge and Petruska (1962) have explained this banded structure as being due to a regular staggered arrangement of molecules in which the dark bands represent gaps between adjacent molecules. This is most easily understood by reference to the schematic model shown in Figure 2.1, and the molecular geometry diagram in Figure 2.2. The explanation for the ordered arrangement of molecules in the fibrils appears to lie in the side-chain groups of the amino acid residues (Hulmes *et al.*, 1973). These can interact with side-chain groups on an adjacent molecule and that spatial arrangement which allows the maximum number of favourable interactions will be the most stable thermodynamically. These pairs of interacting amino acids apparently occur 64 nm apart in the sequence

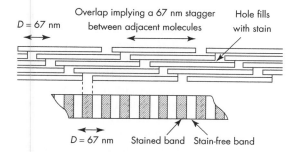

D = 67 nm   Overlap implying a 67 nm stagger between adjacent molecules   Hole fills with stain

D = 67 nm   Stained band   Stain-free band

Figure 2.1   Schematic diagram of the packing arrangement of tropocollagen molecules in the native fibril, according to the model of Hodge and Petruska (1962).

and this is true, not only for ionic attraction between, for example, the positively charged side chain of arginine and the negatively charged aspartic acid residue, but also for the hydrophobic bonds formed by eliminating water from between amino acids.

In addition to these bonds there are a smaller number of stronger covalent cross-links between chains in adjacent molecules (intermolecular cross-links) and also between chains within the same tropocollagen molecule (intramolecular cross-links). Evidence for the existence of various types of covalent bonds in native collagen has been reviewed by Harding (1965). Bonds involving ester, carbohydrate, aldehyde, ε-amino and γ-glutamyl groups were considered. Subsequently, the principal intermolecular and intramolecular cross-links have been shown to be those involving aldehyde groups formed from lysine residues (Bailey *et al.*, 1974; Light and Bailey, 1979). The initial stage in the formation of intermolecular bonds is the oxidative deamination of lysine residues (in the case of skin collagen) or hydroxylysine residues (in the case of bone collagen) located in the non-helical, end-chain regions of the molecules by the enzyme lysyloxidase. The aldehydes produced then condense with side-chain amino groups of hydroxylysine residues situated in the helical regions of neighbouring molecules, forming divalent cross-links. Intramolecular cross-links are established by the aldol condensation of lysine aldehyde groups on adjacent chains. These divalent cross-links should be considered as intermediates since, with time, they are converted to multi-

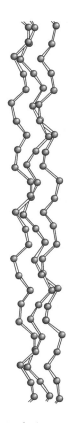

Figure 2.2   Diagram of the tropocollagen molecule, showing only the α-carbon atoms in the chains (reproduced with permission from the Pharmaceutical Press).

valent cross-links capable of linking together several sites (Bailey and Paul, 1998). These bonds are important in conferring high tensile strength and resistance to chemical attack on the collagen fibres.

Conversion of collagen to gelatin

The triple helix structure of tropocollagen can be destroyed (denaturation) by the application of heat (Flory and Weaver, 1960) or by the use of compounds which destroy hydrogen bonds (Steven and Tristam, 1962), with resultant conversion to gelatin. It has been shown that a thermally labile domain, which is deficient in hydroxyproline residues and is located near the carboxy terminus of the molecule, initiates the denaturation process, allowing the triple helix to unzip along its length (Miles *et al.*, 1995). The tropocollagen–gelatin transformation has been extensively studied and reviewed (Jones, 1989a). Denaturation involves breaking only the hydrogen bonds and those hydrophobic bonds that help to stabilise the collagen helix. This is followed by the disentanglement of the chains and dissociation into smaller components with a random coil configuration. Loss of helical structure results in a marked fall in intrinsic viscosity, an alteration in specific optical rotation (from about -400 to $140°$) (Burge and Hynes, 1959), a reduction in the angular dependence of light scattering and an increase in partial specific volume. Thermal denaturation typically occurs at about 40°C, at pH 7, but this temperature varies with the collagen source, the pH and the ionic strength. It is convenient to define the transition in terms of denaturation temperature, T_D, which is the temperature at which the viscosity, or optical rotation, falls by 50% of its original (stable) value in 30 min.

The value of T_D for different collagens correlates with the total imino acid (proline and hydroxyproline) content. For example, fish collagens with their relatively low imino acid contents exhibit T_D values in the range 10–30°C compared with values of 36–40°C for mammalian collagens (Jones, 1989a). Denaturation occurs more readily as the pH is reduced but the effect is somewhat influenced by the acid used. Lower values result from the addition of neutral salts at constant pH. Not surprisingly, compounds that destroy hydrogen bonds, such as urea, guanidine and potassium thiocyanate, can bring about denaturation in the cold.

The products of the denaturation of tropocollagen depend upon whether cross-links remain between the three component protein chains and the situation is further complicated by the fact that the three chains are not identical. Single-chain species (α-type) of about 100 000 Da and two-chain species (β-type) of about 200 000 Da are the main components, although a disordered form of tropocollagen itself (γ-type), which has a molecular weight of about 300 000 Da has been found (Jones, 1989a). Its links can be broken by alkali to give the α-species. Chromatographic separations on carmellose have shown that two

α and two β components exist. The two α components, designated α_1 and α_2, differ in amino acid composition. The α_1 form contains smaller amounts of histidine, hydroxylysine, tyrosine and those amino acids possessing large hydrophobic side chains, but is richer in hydroxyproline.

The β components consist of dimers formed from two α_1 chains (β_{11}) or an α_1 and an α_2 chain (β_{12}), whilst the γ-component is made up of two α_1 chains and one α_2 chain and is designated γ_{-112}. The proportion of α to β components varies with the collagen source and the method of solubilisation. Cod fish soluble collagen has also been extensively studied (Piez, 1964) and is unusual in that the three α-chains are all different, with identification of an α_3 chain.

The amino acid sequence of the α-chain has been pieced together from various studies on specific cleavage of tropocollagen using enzymes or cyanogen bromide (Hulmes *et al.*, 1973). Most of the chain is composed of triplets of the form Gly-X-Y, with Gly-Pro-Y occurring frequently and in particular Gly-Pro-Hypro. The terminal regions of the α-chain do not show this triplet sequence and cannot therefore possess a helical structure.

The amino acid compositions of gelatins are quite similar to those of the parent collagens but consistent slight differences do occur. In particular, the gelatins are slightly richer in the abundant amino acids and relatively poorer in the rarer residues than the corresponding collagens. This is more obvious for alkaline processed gelatins and is attributed principally to the selective loss of peptides deficient in hydroxyproline but containing tyrosine (Eastoe and Leach, 1977). Alkaline processing can also lead to partial conversion of arginine residues to ornithine. Amino acid analyses of several different types of gelatin are illustrated in Table 2.1.

Commercial manufacture of gelatin

Boiling bones or skins in water results in a low yield of impure gelatin with poor physical and organoleptic properties. Commercial processes for converting collagen stock into gelatin are designed with the object of achieving the maximum yield of gelatin consistent with

Table 2.1 Composition of mammalian gelatins (values are given as numbers of residues per 1000 total residues)

	Pigskin[a]	Cattle hide[b]	Cattle bone[b]	Cod skin[c]	Tilapia skin[d]
	Acid	Alkali	Alkali	Autoclaving	Acid
Alanine	111.7	112.0	116.6	107.0	120.6
Glycine	330.0	333.0	335.0	345.0	340.0
Valine	25.9	20.1	21.9	19.0	17.0
Leucine	24.0	23.1	24.3	23.0	23.2
Isoleucine	9.5	12.0	10.8	11.0	8.0
Proline	131.9	129.0	124.2	102.0	119.0
Phenylalanine	13.6	12.3	14.0	13.0	12.2
Tyrosine	2.6	1.5	1.2	3.5	2.8
Serine	34.7	36.5	32.8	69.0	37.1
Threonine	17.9	16.9	18.3	25.0	26.3
Cystine				<1	
Methionine	3.6	5.5	3.9	13.0	9.4
Arginine	49.0	46.2	48.0	51.0	54.3
Histidine	4.0	4.5	4.2	7.5	9.4
Lysine	26.6	27.8	27.6	25.0	24.7
Aspartic acid	45.8	46.0	46.7	52.0	46.4
Glutamic acid	72.1	70.7	72.6	75.0	66.5
Hydroxyproline	90.7	97.6	93.3	53.0	75.7
Hydroxylysine	6.4	5.5	4.3	6.0	7.5

[a] Eastoe (1967).
[b] Eastoe (1955).
[c] Piez and Gross (1960).
[d] Croda (2001).

commercially acceptable values of properties such as gel strength, viscosity, colour, clarity and taste. Manufacture involves the removal of non-collagenous material, conversion of collagen to gelatin, purification and recovery of gelatin in a dry form. The details of the process vary depending upon the nature of the raw material used; the precise operating conditions adopted by manufacturers are, in many instances, commercial secrets. However, a study protocol on the deactivation potential of various stages of bone gelatin manufacture has established that all GME manufacturers of this type of gelatin have adopted the same minimum conditions for 'key stages' of manufacture (GME, 1999). Moreover, legislation introduced to minimise the risk of BSE contamination in edible gelatin imposes certain minimum conditions that must be met. The basic operations involved in pharmaceutical gelatin manufacture are illustrated in the flow chart in Figure 2.3.

Collagenous raw material and its pretreatment

It is possible to utilise a fairly wide range of collagenous raw materials for conversion into gelatin. In practice, the choice has been largely restricted, for economic, processing and quality reasons, to cattle bone, pig bone, cattle skin (hide) and pigskin. Typical compositions for these materials, in terms of water content, grease content and gelatin yield, have been reported (Hinterwaldner, 1977).

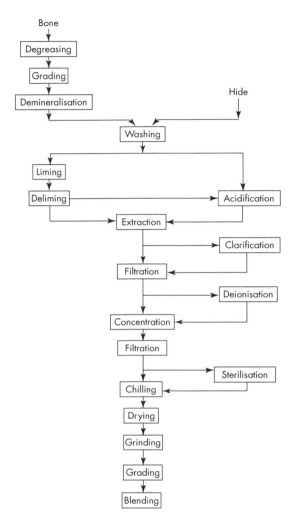

Figure 2.3 Flow chart of the commercial process for the conversion of collagen to gelatin (reproduced with permission from the Pharmaceutical Press).

Pigskin may be received in the form of fresh skins but, more commonly, is supplied in a frozen state to increase its stability during shipment and storage. Cattle hide raw material is a by-product from tanneries and is most commonly received in a wet, limed condition and represents the inner non-grain layer, or corium, known commercially as 'splits'. Hide may also be obtained in the dry state and can also vary in type and age, from thin calfskin splits to thick, mature buffalo hide. These forms may behave quite differently in terms of the

processing required for their conversion to gelatin. Traditionally, bone for gelatin manufacture could be either fresh bone from slaughtered animals, which has a high moisture and fat content and probably some adhering meat and sinew, or so-called 'exotic bone' representing the sun-dried material obtained primarily from India. The latter type of bone has been naturally cleaned and generally shows a much lower level of attached tissue, but still contains a considerable quantity of impurity in the form of dried blood. Exotic bone is no longer used for edible or pharmaceutical gelatin manufacture owing to European Commission (EC) health requirements for the raw material to be exclusively from healthy slaughtered animals (Decision 1999/724/EC).

The composition of bone varies with the species and the age of the animal from which it is obtained and also from one part of the skeleton to another (Johns, 1977). The outer and inner portions of the bones give rise to 'hard' (or compact) and 'soft' (or spongy) pieces, respectively. Data indicating the possible variation in overall composition of fresh, whole bone have been reported, but, despite the difference in physical structure, the chemical composition of bovine compact and spongy bone is very similar.

Other raw materials may be used and concern over the potential for BSE contamination in bovine raw material has led to growing interest in fish skins and, to a lesser extent, also in poultry bone as alternatives. Fish skins may be supplied fresh and in the wet state but more commonly are received in frozen form from fish processing plants or as dried skins. Availability of fish skins is limited since the major proportion of fish for human consumption is not filleted. The species of fish, in particular whether it is classified as 'warm water' or 'cold water', influences both the processing required and the properties of the final gelatin.

Pretreatment of raw materials is necessary. This refers to those procedures necessary to clean the material and to remove inorganic and organic matter which would otherwise have an adverse effect on the subsequent 'chemical conditioning' of the collagen and on gelatin extraction.

Bone requires a very different pretreatment from hide and skin, principally because of its high inorganic content. In all pretreatment processes the aim is to maintain the collagen in an 'undamaged' form to prevent loss of potential gelatin yield or quality.

Bone crushing

Fresh bones have to be sorted to remove inferior or unacceptable raw material and foreign matter. This is still done by hand with the assistance of electromagnets for the removal of tramp iron as the bone is fed on the conveyor belt to the bone cutter. Various types of bone cutter can be used; the design may involve either a true knife-cutting action or a hammer blow. The important parameters are, first, that the bone pieces should be sufficiently small to minimise the time spent in the degreasing and demineralisation stages and, second, that they should not be so small that they create problems in subsequent wet processing, for example poor drainage rate or loss caused by fines passing solids-retaining screens. Generally, the upper size limit will be about 20 mm and the lower about 3 mm. Apart from the size of the bone pieces, their shape can also be important in influencing fat removal and again this is a function of the design of the cutter.

Degreasing

Fresh bone contains, typically, about 35% moisture and 15% fat. This fat must be removed or it will cause processing problems. For example, during the liming stage, fat or calcium soaps formed from it, can clog the capillary spaces in the demineralised bone (ossein) and impede the penetration of alkali. Fat, which appears in the gelatin liquor during extraction, may become partially emulsified, producing poor-clarity liquors that are difficult to filter. More important still is the fact that gelatins with high grease content may show localised non-wettability, in 'windows' in hard gelatin capsules.

It is important to protect the collagen in the bone from thermal denaturation during degreasing. The quality of the removed fat, which is a valuable by-product, is also important. For these reasons, many of the degreasing processes that have been used in the past were unsatisfactory. Current practice is mechanically to agitate bone

and hot water together in a tank using a propeller or screw conveyer. A supply of fresh crushed bones and clean hot water enters the tank at one end and a stream of treated bone and an emulsion of fat and small particles leaves the tank at the other end. The bone chips and fat emulsion are separated and then the bone is treated in a fast stream of hot water to remove soft tissue, by centrifugation or by gravity, prior to hot air drying in a rotary drier. GME (1999) specify minimal operating conditions for a water temperature of 75°C and a contact time of 15 min. Inlet air to the dryer is typically about 400°C and drying time is about 30 min, during which time the bone reaches a temperature of about 85°C. The degreased bone chips typically contain less than 3% fat.

Grading of bone

The sinew component of degreased bone has a very different physical form from the hard and soft bone components; it is important to separate it, in order that it may be treated differently in subsequent processing to avoid losses in the yield and quality. Complete separation and processing of sinew from bone is not really a practical proposition. It is usual to separate the degreased bone into hard, soft and sinew fractions in which:

- the hard bone fraction is virtually free from sinew but will contain a proportion of soft bone;
- the soft bone fraction contains a proportion of both hard bone and sinew;
- the sinew fraction contains a proportion of soft bone but a negligible amount of hard bone.

Separation into these fractions can be effected in the wet state but is most easily effected after the degreased bone has been dried. Differences in specific gravity of the three components are utilised in their separation on a vibrating inclined screen. Improved separation can be achieved by a second screening.

Demineralisation

The inorganic phase of bone is composed principally of calcium phosphate, together with carbonate ions, although it is still not clear whether the major component is hydroxyapatite, tricalcium phosphate hydrate or octacalcium phosphate. Minor constituents, totalling about 2% of the inorganic matter, include magnesium, sodium and potassium chlorides and fluorides, together with trace elements at levels below 100 ppm. Demineralisation of degreased bone is necessary to free the collagen before it can be conditioned and extracted as gelatin. This is achieved by treatment with dilute hydrochloric acid in counter-current flow, so that the fresh bone will meet almost spent acid, while fresh acid is reacted with the bone that is about to leave the process. The main reaction taking place can be considered to be dissolution of tricalcium phosphate to form the monocalcium salt:

$$Ca_3(PO_4)_2 + 4HCl \rightarrow Ca(H_2PO_4)_2 + 2CaCl_2$$

The concentration of acid and the time of treatment are varied according to the type of bone being handled and its grist size. Excessive acid concentrations or over-long treatment times can result in loss of yield and eventual gelatin quality through hydrolysis of the collagen and therefore, in practice, complete demineralisation is not attempted. Temperature control is also important, since heat is generated in the reaction and increased temperature accelerates the rate of collagen hydrolysis. It is usual to employ refrigeration plant during the summer months in warm climates, to maintain a demineralisation temperature of 10–15°C. In general, 4–5% hydrochloric acid is used, with treatment times of about 4–6 days. GME (1999) specify a minimum period of 48 hr and not less than 4% hydrochloric acid (pH <1.5) for the final stage of bone demineralisation. The demineralised bone, ossein, normally shows a residual ash content of 1–2%.

An important by-product from the demineralisation process is dicalcium phosphate, which is obtained by treating the spent acid leach liquor with lime to precipitate dicalcium phosphate at a pH in the region of pH 4.2–5.0. After concentration, the slurry is filtered, washed and dried to give a white free-flowing powder, which may be used in animal feeds and fertilisers or in other technical applications.

Hide

Splits and pieces are normally first given a superficial wash to remove any surface contamination and then fed to a cutter in order to reduce the material to convenient size to allow the material to be pumped or 'flumed' with water. Dry hide, which is an extremely tough material, is sometimes given a weak caustic presoak to soften it before cutting. If the hide is to be acid extracted without conventional liming treatment it must be given a more thorough washing and it is sometimes given a short pretreatment in dilute caustic solution containing hydrogen peroxide prior to neutralisation and water washing. These operations are generally performed in paddle washers which ensure effective movement between the hide pieces and the solution.

Pigskin

Where pigskin is received in a frozen state, the only pretreatment necessary is to thaw the material and cut and wash. Despite its high fat content, no attempt is made to remove the fat at this stage because it is most convenient to separate it during the subsequent acid extraction of gelatin. Any alkaline pretreatment must be avoided to prevent the possibility of formation of soaps.

Fish skins

These are pretreated in a similar manner to pigskin, the frozen skins being thawed, cut and washed with water.

Conditioning of collagen

Liming process

The function of the liming process is to 'condition' the collagen so that gelatin with the desired physical properties is obtained with a good yield on subsequent extraction at moderate temperature and near neutral pH. Liming brings about both physical and chemical changes in the collagen (Jones, 1989a). The amide groups of the glutamine and asparagine residues are converted into carboxyl groups with liberation of ammonia, resulting in a shift in the isoelectric point of the collagen. However, the deamidation reaction is incidental to the collagen-to-gelatin conversion, the change in isoelectric point being practically complete after 10–20 days, whilst conditioning reactions leading to increased yield of gelatin continue for a much longer time. The gelatin derived from limed collagen normally has an isoelectric point of pH 4.8–5.2.

Arginine residues may be deguanidated to form ornithine residues and urea, but the reaction is slow and becomes important only in the later stages of liming. A 34% conversion of arginine to ornithine has been reported for extensively limed calfskin, whereas 3% conversion is a more normal figure for conventionally limed collagen. Ornithine has a side-chain amino group, so that the formation of ornithine is accompanied by an apparent decrease in chain average molecular weight where generation of free amino groups is attributed to peptide bond hydrolysis. Some peptide bond hydrolysis undoubtedly does occur but it has been estimated that this represents less than one bond per chain. The peptide bonds most susceptible to hydrolysis during alkaline processing are those involving glycine, serine, threonine, aspartic acid and glutamic acid at the amino-group terminus, and also these same amino acid residues, plus alanine and phenylalanine, at the carboxyl group end. The bonds most stable to hydrolysis involve proline and hydroxyproline.

Perhaps the most important function of liming is the destruction of cross-links. Alkali-labile bonds almost certainly exist in the non-helical, telopeptide regions where cross-links occur, but not all cross-links are generally destroyed, as evidenced by the existence of very high molecular weight branched chain gelatin molecules. Indeed, complete removal of cross-links may not necessarily be desirable since these very high molecular weight molecules may be important in determining the gel setting rate.

Soaking collagen in acid (e.g. 9% HCl) prior to liming significantly reduces the liming period. However, if this 'dual soak' process is reversed and the acid treatment is given after liming, the same increase in yield is not observed. This has been interpreted as indicating that two types of insolubilising bonds are present in the collagen, one type being labile in acid (or very strong alkali such as 5% NaOH) and the other being labile only in lime. When intact, the acid-labile bonds apparently protect some of the alkali-labile bonds from hydrolysis during liming. Another important role of the liming process is the removal of foreign proteins and mucoproteins, which exist as impurities in the collagen and are soluble at alkaline pH. Collagen fibres, after liming, are indistinguishable from the native fibres under the electron microscope, except that they are swollen and the internal adhesion of each fibril is reduced by rupture of certain intermolecular links. Swelling of collagen during liming proceeds in two stages: the initial rapid equilibrium swelling is complete within about three days and slower swelling, indicative of structural breakdown, proceeds during the entire liming period.

The collagenous raw material may be ossein or cattle skin. The sinew fraction of ossein is not used because of its susceptibility to physical and chemical breakdown during liming. Normally, the 'hard fraction' of ossein is used, although the soft fraction can also be utilised provided it has been sufficiently well segregated from sinew to avoid losses in yield.

Liming is carried out in tanks or pits, or occasionally churns, where the collagen stock is soaked in slaked lime slurry (2–5% concentration) for periods of typically 60–120 days. The length of time in lime depends upon such factors as the nature of the raw materials, temperature, availability of pits and the properties of the gelatin required. Ideally, the liming pits should be sited indoors where some control of temperature can be exercised to prevent marked fluctuations. Typically, liming temperatures of 14–18°C are used. At regular intervals the contents of the pits should be aerated to ensure distribution of lime slurry throughout the pits and to discourage the growth of anaerobic bacteria. The lime slurry may be replaced by fresh slurry several times during the conditioning period, to assist in the removal of impurities.

Apart from its low cost, the use of lime has the advantage that, provided undissolved lime is always present, its solubility acts as a built-in regulator of the alkalinity (pH about 12.5). It also tends to swell collagen to a lesser degree than sodium hydroxide at the same pH. It is possible to accelerate the conditioning process by 'sharpening' the lime liquor by the addition of sodium hydroxide or, at least academically, sodium carbonate, sodium sulfide, hydroxylamine, or calcium chloride and methylamine. By increasing the concentration of sodium hydroxide to about 1–2%, it is possible to condition the material in 10–40 days without using any lime. However, this shorter and consequently more drastic pretreatment is more difficult to control to give a uniform degree of conditioning and usually leads to a greater loss of yield into solution. Part of the 'over-conditioning' is associated with the higher degree of swelling in caustic solution compared with lime slurry, although this can be minimised by the use of a swelling restrainer such as sodium sulfate. This formed the basis of the so-called Sulphate Process (Ward, 1953) by which it was found possible to alkali-condition collagen in 5 days using 5–10% sodium hydroxide solution containing 20% sodium sulfate as a swelling restrictor.

At completion of the liming treatment, 'deliming' is effected by washing with water in a log-washer or paddle-washer for about 24 hr, preferably after solid surface lime has been mechanically removed with water. It is then necessary to neutralise the raw material, which at this stage is generally at pH 9–10, with acid. Various acids may be used but, most commonly, these are hydrochloric, sulphuric or phosphoric. In practice it may be advantageous to over-acidify the material slightly and wash back to the desired pH. However, where collagen has been extensively limed or caustic treated, care must be taken not to acidify to the point where the collagen becomes excessively swollen and fragile. The final pH of the collagen stock is normally required to be in the pH range 5.0–6.5, although more acidic pHs are sometimes used.

Acid process

An alternative to the liming process is to acidify the collagen for subsequent extraction at low pH. It is sufficient simply to bring the collagen to the required equilibrium pH and, since this is normally achieved within 24 hr, acid conditioning offers the advantage of very much shorter processing time compared with the liming process. However, if reasonable gelatin yields and qualities are to be achieved, acid processing is generally restricted to soft bone ossein, sinew, pigskin, calfskin and fish skins. It has also been applied to sheepskin and mature cattle skin, but the age of the animal is a significant factor and a maximum age of 2–3 years for cattle skin has been suggested as being suitable for acid processing. Skins from pigs up to 18 months old give higher gelatin yields and qualities than skins taken from 30-month-old pigs. Typically, pigskin is cut into pieces about 10 cm × 10 cm and washed with water to remove superficial fat. It is then treated with mineral acid solution (normally sulphuric acid or hydrochloric acid) to maintain the pH at about pH 1.8 for at least 5 hr, before washing with water to remove excess acid and to provide a suitable extraction pH of about 4. Cattle skin, soft bone ossein and sinew are less easily extracted than young pigskin and it is usual to adjust these materials to a lower final pH within the range 2–3.5.

Acid conditioning and extraction yields a gelatin with an isoelectric point well above pH 5. This is because a proportion of the glutamine and asparagine residues remain in the amide form, giving rise to isoelectric points in the region of 7.0–9.5 for acid pigskin gelatin or acid fish skin gelatin and pH 6.0–8.0 for acid ossein gelatins. In contrast to the liming process the 'conditioning effect', in terms of breakage of the cross-links and peptide bonds prior to extraction, is minimal. Studies on the titration curves of acid precursor pigskin gelatin and native collagen have shown that peptide bond hydrolysis is the more important reaction in the acid extraction of gelatin (Jones, 1989a).

Extraction and purification of gelatin

The conditions of pH and temperature for extraction are, to a large extent, dictated by the nature of the raw material and the pretreatment it has received, but extraction normally takes place under either acid or neutral conditions and at the minimum temperature needed to give a reasonable extraction rate and a high yield of gelatin. For a given raw material, the extraction rate is greater under acid conditions but, at the same time, thermal degradation of gelatin is considerably more rapid than at neutral pH.

Acid extraction

For materials which have received no alkaline pretreatment but simply acid conditioning, such as soft bone ossein, sinew, pigskin and fish skin, it is necessary to extract at acid pH to obtain reasonable extraction rates and good yields, without resorting to very high temperatures. Pigskin is generally extracted at pH 4 or thereabouts, in a traditional batch-by-batch fashion. The material is loaded into extraction pans constructed from acid-resistant stainless steel, which have false bottoms covered with stainless steel mesh through which liquor can drain. The pans are preferably fitted with lids, lagged to reduce heat loss and fitted with suitable heating equipment. The general procedure is to load the material, allow it to drain, cover it with hot water and then heat the liquor by means of an internal heater or by circulation through an external heat exchanger until the required temperature is achieved. Excessive agitation of the liquor can result in emulsification of the fat and should be avoided. When the gelatin concentration reaches 4–10% the liquor is drained and further 'runs' are made in a similar manner, using progressively high temperatures. Fish skins can be extracted under similar conditions.

Ossein and sinew are less readily extracted than pigskin. The extraction is normally carried out at pH 3.0–3.5, although a pH as low as 2.0 may be used. The secret of successful acid extraction is the rapid removal of the gelatin from the ossein after it has passed into solution in order to reduce thermal hydrolysis and consequent deterioration in physical properties. The batch process has largely been replaced by a semi-continuous extraction process. Hot water, at a controlled temperature and flow rate, is continuously run on to the ossein at the top of the pan, whilst gelatin liquor at a concentration of 3–6% is continuously run off from the bottom and its pH neutralised as soon as possible. The extraction

temperature is raised during the process, to maintain the extraction rate, from about 50°C at the start to about 75°C at the end. It is generally recognised that acid extraction results in a gelatin with a high gel strength/viscosity ratio compared with neutrally extracted limed gelatin. Some control over this ratio can be exercised, however, by selection of the extraction pH within the acid range. In general, high gel strength is favoured by low pH whilst higher viscosities are more easily achieved at a higher pH. As extraction proceeds, the gelatin viscosity generally increases (except for the final extracts, when it may fall again), whereas the gel strength tends to decrease. This is due, in part, to the fact that the pH tends to increase as the extraction progresses.

Limed hide or limed ossein can be extracted at low pH and such commercially produced gelatins generally have lower viscosities than neutrally extracted material. However, this lower viscosity arises through secondary hydrolysis of the gelatin solution rather than from breakage of specific bonds in the collagen. High viscosities can be obtained by extraction at pH as low as 2.0 using continuous extraction under carefully controlled conditions. Similarly, fish skins may be subjected to alkaline conditioning prior to acid extraction.

The water used for extraction may need to be deionised if it has too high a content of dissolved solids and, more particularly, if it contains a high level of nitrate ions which can affect the performance of the gelatin for particular applications. Further consideration of the importance of nitrate ions is given later in this chapter.

Neutral extraction

Limed collagen stock is normally extracted at a neutral or weakly acid pH. The extraction pH may vary from 5.0–8.0 but is most commonly between 5.0 and 6.5. The gelatin is obtained as a series of extracts with an initial temperature of 50–60°C, increasing to boiling point for the final extract. Liquor concentrations of 4–10% are achieved, the rate of extraction being a function not only of temperature but also of the nature of the raw material and the extent of alkaline conditioning. The first fraction of the total gelatin is obtained simply by the thermal breakdown of the hydrogen bonds, which stabilise the helical struc- ture of well-conditioned regions of the collagen. This gelatin is simply 'melted out', with no further rupture of covalent linkages. Later extracts are obtained as a consequence of further breakdown of collagen resulting from the con- ditioning action of the prevailing pH and tem- perature conditions. Neutral extraction tends to give gelatins with lower gel strength to viscosity ratios than acid extraction. High viscosities are favoured by neutral extraction of collagen that has been limed for a long period.

As extraction progresses, the gelatin gel strength generally decreases, although the first extract is, anomalously, often lower in gel strength than the second extract. The viscosity tends to increase for successive runs until the final one, which is near boiling temperature. The 'last run' gelatins generally have poorer physical properties and are darker in colour and stronger in odour, reflecting the more extensive heat treat- ment received during extraction.

Filtration and clarification

The dilute gelatin liquor from the extraction pans may contain suspended collagen particles, fat globules and foreign proteins, if these have not been removed during the conditioning stage. Preliminary coarse separation can be effected by centrifugation, filtration through cloth or stain- less mesh or passage through sludge separators. The traditional way of filtering gelatin liquors, which is probably still the most effective, is to pressure-filter through compressed pads of cellu- lose pulp, which have previously been sterilised with steam. At the end of filtration, the pulp pads are broken up, washed with boiling water to remove gelatin and insoluble matter and then reformed in a press for re-use. Self-cleaning types of filter are also in use, based on the principle of filtration through diatomaceous earth coated on twilled stainless steel support screens. These round support elements are normally mounted horizontally on a central hollow axial shaft in a closed vertical pressure vessel. Vacuum filtration, where gelatin liquor is sucked through a filter bed

on the outside surface of a rotating drum filter, may also be used. In this case some filter aid is also suspended in the liquor, so that the surface of the filter layer on the drum is continuously scraped off and reformed.

Certain gelatin processes produce liquors that, after filtration, still have poor clarity owing to non-collagenous proteins, which have been extracted from the raw material, particularly in the last extraction runs. In such cases it is usually possible to improve the clarity by 'chemical clarification' in which a flocculent precipitate is formed *in situ*, to adsorb finely divided suspended solids and turbidity-forming colloids. Suitable precipitants should not only be good adsorbents but also separate rapidly under gravity or on centrifugation. Various systems for forming floc- culent precipitates of inorganic salts in the gelatin liquor have been used. These include dicalcium phosphate, calcium carbonate and alu- minium hydroxide flocs. Such methods can be particularly useful for acid-extracted gelatin liquor since they provide the possibility of neu- tralising the acid without significantly raising the ash content of the gelatin. For example, if phos- phoric acid is used to condition the collagen for extraction, then by adding lime slurry to gelatin liquor, the pH of the liquor can be raised to 5.0–6.5, depending upon the concentration of phosphate ions, thereby precipitating dicalcium phosphate. The addition of a small amount of aluminium sulfate during precipitation can also assist in producing a more rapidly separating pre- cipitate. In general, chemical clarification is more easily and reproducibly applied to acid-processed gelatin liquors, because alkali-processed gelatin liquors contain soaps that may interfere. It is common practice to give a 'polishing filtration' to the concentrated gelatin liquor, particularly where there is a risk of salts precipitating from the solution during evaporation.

Deionisation

In general, the ash content of even a poor alkali- processed gelatin will be below the 2% maximum limit permitted by the most stringent of the various pharmacopoeias. However, ash levels below 1% are frequently specified by individual pharmaceutical gelatin users and in such cases it may be necessary to deionise the gelatin. Concern about specific ions such as iron, phosphate, sulfite, nitrate or nitrite may also be a reason for resorting to ion exchange. Moreover, because of the ability of ion-exchange to contribute to the overall clearance factor for potential BSE contamination, GME specify deionisation as part of the standard process for bone gelatins.

In the case of acid-extracted gelatins, it is generally necessary to raise the pH of the liquor after extraction from around 3.5 to around 6.0 before further processing. This would bring about an increase in ash content, unless an insoluble, filterable precipitate is formed, or ammonium hydroxide is used as the alkali. Acid pigskin gelatin is frequently offered at pH 4.0–4.5 to avoid this problem, whilst certain acid processes use sulphur dioxide as the conditioning acid, relying on its volatility to increase the pH during evaporation. In most cases, however, acid-extracted gelatins for pharmaceutical applications are treated by ion exchange to raise the pH without increasing the ash content, yielding products with ash contents as low as 0.1%.

Gelatin liquor is normally deionised after filtration and before evaporation, because of the problems of handling the viscous concentrated solutions. Mixed-bed ion-exchange resin columns are theoretically more effective than separate anion and cation resin columns and also offer the advantage that the pH of the liquor is not greatly altered, thus reducing the risk of hydrolysis. However, separate anion and cation columns are widely used because of their relatively simple operation and, particularly, their regeneration. Whether anion exchange should precede or follow cation exchange will depend upon the gelatin liquor. For example, gelatin liquors that have been chemically clarified with dicalcium phosphate, should be cation exchanged first, otherwise precipitation of tricalcium phosphate can occur in the anion resin column as the pH increases above about 7.5. Acid-extracted gelatin liquors, on the other hand, are better anion-exchanged initially since cation exchange is, first,

less effective at acid pH and, second, would make the liquor even more acidic, with greater risk of degradation. Since extremes of pH are generally experienced at some stage in the ion-exchange process it is important to avoid high liquor temperatures and to pass the liquor through the columns as rapidly as possible.

Concentration

Because gelatin is susceptible to thermal hydrolysis it is necessary to keep the evaporation time and temperature to a minimum and therefore vacuum is applied. Various types of evaporator are used including tubular, plate, thin film, centrifugal and 'flash' evaporators (Hinterwaldner, 1977). The plate and tubular types are normally operated in a double- or triple-effect arrangement for reasons of heat economy. It should be possible to concentrate the gelatin solution to a final level of 20–25% with a liquor temperature below 55°C in the final effect. High concentrations, in particular for high-viscosity gelatins, can lead to the risk of localised overheating caused by the reduced circulation rate but, because evaporation is a cheaper operation than drying, there is an incentive to reduce the water content as far as possible by evaporation. Thin film, centrifugal or flash evaporation are frequently used to increase the concentration further. Flash evaporation involves rapidly increasing the temperature of the gelatin solution up to 120–145°C in a plate heat exchanger, followed by expansion into a vacuum separator, resulting in rapid cooling of the liquor. The liquor can be recycled through the system until the required concentration is achieved.

Sterilisation

Since the addition of preservatives to gelatin is not permissible (with the possible exception of limited amounts of sulphur dioxide) and yet gelatin will support bacterial growth, it is necessary to operate a regular and rigid programme of plant sterilisation. Monitoring of microorganisms at various stages of gelatin manufacture, from extraction through to the dry gelatin, is common practice. Flash evaporators are useful in the sterilisation of gelatin since the short-time, high-temperature sequence is effective against bacteria, whilst gelatin degradation is minimal.

Where heat sterilisation is required without the necessity for concentrating the gelatin, the liquor can be flashed to high temperature, 140°C, for a few seconds using live steam followed by expansion into vacuum and cooling. The steriliser is suitably placed immediately prior to the chilling and drying operation. GME specify minimum sterilisation conditions of 138°C for 4 s for all bone gelatins.

Chilling and drying

Commercially, dry gelatin is obtained by chilling the concentrated liquor to form a gel, which is then air dried. The rate-determining stage in the removal of moisture from the gel is the diffusion of water from the interior of the gel to the surface where it is evaporated. If evaporation is too rapid then surface skinning or 'case hardening', can occur. To reduce the risk of this, the air temperature is kept low to begin with and is slowly increased during the drying cycle. The final moisture content of the gelatin should be in the range of 8–13%. Lower moisture contents than this can cause poor dissolution or even partial insolubility of the gelatin.

Gelation may be achieved using a slowly rotating, brine-cooled, stainless steel drum, or an endless stainless steel chilled band, or a scraped-tube, brine-cooled, heat exchanger in which the gelatin is simultaneously chilled and extruded through a noodling or dicing head. The continuous gel sheet which is taken off the drum or band chiller is fed to a cutter or mincer where it is reduced to 'cubes' or 'crumbs'.

The drying stage may be performed either as a continuous or a batch operation. The continuous drier consists of a stainless steel mesh band moving through a series of controlled temperature and humidity zones in the drying tunnel.

The drying air is filtered to remove airborne particles and may be scrubbed to remove gaseous impurities such as sulphur dioxide. Such a band drier is most conveniently fed with noodled gelatin from a scraped-surface heat exchanger. The total drying time is typically less than 4 hr and the air temperature in the final zone is around 75°C. In a typical batch-drying process the gel pieces are supported on a stainless steel screen and exposed to an upward draught of air of controlled temperature and humidity. Movement of the gelatin particles may be effected by rotating 'rakes'. Rotating drum dryers in which the cylindrical wall is permeable to the drying air are also in use. Maximum drying temperatures of about 75°C are employed and the total typical drying time is normally 6–12 hr. A typical capacity for a batch drier would be 750–1000 kg of dry gelatin. Other drying processes such as the fluidised bed method have been tried but have generally found little favour. Spray drying is costly because it is necessary to dry comparatively dilute solutions and because it produces low bulk density, fluffy powder. Drum drying can result in significant thermal degradation of gelatin and produces a very fine powder on milling. Both spray drying and drum drying do offer, however, advantages in the production of cold-water dispersible gelatins by virtue of drying occurring directly from the sol to the dry state, without intermediate gel formation.

Gelatin produced from cold-water fish skins is unusual in that it will not form a gel under normal processing conditions and it is necessary to resort to the more expensive techniques of spray drying or drum drying.

Grinding, sieving and blending

The gelatin from the drier may be in the form of broken sheet, cube, crumb or noodle, depending upon the particular method of drying used. It may be supplied to the customer in these forms, but is usually milled and then classified into specific particle size ranges by sieving or air classification. The particle size achieved by milling will depend partly upon the hardness of the gelatin, which is a function of the grade and moisture content, and partly on the design of the mill. A coarse powder is given by a 'knife-type' mill, whilst finer powder results from the use of a hammer mill. Commercially, standard gelatin grades are achieved by blending individual gelatin batches, which have previously been tested for compliance with physical, chemical and microbiological standards.

Gelatin manufacturing and its effect on transmissible spongiform encephalopathies (TSE) infectivity

Bovine bones and hides are considered to be non-infectious with respect to the BSE agent. Nevertheless, a potential BSE risk arises from the possibility of contamination with infected specified risk material (SRM), such as brain and trigeminal ganglia, spinal cord, vertebrae and dorsal root ganglia. Concern has focused on bovine bone, since it is considered that there is little or no risk of contamination of hide with SRM. A proactive approach to investigating the potential of the gelatin process to remove or destroy potential TSEs in bone raw material has been taken by the GME since the early 1990s, with studies being funded at several different independent research institutions that have the relevant expertise. Several key stages in bone gelatin manufacture appeared to have potential for TSE removal/inactivation. These included degreasing, acid demineralisation, alkaline treatment, filtration, ion exchange and sterilisation.

The University of Goettingen undertook a study to show to what extent the central nervous system (CNS), which is the carrier of BSE, is removed from incoming raw bones for the production of gelatin. It was shown that no nervous tissue could be detected in bones that had been degreased by vigorous agitation in water at 80–85°C for 15 min, using enzyme linked immunosorbent assay (ELISA) and immunoblot techniques. However, this did not provide specific data for the degree of removal of CNS and so a special production scale degreasing trial was run exclusively using heads from healthy cattle. It was found that about 98–99% of the specific marker nervous tissue proteins were removed by degreasing (Manzke et al., 1997). A study of the effect of acid treatment, alkaline treatment and a combination of the two, on infectivity reduction, using a mouse-adapted scrapie strain (ME7) was undertaken by Inveresk Research Institute. The acid treatment involved exposure to 4% hydrochloric acid for 48 hr at 11°C. Alkaline treatment involved exposure to saturated lime solution at 15°C for periods of 20, 45 and 65 days, respectively. The final results showed a reduction factor of log_{10} 1 for acidulation, log_{10} 2 for liming and the effect of varying time within the range 20–65 days was insignificant (Inveresk Research, 1998a) and a factor of log_{10} 3 for the combined treatments (Inveresk Research, 1998b) indicating that the individual treatments are cumulative in their effect.

In 1999, GME initiated a comprehensive study on bovine bone gelatin under the auspices of the European Commission and within the framework of the Commission's BSE Research Program (GME, 1999). The objective was to assess the inactivation/removal potential for the BSE agent of the entire gelatin manufacturing process. In addition, the effect of further individual process steps, such as filtration, ion exchange and flash sterilisation were studied. This was achieved by scaling down the model processes and validating that they were representative of industrial processes used by European gelatin manufacturers. Both the lime process and the acid process were studied using a mouse-adapted BSE strain (301V) and a hamster-adapted scrapie strain (264K), which were used in parallel for much of the study. TSE infectivity was used to spike raw bone (vertebrae) by injection into spinal cord and dorsal root ganglia, at a level which represented about 10^5 times that expected from having one infected and undetected animal in a daily gelatin production batch. No residual infectivity was found in either the acid or limed gelatins, indicating a clearance factor of at least log_{10} 5. The sum of the clearance factors for the individual process steps gave higher clearance factors, indicating that the effect is cumulative but not necessarily additive (GME, 2001). The Scientific Steering Committee of the European Commission has acknowledged a clearance factor of at least $\log_{10}$ 4.8 for acid gelatin and at least $\log_{10}$ 4.9 for limed gelatin in its updated opinion on the safety with regard to TSE risks of gelatin derived from ruminant bones or hides, published in October 2002. A treatment of ossein, following demineralisation, with 0.3 M sodium hydroxide for 2 hr at pH 13 at ambient temperature, followed by washing and acidulation to pH 2, was found to result in an overall clearance factor for acid gelatin of above $\log_{10}$ 5.4.

BSE legislation relating to gelatin

In the EU, bovine-derived pharmaceutical gelatin, in common with other ruminant animal-derived pharmaceutical raw materials, is currently subject to Commission Directive 2001/83/EC. This requires evidence of compliance with the joint Committee for Proprietary Medicinal Products (CPMP) and Committee for Veterinary Medicinal Products (CVMP) 'Note for guidance on minimising the risk of transmitting animal spongiform encephalopathy agents via human and veterinary medicinal products' (CPMP/CVMP, 2001). This evidence is provided by means of a general 'Certificate of Conformity' issued by the European Directorate for Safety of Medicines (EDQM) to the gelatin manufacturer, as prescribed in general chapter 5.2.8 in the European Pharmacopoeia (PhEur). This Certificate is issued based on a detailed dossier submitted by the manufacturer on all aspects of gelatin manufacture relative to TSEs and also requires a quality system such as ISO 9000 together with hazard at critical control points (HACCP) to be in place. Conditions for gelatin 'compliance' have evolved based on expert opinion on TSEs (e.g. FDA, 1997; WHO, 1997; OIE, 1999; SSC, 1999, 2002a, 2002b) and new data on geographical risks and on the removal/deactivation effect of the gelatin process.

Legislation makes a significant distinction between risks posed by bovine bone gelatin and bovine hide gelatin. The 'Note for Guidance' lists a number of parameters which contribute to the safety of bone gelatin, such as geographical sourcing of raw material, removal of the SRM, the manufacturing method and quality and product-

safety systems, together with traceability procedures and supplier audits (CPMP/CVMP, 2001). In contrast, the only parameter specified for bovine hide gelatin is 'cross-contamination with possible infectious material should be avoided'.

The United States Food and Drug Administration (FDA) has made less of a distinction between the UK and the rest of Europe, in terms of the geographical risk presented by bovine raw material. The US Department of Agriculture (USDA, 1998) not only classified as 'BSE countries' Belgium, France, Luxembourg, Netherlands, Oman, Portugal, Republic of Ireland, Switzerland and the United Kingdom but also imposed import restrictions on meat and edible products from the following countries: Albania, Austria, Bosnia-Herzegovina, Bulgaria, Croatia, Czech Republic, Denmark, Federal Republic of Yugoslavia, Finland, Germany, Greece, Hungary, Italy, the former Yugoslav Federal Republic of Macedonia, Norway, Poland, Romania and the Slovak Republic. In contrast, the FDA continues to maintain the view (FDA, 1997) that, 'At this time, there does not appear to be a basis for objection to the use of gelatin produced from bovine hides and bones in FDA-regulated products for human use if the gelatin is produced in the United States from US-derived raw materials.'

Around 1998, the Scientific Steering Committee (SSC) of the European Commission started to develop a methodology for arriving at assigning to a country a 'geographical risk of BSE' (GBR). This lead to an Opinion in July 2000, which was subsequently updated in January 2002 (SSC, 2000, 2002b). GBR is a qualitative indicator of the likelihood of the presence of one or more cattle being infected with BSE in a country. It is based on a consideration of the risk that if the BSE agent was imported into the country and, then if introduced into the food chain, whether it would have been recycled and amplified or eliminated by the BSE/cattle system of that country. GBR categories range from I to IV: with I representing 'highly unlikely', II 'unlikely but not excluded', III 'likely but not confirmed or confirmed, at a lower level', and IV 'confirmed at a higher level'. Of the 50 countries that have been classified (2002b), two countries, the United Kingdom and mainland Portugal have been rated as 'IV' and 16 countries rated as 'III'. The United States and Canada have been classified as 'II', whilst the majority of western European countries such as Belgium, France, Germany, Italy, the Netherlands and Switzerland have been classified as 'III'.

The CPMP/CVMP (2001) 'Note for Guidance' states that the preferred method for the manufacture of gelatin is the alkaline process. This is due to the additional clearance factor from the alkaline treatment step. This is also reflected in Decision 1999/724/EC, which specifies that category III (low BSE risk) ruminant bone must be subjected to an alkaline treatment with saturated lime solution (pH >12.5) for a period of at least 20 days. Additionally, the following minimum conditions must be met for other key stages of manufacture: the bone is finely crushed and degreased with hot water; demineralised with dilute hydrochloric acid, at minimum concentration of 4% and pH <1.5, over a period of at least two days; undergoes a sterilisation step of 138–140°C for 4 s. Provision is made for the use of an equivalent process approved by the Commission after consultation with the appropriate Scientific Committee. The acid process is consequently restricted to bone from category I and II countries.

Molecular weight of gelatin

Because gelatins are obtained by denaturation of collagen, a heterogeneous product normally results, with a range of molecular weight species. In such a polydisperse system the average molecular weight attributed to a gelatin will depend upon the method of determination. Many methods are available, osmometry, viscometry, ultracentrifugation, end-group analysis, gel filtration or permeation chromatography, size exclusion high pressure liquid chromatography (SEHPLC) and gel electrophoresis and there is an extensive literature on this topic (Jones, 1989a). Only light scattering is capable of providing absolute values for molecular weight, but gel electrophoresis and more particularly, SEHPLC, are proving to be the most useful and widely adopted techniques. They give good

resolution of molecular species and also allow calculation of values of weight average molecular weight (M_w) and number average molecular weight (M_n). SEHPLC has the advantage of producing more rapid results. In this case, it is necessary to produce a calibration curve relating molecular weight to elution time. The choice of molecular weight standards can influence the values of M_w and M_n for gelatin. Some workers use non-gelatin protein standards or polystyrene sulfonate standards but it is preferable to use a gelatin standard for which the α, β and γ peaks are identifiable.

The weight average molecular weight of commercial gelatins may vary from about 20 000 to 200 000 Da, but very much higher molecular weights (in excess of 10^6) have been reported for gelatin fractions separated by alcohol coacervation or gel electrophoresis. These high molecular weight species have been shown not to be due to aggregation. Number average chain weights in the range 40 000 to 80 000 Da have been reported for high-grade commercial gelatins but other reports suggest that these estimates may be on the high side (see later). Values of number average molecular weights also falling within this range have been reported for hard capsule gelatin (Robinson et al., 1975a). Comparative molecular weight data obtained by light scattering (weight average molecular weight) and end-group analysis (number average molecular weight) demonstrated that high molecular weight fractions from both acid and limed gelatins are branched structures (Courts and Stainsby, 1958). More recent work suggests that the degree of branching may be even greater than was originally thought, owing to limitations of the early technique and the presence of chains that do not possess any terminal amino groups (Stainsby, 1977a).

The lower intrinsic viscosities exhibited by acid-processed gelatins compared with lime-processed gelatins of comparable molecular weight have been interpreted as indicating a more compact, more highly cross-linked structure for these so-called acid gelatins (Robinson et al., 1975a). The average molecular weight and the molecular weight distribution are functions of the collagen-to-gelatin conversion process and they have an important effect on the physical

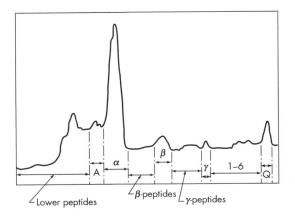

Figure 2.4   Molecular weight distribution of a hard capsule grade, limed ossein gelatin (Bloom strength, 250 g), determined by polyacrylamide gel electrophoresis (2.5% gel) (reproduced with permission from the Pharmaceutical Press).

properties of the gelatin. A typical molecular weight distribution for hard capsule grade limed ossein gelatin determined by gel electrophoresis is illustrated in Figure 2.4 and the molecular fractions details are given in Table 2.2. This shows the prominent α-peak characteristic of limed gelatins which is a consequence of the selective breaking of cross-links.

Acidic and basic properties

Several of the amino acid residues of gelatin possess ionisable groups (carboxyl, phenolic, amino, guanidino and imidazole), which are distributed along the length of the molecule. The acidic and basic side-chain groups together with the terminal amino and carboxyl groups, enable gelatin to adopt a different net charge, which may be either negative or positive, depending upon the pH of the solution. The numbers of these groups can be determined from titration curves; data for both acid- and lime-processed gelatin are summarised in Table 2.3. The difference in the number of free carboxyl groups present in acid and limed gelatins is due to the differing degrees of deamidation of glutamine and asparagine residues. The isoionic point is the pH of a solution of a protein that contains no non-colloidal ions other than hydrogen or hydroxyl. In the absence of other ions, this pH corresponds to there being zero average net charge on the molecule and coincides with the isoelectric point for gelatin. The isoelectric point is the pH of a gelatin solution in which no net migration of the protein is produced by application of an electric field; it is affected by the ionic strength of the solution. The isoionic point is most conveniently

Table 2.2   Molecular weights and relative proportions of molecular species separated by gel electrophoresis of limed ossein gelatin (reproduced with permission from the Pharmaceutical Press)

Molecular species	Molecular weight	Fraction (%)
<A[a]	<86 000	39.8
A[a]	86 000	8.3
α[b]	95 000	14.4
β-peptides[b, c]	145 000	6.8
β	190 000	5.2
γ-peptides[b, d]	237 000	9.3
γ[b]	285 000	1.3
1–6[e]	380 000–855 000	11.7
Q (microgel)[f]	$1.5 \times 10^6 - 1.5 \times 10^7$	3.2

[a] A and <A are molecules of progressively lower weight than α-molecules.

[b] The molecular species α, β and γ are as defined in the text.

[c] β-peptides are molecules intermediate in size between α and β molecules.

[d] γ-peptides are partially hydrolysed γ-molecules.

[e] Molecules designated 1–6 are species identifiable as greater in molecular weight than γ-molecules and are oligomers of the α-chain.

[f] Molecules designated Q (known as microgel) are polymerised species of extremely high molecular weight.

determined by deionising a dilute (2–5%) gelatin solution using a mixed-bed ion-exchange column, whilst the isoelectric point can be determined by observing the pH for which there is zero electrophoretic mobility. The changes in ionisation with pH affect the shape of the gelatin molecules in solution and the interactions between them, and account to some extent for the pH-dependence of the physical properties. The isoionic pH is of practical importance since it coincides with extreme values for many of the solution and gel properties of gelatin. It is generally accepted that the isoionic point of lime-processed gelatin falls within the range 4.8–5.2 whilst acid-processed gelatins exhibit values of 6.0–9.4, although it should be noted that the value does not unequivocally differentiate between acid-extracted and neutral-extracted gelatins. Isoelectric focusing techniques applied to gelatins have shown that a distribution of isoelectric point values occurs, not only in acid-processed gelatins but also in limed gelatins (Maxey and Palmer, 1976; Li-juan *et al.*, 1985; Toda, 1985). For limed gelatins, the range was 0.5–0.7 pH units. A gelatin exhibiting an average isoelectric point of 4.85 showed components ranging from about 4.6 to 5.1. Acid-processed gelatins showed a wider distribution (about 2 pH units), as might be expected, owing to more widely differing degrees of deamidation of the molecules. The distribution is not merely a characteristic of commercial blended gelatins since a pure α-component of gelatin gave a similar distribution to that of limed gelatin. Hydrolysis of limed gelatin by enzymes or alkali results in a reduction of the average isoelectric point to as low as pH 4.6 in the case of alkaline hydrolysis. This is due to the basicity of the α-amino groups liberated on hydrolysis differing from that of the ε-amino groups, and the acidity of the α-carboxyl groups differing from that of the side-chain carboxyls. In addition, conversion of arginine residues to less basic ornithine residues will contribute to the reduction of the isoelectric point value.

Gel strength

As a protein, gelatin has the unique ability to form a thermally reversible gel. The sol/gel and gel/sol transformations occur very readily when the temperature is changed over a comparatively small range. For any given gelatin, gel strength depends upon the gelatin concentration, pH, temperature and maturing time. Not all gelatins form a gel under the conditions of the Bloom test. Cold-water fish gelatins have zero Bloom, even when they have comparable average molecular weights and viscosities to high Bloom mammalian gelatins. This would appear to make them unsuitable for capsule production, although gels are formed at higher concentrations and/or lower temperatures than the conditions of the Bloom test. WO 99/33924 (1999) describes the manufacture of hard capsules from non-gelling fish gelatin using carrageenan as a gelling agent. 'Non-gelling' hydrolysed mammalian gelatins are also available, but these have low average molecular weights.

Commercially, gelatin gel strength is determined by a standard, but arbitrary test, the 'Bloom' gel strength test, which measures the force required to depress the surface of a 6.67%

Table 2.3   Content of acidic and basic groups in gelatin (Kenchington and Ward, 1954)

Group	pH Titration range (40°C)	Concentration (mmol g^{-1} protein)	
		Acid processed	Lime processed
Carboxyl	1.5–6.5	0.85	1.23
Amide	–	0.35	0.03
α-amino + imidazole	6.5–8.0	0.06	0.07
ε-amino	8.0–11.5	0.42	0.42
Guanidino	–	0.49	0.48

w/w gel, matured at 10°C for 16–18 hr in a standard bottle, by a distance of 4 mm using a flat-bottomed plunger 12.7 mm in diameter. The original Bloom gelometer, which was named after its inventor in 1925, uses a force applied in the form of a stream of lead shot and the weight, in grams, is termed the 'Bloom' strength. In the early 1960s a semi-automated version of the Bloom gelometer, the 'Boucher Electronic Jelly Tester', became commercially available and its use was recognised in British Standard 757:1975 'Sampling and Testing of Gelatin'. In this instrument the plunger is suspended from one end of a beam mounted on a torsion wire. Force is applied to the plunger by torque on the wire. The instrument is calibrated to give 'Bloom' values. More recently, further-improved gelometers such as the LFRA Texture Analyser (Stevens & Sons Ltd, London) and the Texture Analyser TA-XT2 (Stable Micro Systems Ltd, London) have been developed and are now widely used.

The procedure used in the USA for the Bloom test (AOAC, 1984) used to differ from the BS method in the shape of the plunger. The BS plunger had its bottom edge rounded to a radius of curvature of 0.44 mm whilst the AOAC plunger had a sharp edge. This affects the Bloom result to the extent that the AOAC Bloom is about 3% higher than the BS Bloom. To overcome this anomaly and to standardise testing in Europe and North America, the GME initiated a switch to using the AOAC plunger from July 1998. BS 757: 1975 is no longer an active Standard, in so much that it is no longer reviewed and updated. GME (2000) has produced an updated Bloom test procedure in its Gelatin Monograph.

Whilst Bloom gel strength is widely accepted as a measure of quality, gelatins are rarely used under conditions which even approximate to those of the test procedure. In hard capsule manufacture, for example, the gelatin gel coating the stainless steel pins is formed in a matter of seconds from an approximately 30% w/w solution on cooling from about 50 to 25°C. In the manufacture of soft capsules by the rotary die process, a concentrated gelatin solution is rapidly gelled on a chilled drum, but in this case the situation is complicated by the inclusion of a plasticizer, normally glycerol.

Nature of the gelatin gel

When an aqueous solution of 'gelling' gelatin above a certain minimum critical concentration is cooled below 40°C, a three-dimensional gel network is formed. Gelation is accompanied by changes in optical rotation (Jones, 1989b). These changes have been interpreted as indicating the partial reordering of gelatin molecules into the collagen helical structure. The transformation can be considered as a three-stage process, the first stage being the intramolecular rearrangement of imino acid-rich chain segments of single-chain gelatin molecules, so that their configurations are similar to those of the same segments in the collagen structure. This is known as the 'collagen fold' and is responsible for the observed increase in optical rotation. The second stage requires the association of separate chains in 'collagen-fold form' at these regions to form a three-dimensional network. These network junction points formed by association of two or three ordered segments (from two or more individual chains, respectively) create the 'crystallites' indicated by X-ray diffraction studies on gelatin gels and are linked by the non-helical regions of the gelatin chains. The third stage involves the stabilisation of this structure by lateral interchain hydrogen bonding within the helical regions. The formation of the gelatin gel can therefore be considered

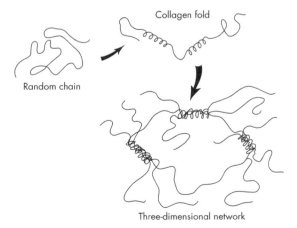

Collagen fold

Random chain

Three-dimensional network

Figure 2.5 Schematic representation of the formation of a three-dimensional network in gelatin gels (reproduced with permission from the Pharmaceutical Press).

as the production of a three-dimensional network of gelatin molecules with water entrapped in the meshes. Development of the equilibrium gel rigidity occurs through the propagation of collagen folding in the junction regions and by the formation of new, but less stable links, resulting in a fine network within the coarse network. This gel structure for gelatin has been dramatically demonstrated by freeze-etching electron microscopy studies. A schematic view of gel formation is shown in Figure 2.5. Not all gelatin molecules contribute to the gel network. Ultracentrifugation studies on gelatin gels have demonstrated the existence of a non-gelling or weakly gelling 'sol' component. This sol component can also be extracted by dialysis of the gel; the proportion of the total gelatin existing as sol decreases with maturing time and decreasing temperature. The sol component, which can represent as much as 40% of the gelatin in a 2% gel at 20°C, depending on the gelatin used, is more polydisperse and has a lower average molecular weight than the gel component. It also exhibits reduced ability for collagen folding. Moreover, differences in amino acid composition and N-terminal amino acids for the sol and gel fractions have been detected. The evidence is that the sol fraction, at temperatures below about 21°C, consists mainly of α_2 chains or their degraded fragments. It has also been observed that the α_2 chains are not found in the sol phase in gelatin containing no molecules smaller than the α chains.

Effect of molecular weight

It was established over 50 years ago that there was a linear relationship between the square root of the rigidity modulus G and the weight average molecular weight (M_w) for a series of low molecular weight (33 000–72 000 Da) gelatins prepared from a high molecular weight lime-processed gelatin by progressive thermal hydrolysis (Jones, 1989b). Other workers have shown that above a certain critical average molecular weight the rigidity modulus was practically independent of molecular weight, provided the temperature of the gel was sufficiently below the melting point (Jones, 1989b). The rigidity plateau differed for gelatins of different origins, but this structural feature (called the 'rigidity factor') could not be related to intrinsic viscosity or to the chain average molecular weight. Another factor which influences the position of the plateau is the degree of racemisation of the amino-acid residues in the protein chain. It is known that racemisation can occur at alkaline pH and this can impair the ability to form the collagen fold. The very high molecular weight components (i.e. greater than that of the α-chain), which of necessity must be multi-chain molecules, exhibit a reduced tendency to form useful gel network links despite their greater tendency to form the collagen fold. It has been suggested that this reflects a tendency for intramolecular aggregation to occur, which would effectively reduce the number of collagen folds available for network formation. As the temperature approaches the melting point of the gel the 'critical molecular weight' increases and the rigidity becomes more dependent upon molecular weight. The rigidity of commercial gelatins possessing a broad distribution of molecular weights depends upon the proportion of molecules with molecular weights below the critical value.

Effect of temperature and time

The strength of a gelatin gel increases rapidly during the first few hours and thereafter more slowly, at any given temperature (Saunders and Ward, 1958). It does not reach a constant value, although it has been shown that an approximately constant rigidity can be achieved within 5 hr at the maturing temperature if the gel is first prechilled and aged at a much lower temperature. This illustrates the fact that rigidity is not simply related to the maturing temperature and time but also depends upon the thermal history. Prematuring the gel at a higher temperature than the final maturing temperature results in a higher gel strength than if the gel has been chilled directly to the maturing temperature. These observations can be explained on the basis that rapid cooling of a gelatin solution results in a less ordered arrangement of gelatin molecules because they have been 'frozen' in the configuration that existed in the solution state. This leads to only limited development of network junctions and is called a 'fine' gel network. In contrast, slow cooling leads to the formation of a 'coarse' network in which more highly organised links have had a chance to form (Stainsby, 1977b).

For a given concentration, the rigidity of a gelatin gel decreases with increasing maturing temperature. An approximately linear relationship between temperature and G/c^2 (where G is the rigidity modulus and c is the concentration) was found for low molecular weight gelatins over the range 0–30°C, but the relationship has not proved to be universal for all gelatins. Different types of gelatin show varying dependencies of rigidity upon temperature. Moreover, for gelatins of similar type (i.e. similar raw material and processing) low Bloom strength gelatins exhibit greater sensitivity to temperature.

Effect of gelatin concentration

The rigidity modulus (G) of a gelatin gel is a function of its concentration (c) and, in general, G is approximately proportional to the square of the concentration (Jones, 1989b). Not surprisingly, in view of the differing contribution to the gel network expected for each of the different molecular species present to varying extents in commercial gelatins, deviations from the square law have been found, particularly for very dilute gels, less than 2%, for very concentrated gels, 20–30%, and for highly degraded gelatins. The deviation can mean that the exponent is greater or less than 2, and it has been concluded that a general relationship is unlikely to apply to all gelatins (Ward and Saunders, 1958). An average value of 1.7 has been found most useful for calculating Bloom gel strength from gel strengths determined at concentrations differing from the standard 6.67%, for 'commercial gelatins' containing moisture and ash (Kramer and Rosenthal, 1965).

Effect of pH

The extent to which the strength of a gelatin gel is affected by pH varies with the type of gelatin (e.g. acid- or alkali-processed) and also with the gelation conditions. Early work on an alkali-processed gelatin at 10% concentration, following a low-temperature maturing period, indicated that rigidity was independent of pH over the range 4.4–9.0. When the concentration was reduced to 1.5%, the pH range for constant rigidity was reduced to 4.4–6.7. Later work on an acid-extracted limed gelatin at a concentration of 2.7% and a constant ionic strength of 0.15, using rather weak gels obtained by maturing at 25°C, showed a greater dependence upon pH. Over a similar pH range, an increase in rigidity was observed as the pH increased to 9, at which point the gel strength was maximal. A more marked dependence of rigidity on pH was, however, detected below about pH 4 and above pH 10. Acid-processed gelatins show a more pronounced reduction in rigidity below pH 4.5 than do limed gelatins. This is a reflection of the greater degree of hydration of the gelatin molecules at pH values well away from the isoelectric point.

Effect of blending

Because of the relationship between G and c^2, the rigidity of a mixture of several gelatins can be calculated from the equation:

$$\sqrt{G_{\text{blend}}} = a_1\sqrt{G_1} + a_2\sqrt{G_2} + a_3\sqrt{G_3} + a_n\sqrt{G_n}$$

where a_1 is the fraction of gelatin of rigidity G_1 in the blend, and so on.

This general relationship is useful for blending gelatins of similar types that are not widely different in gel strength and where the exponent is fairly close to 2. However, synergistic increases in gel strength can be achieved by blending together acid- and lime-processed gelatins, when the pH of the system is such that the gelatins are oppositely charged. This will normally be the case in the region pH 5.0–6.5. This synergism has been demonstrated for 'hard capsule grade' acid ossein and limed ossein gelatins which were blended together in various proportions, the rigidities then being measured over the concentration range 5–50% w/v (Robinson *et al.*, 1975a). Maxima in the rigidity–concentration plots occurred when the blend contained about 40–50% of limed gelatin. For the 50% w/v gels the increase in the rigidity modulus was 15-fold. In contrast, at low concentrations (3.5% or less) a decrease in gel strength occurred owing to coacervate formation.

Effect of additives

The effect of added glycerol on gelatin rigidity has been quite widely studied (e.g. Carless and Nixon, 1970) because of the importance of the gelatin–glycerol system in the manufacture of soft capsules and suppositories. Addition of glycerol significantly increases gelatin gel rigidity: for gels containing 4–15% gelatin and up to 40% w/w of glycerol a general relationship for rigidity modulus has been derived (Nixon et al., 1966):

$$G = a + bZ^2 + (c + dZ)g$$

where Z is percentage gelatin concentration, g is percentage glycerol concentration and a, b, c and d are constants whose values depend upon the grade of gelatin. Other polyhydric alcohols such as sorbitol, and sugars, such as sucrose (Jones, 1977) and maltodextrins (Jones, 1977) can increase gel strength. However, at high additive levels and high total solids the rigidity can start to fall again. Also, with maltodextrins of increasing molecular weight or polyethylene glycols of increasing degrees of polymerisation, the 'critical concentration' of the additive decreases with increasing molecular weight (Holmes et al., 1986). Ethanol produces an increase in rigidity, provided that the concentration is low enough to avoid precipitation of the gelatin from solution. Formaldehyde cross-linking also increases the rigidity. Reduction in rigidity can be effected by the addition of electrolytes. The addition of sodium chloride at a concentration of up to 17.5% in a dilute gel produced a decrease in rigidity, which was linear with the square root of the ionic strength (I). The following equation was obeyed:

$$G = G_0(1 - 0.77I^{1/2})$$

where, G_0 is the rigidity of the gel without additives. Compounds that destroy hydrogen bonds, such as urea, phenol, ammonium nitrate, lithium bromide and potassium isothiocyanate at high concentration, can completely suppress gelation.

Viscosity

Commercially, gelatin viscosity is routinely measured at 6.67% w/w concentration at 60°C using either an Ostwald or a pipette viscometer. For the gelatin used in hard capsule manufacture, measurements are frequently also made at 12.5% w/w. A relationship between viscosities at these two concentrations has been established (Stainsby, 1958, 1977a). Over a wider range of concentration, i.e. 2–60%, a slight S-shaped deviation from the linear relationship between viscosity and concentration, $\log \eta = kc$, has been observed in the temperature range 50–90°C (Finch and Jobling, 1977). Over the restricted concentration range of 10–50% the logarithmic relationship is quite adequate. Above the setting point, viscosity decreases exponentially with increasing temperature (Croome, 1953; Finch and Jobling, 1977).

The effect of pH in concentrated solutions is similar to that for dilute solutions, with the viscosity being lowest at the iso-ionic point and highest at pH 3 and 10.5, respectively. However, as the gelatin concentration increases, the effect of pH becomes less significant, particularly in acid solutions (Stainsby, 1977a). This is due to the inevitable increase in ionic strength with increasing gelatin concentration, owing to the presence of the counterions associated with the gelatin. For any fixed gelatin concentration in the approximate range 6.67–50% and any fixed temperature in the range 5–90°C, the viscosity of a blend of several gelatins can be calculated from the relationship:

$$\log \eta_{blend} = a_1 \log \eta_1 + a_2 \log \eta_2 + a_n \log \eta_n$$

where a_1 is the weight fraction of gelatin in the blend of viscosity η_1 when tested at the same fixed concentration and temperature. This relationship does not apply to blends of acid and limed gelatins, since synergistic increases in viscosity, analogous to the gel strength increases referred to earlier, may occur. Blends of hard capsule grade limed ossein and acid pigskin gelatins in varying proportions exhibit marked viscosity increases. Acid ossein gelatin produces less marked synergistic viscosity increases with limed ossein gelatin (although in all cases the viscosity is higher than the calculated value) and the viscosity–blend curve is much smoother than the corresponding curve for limed ossein/acid pigskin blends. Acid ossein gelatin also produces a smooth curve for blending with either pigskin gelatin alone or limed ossein/acid pigskin mixtures. At temperatures above the setting point, but below about 43°C, depending upon gelatin concentration and viscosity, the viscosity can increase with time and become non-Newtonian in character. This is due to the association of the gelatin molecules to form aggregates. It has also been shown that, for a 50:50 blend of hard capsule grade acid ossein gelatin and limed ossein gelatin in 30% aqueous solution at 35°C, the rate of development of increased viscosity is very much greater than for either of the component gelatins alone (Robinson et al., 1975a).

Degradation of gelatin

Gelatin in solution is susceptible to thermal or enzymatic hydrolysis, leading to a reduction in average molecular weight and a change in many of the useful properties (Jones, 1989b). The extent of thermal hydrolysis is a function of temperature, time and pH, and is minimal at near-neutral pH. Gel strength, viscosity and chain average molecular weight have each been used to monitor hydrolysis; acid-processed gelatins are more resistant to acid and more susceptible to alkaline hydrolysis than lime-processed gelatins. This difference is more obvious in the effects on gel strength than on viscosity. Studies of the terminal amino residues created during hydrolysis show that the pH dependence of peptide bond scission varies for different N-linked amino acid residues. Peptide links involving the amino groups of serine and threonine are particularly susceptible to both acid and alkaline hydrolysis, whereas those involving glutamic acid peptides are very stable. Aspartic acid peptides are particularly susceptible to acid hydrolysis. Thermal degradation of high-grade gelatin results in a more rapid reduction of viscosity than of Bloom gel strength, particularly in acid. A general formula for calculating the loss in Bloom strength on exposing gelatin solutions to various pH values in the range 3–9 and temperatures up to 100°C has been reported for acid pigskin (Tiemstra, 1968), but it is not directly applicable to limed gelatins on acid hydrolysis.

One practical implication of the susceptibility of gelatin to thermal degradation is that gelatin solutions cannot be autoclaved as a means of sterilisation without suffering deterioration in physical properties. Dissolution temperatures of 55–70°C can safely be used for gelatins that have pH values in the normal 'commercial range' of 5.0–6.5, and solutions can be maintained at 45–60°C for several hours without significant change in physical characteristics, provided that the pH is maintained in this range. At temperatures of about 40°C, the rate of thermal hydrolysis is negligible, but the rate of bacterial or enzymatic degradation increases whenever contamination by viable bacteria or enzymes exists. Gelatin in solution has a random configuration; consequently gelatin peptide segments can conform to the configurations of the active centres of enzymes with the result that gelatin is susceptible to a range of proteolytic enzymes. Gelatin hydrolysis using papain, pepsin, chymotrypsin, trypsin and commercial bacterial enzymes has been studied (Courts, 1955; Jacobson, 1976).

It is usual for specifications for hard capsule gelatins to include a test for 'degradation rate', since the stability of solution viscosity at 45–60°C over a number of hours is an important characteristic. This is commonly, but arbitrarily, carried out by measuring the percentage drop in viscosity of a 12.5% w/w solution at 60°C over 17 hr. Under these conditions, a viscosity drop of more than 20% suggests bacterial or extracellular enzymatic action. It is possible to distinguish between the action of enzymes and bacteria by measuring viscosity degradation rates at 40°C in the absence and presence of preservatives. The initial rate of thermal hydrolysis is greater for higher viscosity gelatins.

Setting point and setting time of gelatin gels

The gelation temperature, or more particularly the time for the onset of gelation, is of practical importance in many applications, not least in hard capsule manufacture. Despite its practical interest, there is no universally accepted or adopted procedure for measuring setting time, as there is for viscosity and gel strength. Various methods have been described (Janus et al., 1965; Marrs and Wood, 1972; British Standard 757,

1975; Coopes, 1975; Wainewright, 1977; Itoh, 1985) which basically attempt to determine either the point at which the viscosity of the solution becomes very high, or the development of a particular degree of rigidity after the setting point has been passed.

Procedures range from simple techniques, such as that of British Standard 757 (1975), which involves stirring a 10% gelatin solution with a thermometer while it is being cooled from 45 to 20°C, and noting the point at which the solution no longer drips from the end of the thermometer when it is removed, to rather more sophisticated techniques such as that described by Marrs and Wood (1972). In this method, drops of carbon tetrachloride are released into a vertical column of gelatin solution, at a constant drop rate, and are used to detect the setting point as the system is cooled from $40 \pm 0.10°C$ to $23 \pm 0.02°C$. The viscosity increases rapidly near the point of gelation, when the drops become almost stationary in the solution. This method has been used to show that setting time is related to the content of gelatin molecules with a very high molecular weight, i.e. greater than 500 000 Da (Bartley and Marrs, 1974).

Similar conclusions followed from setting-rate studies using an oscillating hollow cup viscometer (Bohonek and Spühler, 1976). It was also observed that ossein gelatins (whether limed or acid processed) gelled more rapidly than skin gelatins (whether acid pigskin or limed) even though they had comparable contents of very high molecular weight components (Bartley and Marrs, 1974). Although there is no published work on the relationship between setting-rate characteristics and capsule manufacture, one method which would appear to be useful for studies of both hard and soft capsule gelatins is that described by Janus et al. (1965), because it enables short gelation times of concentrated gelatin solutions to be measured. Uniform drops of gelatin solution are placed at short intervals on the polished surface of a level, hollow bar through which water at controlled temperature is circulated. Gelation time is determined by rotating the bar through 90°C and observing the position of the drop that just fails to run. The bar is enclosed in a high humidity environment to prevent skin formation but no doubt the equip-

ment could be adapted to lower fixed humidities comparable with those existing in hard capsule manufacture. Janus used this method to establish that a linear relationship exists between the logarithm of setting rate at 20°C and the logarithm of gelatin concentration, up to 10% concentration. Setting rate is affected by pH; minima in the pH–setting time curves have been reported at pH 4.5–5.5 for lime-processed gelatins and at pH 6 for acid-processed gelatins (Hopp, 1964). Similar results have been reported in a separate study but minima were also found at about pH 9 for both acid pigskin and limed hide gelatin (Marrs and Wood, 1972). Below about pH 4.5 and above about pH 9.5, the setting time increased rapidly. Itoh (1985) found a sharp minimum setting time at the iso-ionic point for limed gelatin, but a broader minimum in the region of the iso-ionic point for acid pigskin gelatin. Extremes of pH resulted in marked increases in setting time, as found by other workers.

Neutral salts such as sodium chloride and potassium chloride have only a minor effect on setting rate. In the pH range 4.5–10.8 there is a slight reduction and outside this range a slight increase in setting rate (Marrs and Wood, 1972). Increasing the gel strength of the gelatin reduces setting time for both acid and limed gelatins (Itoh, 1985). Warm-water fish gelatins show lower setting points and setting rates than mammalian gelatins of similar Bloom and viscosity values. Cold-water fish gelatins may set or not, depending upon the test conditions. Using the setting point test of the Photographic and Gelatin Industries, Japan (PAGI, 1992), which is a variation of the method of British Standard 757 (1975), using 10% gelatin solutions, setting points for hard capsule grade gelatins have been found to be of the order 26°C for mammalian gelatins and 18–20°C for warm-water fish gelatins. Cold water fish gelatins of similar viscosity show setting points in the region of 8–10°C.

Melting point of gelatin gels

The ability of gelatin gels to melt at body temperature has obvious significance for hard and soft capsules. The actual melting point depends on many factors, including concentration, molecular weight, pH, thermal history and additives. As a gelatin gel is warmed, the rigidity decreases and then disappears relatively rapidly, but the melting point and setting point do not coincide, even when the heating and cooling rates are very slow. Indeed, holding gels at temperatures near the melting point can result in higher melting temperatures than for comparable gels heated at a constant rate (Stainsby, 1977b). For example, the melting point of a 6% gel formed at 0°C rose from 31.5 to 34.5°C during storage for one day at 25°C. This demonstrates that the gel network is continually being reorganised to give linkages of greater thermal stability. Increasing the gelatin concentration in the range 1–5% increases the melting point for gelatin gels matured at 0°C, but for high-grade gelatins the increase is quite small once the concentration exceeds 5%. A higher maturing temperature of 25°C, however, gave a maximum melting point at pH 5 for limed gelatin (Jones, 1989b). Although the general term melting point has not been precisely defined for gelatin, it will be apparent that this property, in common with setting point, is influenced by the test procedure and no universally adopted method exists. However, BS 757 (1975) describes a simple procedure which is capable of giving results to within 0.25°C. In practice, it is often of more interest to know the melting point of gelatin films in water or aqueous solution and, in this situation, the so-called melting point is the temperature of dissolution of the gel in water. A useful method has been described (Tabor, 1968) in which the melting point is determined by observing the temperature at which the swollen gelatin film ruptures under the weight of a standard stainless steel ball; this method has been used to study the effect of cross-linking agents.

Properties of gelatin films

Effect of drying conditions

Films formed by drying gelatin from the sol state possess different physical properties from those dried at low temperatures from the gel state. X-ray diffraction studies (Bradbury and Martin, 1952) have shown that hot-dried films give only a broad diffuse diffraction pattern typical of liquids or glassy materials, indicating a relatively simple random coil or amorphous structure. In contrast, cold-dried films give an X-ray pattern with a series of diffracted areas and spots, indicating a high degree of ordering, with triple-helix crystallites orientated in a direction parallel to the plane of the film. It has been estimated that up to 20% of the gelatin molecules in cold-dried films are in the triple-helix crystalline form (Jolley, 1970). The degree of orientation in the plane of the films is reduced as the drying temperature approaches the gel melting point and, as might be expected, thin (about 16 μm dry film thickness) gelatin films which have been rapidly dried have been shown, by optical rotation measurements, to contain more randomly orientated helices than thicker films dried more slowly (Coopes, 1976a, 1976b). The difference in structure between hot- and cold-dried gelatin films is reflected in their mechanical properties and their solubility characteristics in cold water (Kozlov and Burdygina, 1983). Cold-dried films have a greater tensile strength than hot-dried films, when measured at relative humidities (RHs) of 45–85%, and a greater elongation at break at RHs up to 65%. However, at RHs of 75% and 85% the situation is reversed and hot-dried films exhibit marked extensibility (Bradbury and Martin, 1952). Cold-dried films swell without dissolving in cold water, whereas hot-dried films can dissolve spontaneously, at least at their surface. Again, this is consistent with the picture of individual molecules of amorphous gelatin being free to dissolve whilst the ordered structure of the molecules of the crystalline film prevents dissolution. The hot-dried film may not dissolve completely, particularly if it is not very thin, because as the water diffuses into the film the amorphous structure may be converted into a crystalline gel network more rapidly than it can dissolve.

Properties of cold-dried films

The properties of cold-dried gelatin films are of obvious importance in the production of hard capsules. They are intimately related to moisture content and may be affected by pH, additives and RH.

Effect of moisture content

Gelatin of very low (less than 5%) moisture content is too brittle for almost all practical purposes. A higher content of moisture, which acts as a plasticizer, is necessary in hard capsules to avoid problems at the trimming stage during manufacture and in subsequent handling and filling operations. Equilibrium moisture sorption–relative humidity isotherms for gelatin have been obtained by several workers (Sheppard *et al.*, 1940; Bell *et al.*, 1973). In the earlier study, identical isotherms were reported for acid- and lime-processed gelatins below 80–90% RH. A hysteresis effect was apparent whereby the moisture content was influenced by the previous moisture history, higher moisture content being achieved on desorption than on absorption. Equilibrium moisture content decreased with increasing gel concentration in the range 2–15% and was also slightly affected by pH; at 47% RH the moisture content decreased progressively on reducing the pH below 4.9. Hydrolysis of the original gelatin, which reduced its viscosity by a factor of three, had little effect on the moisture content; neither did varying the film thickness. The results of later work, although of a similar order, are generally not in close agreement. For example, literature data for the equilibrium moisture content of gelatin at 44% RH and 25°C are compared in Table 2.4.

If these apparent differences in equilibrium moisture content for various gelatins were real, then they would be expected to be of significance for hard capsules. The work of Melia (1983) indicates that there are only small but, nevertheless, statistically significant differences in the equilibrium moisture contents of gelatins. Whilst uncertainty may exist over the equilibrium moisture values for gelatin, there is evidence that the rate at which gelatins dry also varies with gelatin type. It has been reported that acid pigskin gelatin dries more rapidly than lime-processed gelatins and that this characteristic is important in avoiding brittleness in capsule gelatin (Knox, 1960). Certainly this observation is in agreement with practical experience from hard capsule manufacture, where brittle capsules are produced when the gelatin blend contains in excess of a critical percentage (normally 40–60%) of pigskin gelatin. In contrast, it is generally found that inclusion of acid ossein gelatin into the blend reduces brittleness. However, a detailed study (Melia, 1983) of the drying rate of films cast from acid ossein,

Table 2.4 Equilibrium moisture content of gelatin film at 44% relative humidity and 25°C (reproduced with permission from the Pharmaceutical Press)

Authors	Moisture content (%)	
	Sorption	Desorption
Sheppard *et al.* (1940)	16.5	18.5
Mason and Silcox (1943)	17.8	–
Calhoun and Leister (1959)	–	11.5
Ito *et al.* (1969)	13.5	17.0
Bell *et al.* (1973)	13.4	15.0
Melia (1983)	–	13.46
		(Limed ossein gelatin 1)
	–	13.27
		(Limed ossein gelatin 2)
	–	13.24
		(Acid ossein gelatin)
	–	13.09
		(Acid pigskin gelatin)
Eith *et al.* (1986)	16.0 (20°C)	–
Marshall and Petrie (1979)	11.3 (21°C)	–

limed ossein and pigskin gelatins, over the range 12–85% RH and 20–35°C, failed to demonstrate significant differences between the gelatins. In this work, 25% gelatin solutions were cast as thin gelled films (typically 0.40 ± 0.01 mm thickness) which were then dried at a controlled RH and temperature and fixed air flow, whilst freely suspended from a microbalance to enable moisture loss to be continuously monitored. At 25°C and 44% RH, the drying rate was independent of film thickness over the range 0.24–0.56 mm, indicating that, under these conditions, diffusion of moisture through the film was not a rate-determining step. It is possible that, under different conditions, intrinsic differences in drying characteristics of gelatin may be apparent, for example, where drying occurs from only one side of the film and shrinkage of the film is restricted by the 'backing' and higher airflow rates are experienced. Since the time of this study, commercial instruments capable of measuring dynamic vapour sorption/desorption under controlled RHs have become available which make it much easier to measure isotherms and also the rate of approach to equilibrium.

The tensile strength, hardness and Young's modulus of thin gelatin films, measured as a function of moisture content, indicate a maximum tensile strength at a moisture content of about 10%, decreasing to 90% of the maximum for moisture levels between 10% and 20% (Healey *et al.*, 1974). Other workers (Kellaway *et al.*, 1978a) have compared the mechanical properties of hard capsule grade gelatin films prepared from limed ossein and acid ossein gelatins and 1:1 blends of the two. Little difference in Young's modulus was found, at low stresses, between the acid and limed gelatins, which showed a maximum at 14–16% moisture (non-equilibrium). In contrast, the blend of the two gelatin types showed an increased modulus over the moisture range 20–50%, with a maximum at 22% moisture, although at lower moisture contents the difference was no longer apparent. At higher stresses, when the gelatin films had been allowed to attain equilibrium moisture values in the range 10–20%, the gelatin blend behaved in a more closely similar manner to the individual types, with values of Young's modulus decreasing in the order 1:1 blend, limed and acid gelatin. Below a critical equilibrium

moisture content (13–16%), all three gelatin films exhibited stress relaxation at low strains but, at higher equilibrium moisture values, the stress increased with time owing to contraction of the film. In contrast, more-rapidly dried, non-equilibrium moisture content films did not shrink. This phenomenon was presumed to be related to the orientation of crystallites in the films.

Effect of pH

The effect of pH on the mechanical properties of low moisture content films appears not to have been studied, although the rupture properties of 10% gel strips, matured and tested at 10°C, have been investigated (Marrs and Wood, 1972). Acid and limed ossein gelatins behaved similarly, exhibiting essentially constant values for rupture load and extension at break, over the pH range 4–11. For limed hide gelatins, there was a minimum in the rupture load (or extensibility)–pH curve near pH 7, with maxima at around pH 5 and pH 9.5, which were very much higher than for ossein gelatins.

Effect of additives

Tensile strength measurements on deionised gelatin films at low RH suggest that deionisation increases tensile strength (Norris and McGraw, 1964). Thus one would expect additives to reduce the strength. The addition of calcium chloride to limed hide gelatin reduced the rupture load over the pH range 4–11 for 10% gels (Marrs and Wood, 1972). The effect of the addition of dyes (erythrosine, amaranth, brilliant blue FCF and indigo carmine) and of titanium dioxide on the mechanical properties of thin gelatin films equilibrated to 16% moisture was studied by Robinson *et al.* (1975b). There was little or no effect on Young's modulus and tensile strength, but the dyes all influenced the stress relaxation characteristics of the gelatin. In some cases a low concentration of dye (0.1% w/v) was more effective in reducing the percentage increase in stress (i.e. the degree of contraction) of the films than a higher concentration (1.0% w/v), but amaranth had the greatest effect at the 1.0% level. Similar results were found for both limed and acid ossein gelatins.

The most important additive for soft capsules is glycerol, which exhibits strong molecular inter-actions with gelatin. When added as a plasticizer at low levels (i.e. up to 10% w/w on gelatin) it reduces the affinity of gelatin for moisture (Pouradier and Hodot, 1972). The amount of water sorbed by a mixture of gelatin and glycerol is less than that sorbed at the same RH by the two materials separately. This effect has been attributed to the partial or total blocking of the hydrophilic groups of gelatin by glycerol mol-ecules. The inclusion of glycerol in hot-dried film will transform the gelatin from a comparatively brittle, stiff material to a highly extensible rubber. Measurements of tensile relaxation modulus show that devitrification occurs at about 30% w/w glycerol, while the modulus remains almost inde-pendent of glycerol content in the ranges 0–30% w/w and 60–75% w/w. At 85–95% w/w glycerol, relatively weak gels, which tear easily, are pro-duced (Yannas, 1972).

The viscoelastic properties of thin films formed from gelatin–glycerol–water mixtures typical of those used in soft capsule manufacture have been studied. Stress relaxation measurements showed a decrease in elasticity with increasing ageing time up to 48 hr, followed by a progressive increase. These changes appeared to reflect an initial change to a more amorphous structure followed by a gradual restoration of crystallinity. It has been reported that gelatin films should have a minimum moisture content of about 12% if the inclusion of glycerol (or other low molecular weight plasticizer) is to have a plasticizing action as indicated by measurement of resistance to impact (Kozlov and Burdygina, 1983). However, even at lower moisture levels, glycerol appreciably reduces internal stress when subjected to iso-metric heating.

Effect of exposure to high RH

The water solubility of gelatin capsules and gelatin-bound or -coated tablets may progres-sively change with time in tropical climates where conditions of high temperature and, more particularly, high RH, prevail. It has also been

Table 2.5 Melting points of isoelectric gelatin films after conditioning at 50°C for eight days (Jopling, 1956); reproduced with permission from the Society of Chemical Industry

Gelatin type	% RH	Melting point[a]
Lime-processed ossein	25	31.5
	60	31
	85	>97
	95	86
50% Lime-processed ossein	10	33
	68	34.5
50% Lime processed hide	85	>97
	96	86.5
Acid-processed ossein	25	31
	60	31
	85	93
	95	53.5

[a] 'Melting point' (more precisely the dissolution temperature) of a gelatin film determined by placing the film in a water bath at 20°C, the temperature then being raised at a rate of 1.7°C per minute until the gelatin dissolved. The initial melting point of all the samples was 32°C.

reported that gelatin can become insoluble under comparatively mild conditions (Pankhurst, 1947; Jopling, 1956). Gelatin heated at 45°C and high RH for 5 days lost some of its ability to swell in water and had an increased melting point. At RHs approaching 100%, hydrothermal contraction, similar to that of collagen, occurred (Pankhurst, 1947). Similar results have been reported for films conditioned at 85–95% RH at 50°C. Even at 20°C some reduction in water absorption was observed, but not an increase in melting point. The effect of exposing gelatin film to various RHs at 50°C is illustrated in Table 2.5.

The change in gelatin film properties at high RH has been explained by assuming that suf-ficient water is absorbed by the film to enable weaker bonds involved in gelation to reform to give crystallites of greater strength and stability. The fact that the maximum increase occurs at 85% RH rather than at 100% is due to excessive moisture absorption at the higher level which increases the separation between molecules and makes reformation of bonds more difficult. Hydrothermal contraction of gelatin film, 0.6 mm in thickness, exposed to 80% RH at 21°C represented as much as 2.4% of the film length (Calhoun and Leister, 1959).

Chemical interactions in gelatin solutions

The gelatin molecule possesses a number of side-chain groups capable of reacting chemically with a variety of modifying reagents, producing derivatives with their own particular chemical and physical properties. However, apart from these 'true derivatives' which involve the formation of covalent bonds, gelatin is ampholytic and is capable of ionic interaction with oppositely charged molecules. In addition to these reactions of the gelatin molecule itself, chemical impurities naturally present in the gelatin may be responsible for chemical changes observed in gelatin-containing pharmaceutical systems. As an example, soft capsule walls formed from gelatin with a high iron content may become blackened in capsules containing ascorbic acid. In other cases, the mechanism for the observed chemical change may be unknown and therefore inseparable from the gelatin. Since the impurities present in a gelatin will be a function of its manufacturing process, it is always possible that certain gelatin types from particular manufacturers may exhibit their own peculiar characteristics. Some of the chemical interactions of gelatins which are of significance in pharmacy and, more particularly, to capsules are reviewed in this section.

Reactivity of the gelatin molecule

The reactive side-chain groups of gelatin consist of amino, carboxyl and hydroxyl groups, of which the amino groups are the most amenable to chemical modification and the hydroxyl groups least so. The amino groups arise from lysine and hydroxylysine residues, together with the imidazolium group from histidine and the guanidinium group from arginine. There is only about one α-amino group per 25 ε-amino groups in high-grade gelatin. Reaction generally occurs only with uncharged amino groups, which

means that the highly basic guanidino group is unreactive unless a very alkaline pH is employed. The imidazolium group is also less reactive than the α- and ε-amino groups and there is evidence that its bonds, once formed, are also less stable. A wide range of modifying reagents for gelatin amino groups has been investigated and reviewed (Clark and Courts, 1977) of which acid chlorides and anhydrides, epoxides and aldehydes are typical examples. Reagents that are capable of reacting with two amino groups can cross-link the gelatin to give increased solution viscosity and increased gel melting temperature, even to a point where the gelatin eventually becomes insoluble. Succinylated gelatins have found limited use for soft capsule manufacture where the capsule fill contains components (e.g. aromatic oils) capable of cross-linking gelatin amino groups. By blocking the amino groups, the development of brittle capsule walls is avoided. Succinylated gelatins suffer the disadvantage of exhibiting a faster hydrolysis rate than unmodified gelatin. In contrast, acetylated gelatin shows a hydrolysis rate more comparable with gelatin.

Formaldehyde has been used to cross-link gelatin for the manufacture of enteric hard and soft capsules. It has the disadvantages of a marked pH-dependence of the reaction and difficulty in controlling its rate and extent because it continues in the dry state for a relatively long period. The reaction mechanism is believed to involve the initial formation of hydroxymethylamino groups on lysine and arginine residues, the latter being the rate-determining step. The hydroxymethyl lysine eliminates water to give a cationic imine that can react with hydroxymethyl arginine to form a dimethylene ether bridge. This subsequently reorganises to form a methylene link between amino groups of the original lysine and arginine residues. 'After-hardening' in the dry state appears to be due to slow release of formaldehyde by depolymerisation of polyoxymethylene, which is present in aqueous formaldehyde solution (Davis and Tabor, 1963; Taylor et al., 1978). Formaldehyde also reduces helix formation (Moll et al., 1976).

Ionic interactions of gelatin

Interactions with dyes

The interaction of eight F D & C dyes with acid- and lime-processed gelatins has been examined by measuring changes in the visible spectrum. In solution, all dyes were found to interact with acid-processed gelatin, primarily by electrostatic bonding, although erythrosine, which showed the greatest interaction, also showed hydrogen bonding, hydrophobic bonding and even some evidence of irreversible interaction (Cooper *et al.*, 1973). Lime-processed gelatin had no effect on the dye spectra until the very acid conditions of pH 1, when the gelatin possessed a sufficiently high net positive charge. In the dry film state, interaction was examined in the visible and infrared regions by attenuated total reflectance spectroscopy and was apparent for all dyes and for both gelatins. Erythrosine and Wool Violet 5BN showed the greatest effects. Dyes with the greatest changes in spectral activity had the slowest release rates from gelatin films during dissolution in simulated gastric or intestinal fluid. In a further study (Kellaway *et al.*, 1978b), the behaviour of erythrosine with acid- and lime-processed gelatins in solution, its effect on absorbance and the wavelength for maximum absorbance were examined. By fractionating the gelatins by gel chromatography, a linear relationship was found between gelatin molecular weight and the quantity of bound dye, although the slopes of the plots were naturally very different for acid and limed gelatins.

Lime-processed gelatin has been reported to increase the rate of fading of indigo carmine at 60°C, although there could have been a contribution by inorganic impurities present in the gelatin (Khalil and El-Gamal, 1978). At a concentration of 0.02% w/w, indigo carmine, tartrazine and amaranth reduce gel structure formation in 3% and 6% gelatin gels (Artemova and Usova, 1964). The compatibility of amaranth, sunset yellow FCF and indigo carmine with gelatin is influenced by the sulphur dioxide content of the gelatin, which should contain less than about 60 ppm when these dyes are used (Torrado Valeiras, 1966).

Gas generation in gelatin solutions

A defect that can occur in hard capsules is the appearance in the capsule walls of very fine bubbles. This problem can often be traced back to the presence of gas bubbles in the gelatin solution circulating to the 'dip tank', at the stage in production where the capsules are formed by dipping stainless steel pins into the gelatin liquor. Gas bubbles can also develop after the gelatin solution has been bubble-free for several hours on the machines. The problem has been found to be specific to particular gelatin batches but not to particular gelatin types, and is frequently aggravated by the presence of certain dyes (e.g. amaranth and erythrosine). Gas generation was formerly prevented by relatively large additions of sulphur dioxide, e.g. 500–1000 ppm, when its use as a preservative was permitted. Various theories have been proposed to explain gas generation, including the presence of hydrogen peroxide. Nitrate and nitrite ions present in trace amounts in the gelatin, or derived from the water supply, have been correlated with the incidence of bubbles, particularly when nitrate-reducing thermophilic organisms such as *B. stearothermophilus* or *E. coli* are present in the system. Thermophilic organisms are dependent upon the presence of certain inorganic nutrients, namely calcium and magnesium ions, for their growth. By depriving them of these nutrients the mechanism for bubble formation can be prevented. The addition of citrate ions to gelatin, at levels of 0.1–0.5%, has been found to reduce or prevent bubble problems.

Tests and specifications for gelatin

Standard methods for sampling and testing gelatins for chemical and physical properties are given in British Standard 757 (1975) and methods for the microbiological testing of gelatin appear in British Standard 5349 (1976–7). Updated test methods for physical, chemical and microbiological testing of edible gelatin have been produced by GME (2000), which cover all items in the detailed specification of Commission Decision 1999/724/EC. This specification is, in many ways, more detailed than the gelatin monographs of pharmacopoeias. For example, limits are set for the following trace metals: arsenic, lead, cadmium, mercury, chromium, copper and zinc. However, the need for many of these tests is questionable. The upper limits set reflect safe limits based on assumed maximum daily intake, but are generally far in excess of actual levels that occur in commercial gelatins. This can be illustrated for copper (limit 30 mg kg^{-1}) and zinc (limit 50 mg kg^{-1}), for which actual values are typically <1 mg kg^{-1} in each case.

The gelatin monographs of various pharmacopoeias differ in certain respects. Table 2.6

summarises the gelatin specifications of the European Pharmacopoeia (PhEur), United States Pharmacopoeia (USP/NF) and Japanese Pharmacopoeia (JP). In practice, pharmacopoeial specifications generally represent the minimal requirements; capsule manufacturers' specifications are more detailed and stringent and may include performance tests designed to assess gelatins under conditions that more closely approximate manufacturing conditions. Thus soft capsule gelatins may be evaluated for clarity as concentrated solutions in glycerol/water and hard capsule gelatins may be tested for viscosity at 30% concentration at 50°C. Stringent microbiological standards frequently relate to the

Table 2.6 Pharmacopoeial standards for gelatin

Test	PhEur (2002)	USP 26/NF 21	JP XIV
Identification			
1 Biuret reaction	Positive	–	–
2 Gel test on 5% solution	Positive	–	–
3 Reaction with potassium dichromate	–	Positive	–
4 Reaction with tannic acid	–	Positive	Positive
5 Precipitate with chromium (VI) oxide or 2,4,6-trinitrophenol	–	–	Positive
Clarity, colour of solution	To meet standards	Only slightly opalescent as 2.5% solution	Only slightly opalescent as 2.5% solution colour to meet standard
Odour, 2.5% solution	–	Not disagreeable	Not disagreeable
pH, 1% solution	3.8–7.6	–	–
Gel strength (Bloom)	150–250[a]	–	–
Moisture (% max.)	15	–	15
Ash (% max.)	2	2	2
Sulphur dioxide (ppm max.)	200	40[b]	61.7[c]
Heavy metals (ppm max.)	50	50	50
Arsenic (ppm max.)	1	0.8	1
Hydrogen peroxide (ppm max.)	100	–	–
Test for phenolic preservatives	Negative	–	–
Mercury (ppm max.)	–	–	0.1
Microbiological standards			
Total count (organisms/g max.)	1000	1000	–
E. coli	Absent in 1 g	Absent in 10 g	–
Salmonellae	Absent in 10 g	Absent in 10 g	–

[a] Gelatin intended for use in the preparation of pessaries, suppositories or zinc gelatin.
[b] Maximum 1500 ppm permitted for gelatin used in capsules or for tablet coating.
[c] Maximum 1028 ppm permitted for capsules or tablets.

Table 2.7 Typical specification for hard capsule gelatins

	Limed ossein/hide	Acid ossein	Acid pigskin	Fish
Bloom (g)[a]	200–220	245–270	240–255	220–260
	240–265		260–285	
Viscosity, 6.67%, 60°C (mPa s)	4.2–4.8	3.3–3.7	4.2–4.8	3.5–4.5
Viscosity, 12.5%, 60°C (mPa s)	19.0–20.5	12.5–14.5	19.0–20.5	13.0–20.0
% Viscosity drop max. (18 hr/60°C)	20	20	20	20
Moisture (% max.)	13.0	13.0	13.0	13.0
Ash (% max.)	1.0	1.0	1.0	1.0
pH, 1% solution	5.5–6.0	5.5–6.0	5.2–5.8	5.5–6.0
Iso-ionic point, pH	4.7–5.3	6.0–8.0	7.0–9.4	7.0–9.4
Particle size				
% Passing 4 US mesh	100	100	100	100
% Passing 40 US mesh, max.	5	5	5	5
Sulphur dioxide (ppm max.)	40	40	40	40
Hydrogen peroxide	Absent	Absent	Absent	Absent
Nitrate (ppm max.)	300	300	300	300
Nitrite (ppm max.)	30	30	30	30
Microbiological standards				
Total count (organisms/g max.)	500	500	500	500
Salmonellae in 25 g	Absent	Absent	Absent	Absent
E coli in 10 g	Absent	Absent	Absent	Absent
Pseudomonas aeruginosa in 10 g	Absent	Absent	Absent	Absent
Staphylococcus aureus in 10 g	Absent	Absent	Absent	Absent
Thermophiles	Absent	Absent	Absent	Absent

[a] corrected to 11.5% moisture.

Table 2.8 Typical specification for soft capsule gelatins

	Limed ossein/hide	Acid ossein	Fish
Bloom (g)	150–175	180–210	160–210
Viscosity, 6.67%, 60°C (mPa s)	3.6–4.0	2.7–3.2	3.0–3.5
Moisture (% max.)	13.0	13.0	13.0
Ash (% max.)	1.0	1.0	1.0
pH	5.0–6.0	5.0–6.0	5.0–6.0
Iso-ionic point, pH	4.7–5.3	6.0–8.0	7.0–9.4
Particle size			
% Passing 10 US mesh	100	100	100
% Passing 60 US mesh, max.	5	5	5
Sulphur dioxide (ppm max.)	60	60	60
Hydrogen peroxide (ppm max.)	60	60	60
Iron (ppm max.)	30	30	30

Continued

Table 2.8   *Continued*

	Limed ossein/hide	Acid ossein	Fish
Clarity (45% solution in 30% glycerol)	Clear, no precipitate	Clear, no precipitate	Clear, no precipitate
Microbiological standards			
Total count (organisms/g max.)	500	500	500
Salmonellae in 25 g	Absent	Absent	Absent
E. coli in 10 g	Absent	Absent	Absent
Pseudomonas aeruginosa in 10 g	Absent	Absent	Absent
Staphylococcus aureus in 10 g	Absent	Absent	Absent

requirements of the final customer for the capsules. Because physical properties, such as Bloom strength and viscosity, vary with the moisture content of the dry gelatin, it is common practice to correct these values to a standard moisture level of 11.5%. For hard capsules a correction of ± 5 g Bloom or ± 0.5 mPa s viscosity (at 12.5% concentration and 60°C) for each 1.0% moisture difference from 11.5% is taken as the 'rule of thumb'. Typical specifications, in addition to the standards demanded by the relevant pharmacopoeia, for gelatins used for manufacture of hard and soft capsules respectively are given in Tables 2.7 and 2.8.

References

AOAC (1984). Williams, S. (Ed.), *Official Methods of Analysis*, 14th edition. Arlington, Virginia: Association of Official Analytical Chemists, 429.

Artemova, V. M. and Usova, E. M. (1964). Effect of food dyes on the structure formation in gelatin gels. *Izvestiya Viysshiki Uchebnykh Zavedenii, Pishchevaya Tekhnol.*, (2) 46–48 (Russian), per *Chem. Abs.*, 64, 2659F.

Bailey, A. J. and Paul, R. G. (1998). Collagen a not so simple protein. *J. Soc. Leath. Techn. Chem.*, 82, 104–110.

Bailey, A. J, Robins, S. P. and Balian, G. (1974). Biological significance of intermolecular cross-links of collagen. *Nature*, 251, 105–109.

Bartley, J. P. and Marrs, W. M. (1974). *Electrophoresis of gelatin (part 2)*. The British Food Manufacturing Industries Research Association Technical Circular 582.

Bell, J. H., Stevenson, N. A. and Taylor, J. E. (1973). A moisture transfer effect in hard capsules of sodium cromoglycate. *J. Pharm. Pharmacol.*, 25, Suppl., 96P–103P.

Bogue, R. H. (1922). Introduction: historical and statistical considerations. In: *The Chemistry and Technology of Gelatine and Glue*. New York: McGraw-Hill, 1–12.

Bohonek, J. and Spühler, A. (1976). Photographic gelatin II. In: Cox R. J. (Ed.), *Proceedings of the Royal Photographic Society Symposium*, Cambridge, 1974. London: Academic Press, 57–71.

Bradbury, E. and Martin, C. (1952). The effect of temperature of preparation on the mechanical properties and structure of gelatin films. *Proc. R. Soc. A*, 214, 183–192.

British Standard (BS) 757 (1975). *Methods for sampling and testing of gelatines*. Gr8. London: British Standards Institution.

British Standard (BS) 5349 (1976–7). *Methods for microbiological examination of gelatine*. London: British Standards Institution.

Burge, R. E. and Hynes, R. D. (1959). The thermal denaturation of collagen in solution and its structural implications. *J. Mol. Biol.*, 1, 155–164.

Calhoun, J. M. and Leister, D. A. (1959). Effect of gelatin layers on the dimensional stability of photographic film. *Photogr. Sci. Engng*, 3, 8–17.

Carless, J. E. and Nixon, I. R. (1970). Some applications

of rigidity and yield values in a study of gelatin and laponite gels. *J. Soc. Cosmet. Chem.*, 21, 427–440.

Clark, R. C. and Courts, A. (1977). The chemical reactivity of gelatin. In: Ward A. G. and Courts A. (Eds.), *The Science and Technology of Gelatin*. London: Academic Press, 209–247.

Cooper, J. W., Ansell, H. C. and Cadwallader, D. E. (1973). Liquid and solid solution interactions of primary certified colorants with pharmaceutical gelatins. *J. Pharm. Sci.*, 62, 1156–1164.

Coopes, I. H. (1975). Gel chromatography studies of gelatin. *J. Polym. Sci.*, 49, 97–107.

Coopes, I. H. (1976a). Temperature dependence of the structure-forming process in gelatin films. *J. Polym. Sci., Polym. Symp.*, 55, 127–138.

Coopes, I. H. (1976b). Photographic gelatin II. In: Cox R. J. (Ed.), *Proceedings of the Royal Photographic Society Symposium*, Cambridge, 1974. London: Academic Press, 121–129.

Courts, A. (1955). The N-terminal amino acid residues of gelatin: 3. Enzyme degradation. *Biochem. J.*, 59, 382–386.

Courts, A. and Stainsby, G. (1958). Evidence for multichain gelatin molecules. In: Stainsby, G. (Ed.), *Recent Advances in Gelatin and Glue Research*. London: Pergamon Press, 100–105.

Cowan, P. M. and McGavin, S. (1955). Structure of poly-l-proline. *Nature*, 176, 501–503.

CPMP/CVMP (2001). *Note for guidance in minimising the risk of transmitting animal spongiform encephalopathy agents via human and veterinary medicinal products.* EMEA/410/01 rev. London: The European Agency for the Evaluation of Medicinal Products, May 2001, 11 pages.

Croda (2001). Data sheet on fish gelatin (version June 2001). Croda Colloids: Widnes, Cheshire.

Croome, R. J. (1953). The variation of the viscosity of gelatin sols with temperature. *J. Appl. Chem. London*, 3, 330–334.

Davis, P. and Tabor, B. E. (1963). Kinetic study of crosslinking of gelatin by formaldehyde and glyoxal. *J. Polym. Sci. Part A*, 1, 799–815.

Eastoe, J. E. (1955). Amino acid composition of mammalian collagen and gelatin. *Biochem. J.*, 61, 589–602.

Eastoe, J. E. (1967). Composition of collagen and allied proteins. In: Ramachandran, G. N. (Ed.), *Treatise on Collagen, Vol. 1*. London: Academic Press, 1–72.

Eastoe, J. E. and Leach A. A. (1977). Chemical constitution of gelatin. In: Ward, A. G. and Courts, A. (Eds.), *The Science and Technology of Gelatin*. London: Academic Press, 77–85.

Eith, L., Stepto, R. F. T., Tomka I. and Wittwer, F. (1986). The injection moulded capsule. *Proceedings Pharmaceutical Technology Conference*, Harrogate, UK.

European Pharmacopoeia (PhEur) (2002). 4th edition, Council of Europe, Strasbourg.

FDA (1997). Guidance for Industry: The Sourcing and Processing of Gelatin to Reduce the Potential Risk Posed by Bovine Spongiform Encephalopathy (BSE) in FDA-Regulated Products for Human Use, September 1997.

Finch, C. A. and Jobling, A. (1977). The physical properties of gelatin: B. Mechanical properties of gelatin gels. In: Ward, A. G. and Courts, A. (Eds.), *The Science and Technology of Gelatin*. London: Academic Press, 263–271.

Flory, P. J. and Weaver, E. S. (1960). Helix coil transitions in dilute aqueous collagen solutions. *J. Am. Chem. Soc.*, 82, 4518–4525.

GME (1999). Evaluation of the inactivation/removal effect of the gelatin manufacturing process on TSE infectivity: Protocol prepared by Gelatin Manufacturers of Europe, 18th November 1999.

GME (2000). Standardised Methods for the Testing of Edible Gelatin – Gelatin Monograph, Version 1, 1st July 2000.

GME (2001). Gelatine Process Study Workshop, Brussels, 5 December 2001.

GME (2002). www.gelatine.org (accessed 10 September 2003).

Harding, J. J. (1965). The unusual links and cross-links of collagen. *Adv. Protein Chem.*, 20, 109–190.

Healey, J. N. C., Rubinstein M. H. and Walters, C. (1974). The mechanical properties of some binders used in tableting. *J. Pharm. Pharmacol.*, 26, Suppl. 41P–46P.

Hinterwaldner, R. (1977). Technology of gelatin manufacture. In: Ward, A. G. and Courts, A. (Eds.), *The Science and Technology of Gelatin*, London: Academic Press, 295–314.

Hodge, A. J. and Petruska, J. A. (1962). Recent studies with the electron microscope on ordered aggregates of the tropocollagen molecule. In: Ramachandran, G. N. (Ed.), *Aspects of Protein Structure*. New York: Academic Press, 289–300.

Holmes, A. W., Marrs, W. M. and Boyar, M. M. (1986). Hydrocolloids and processed food. In: *Proceedings of the Third International Conference*, Wrexham 1985. London: Elsevier Applied Science, 245–252.

Hopp, V. (1964). The influence of the physical properties of gelatin on its gelation time. *Leder*, 15, 59–63.

Hulmes, D. J. S., Müller, A., Parry, D. A., Piez, K. A. and Woodhead-Galloway, J. (1973). Analysis of the primary structure of collagen for the origins of molecular packing. *J. Mol. Biol.*, 79, 137–148.

Inveresk Research (1998a). Validation of the clearance of scrapie from the manufacturing process of gelatine. Final Report. Inveresk Project No. 855028, Inveresk Report No. 14682. Tranent, Scotland, 28 pp.

Inveresk Research (1998b). Validation of the clearance of scrapie from the manufacturing process of gelatine: additional stage. Final Report. Inveresk Project No. 855080, Inveresk Report No. 14683. Tranent, Scotland, 28 pp.

Ito, K., Kaga, S. and Takeya, Y. (1969). Studies on hard gelatin capsules. I. Water vapour transfer between capsules and powders. *Chem. Pharm. Bull.*, 17, 1134–1137.

Itoh, N. (1985). Setting time of gelatin. In: Amman-Brass, H. and Pouradier, J. (Ed.), *Photographic gelatin. Proceedings of the Fourth IAG Conference*, 1983, Fribourg, 134–144.

Jacobson, R. E. (1976). The hydrolysis of gelatin by proteolytic enzymes and their use in photographic emulsion preparation. In: Cox, R. J. (Ed.), *Photographic gelatin II. Proceedings of the Royal Photographic Society Symposium*, Cambridge 1974. London: Academic Press, 233–251.

Janus, J. W., Tabor, B. E. and Darlow, R. L. R. (1965). The setting of gelatin sols. *Kolloidzeitschrift*, 205, 134–139.

Japanese Pharmacopoeia (JP) (2001). 14th edition, English Version. Tokyo: Society of Japanese Pharmacopoeia, 923–924.

Johns, P. (1977). Structure and composition of collagen containing tissues. In: Ward, A. G. and Courts, A. (Eds.), *The Science and Technology of Gelatin*. London: Academic Press, 31–72.

Jolley, J. E. (1970). The microstructure of photographic gelatin binders. *Photogr. Sci. Engng.*, 14, 169–177.

Jones, N. R. (1977). Uses of gelatin in edible products. In: Ward, A. G. and Courts, A. (Eds.), *The Science and Technology of Gelatin*. London: Academic Press, 365–394.

Jones, R. T. (1971). Pharmaceutical gelatin: Applications. *Process Biochem.*, 6(7), 19–22.

Jones, R. T. (1989a). Gelatin structure and manufacture. In: Ridgway, K. (Ed.), *Hard Capsules, Development And Technology*. London: Pharmaceutical Press, 13–30.

Jones, R. T. (1989b). Gelatin: physical and chemical properties. In: Ridgway, K. (Ed.), *Hard Capsules, Development And Technology*. London: Pharmaceutical Press, 31–48.

Jopling, D. W. (1956). The swelling of gelatin films, the effects of drying temperature and of conditioning the layers in atmospheres of high relative humidity. *J. Appl. Chem. London*, 6, 79–84.

Kellaway, I. W., Marriott, C. and Robinson, J. A. J. (1978a). The mechanical properties of gelatin films. 1. The influence of water content and preparative conditions. *Can. J. Pharm. Sci.*, 13, 83–86.

Kellaway, I. W., Marriott, C. and Robinson, J. A. J. (1978b). The mechanical properties of gelatin films. 1. The influence of titanium dioxide and dyes. *Can. J. Pharm. Sci.* 13, 87–90.

Kenchington, A. W. and Ward, A. G. (1954). Titration curve of gelatin. *Biochem. J.*, 58, 202–207.

Khalil, S. A. H., and El-Gamal, S. S. (1978). Indigocarmin: factors affecting thermal stability. *Manuf. Chem.*, 49(5), 52–59.

Knox, C. B. (1960). British Patent No. 1 836 082, Gelatine Co. Inc.

Koepff, P. (1985). History of industrial gelatine production (with special reference to photographic gelatin). In: Amman-Brass, H. and Pouradier, J. (Eds.), *Photographic gelatin, Proceedings of the Fourth IAG Conference*, Fribourg 1983, 3–35.

Kozlov, P. V. and Burdygina, G. I. (1983). The structure and properties of solid gelatin and the principle of their modification. *Polymer*, 24, 651–666.

Kramer, F. and Rosenthal, H. (1965). Determination of Bloom of gelatin solutions at non-standard concentrations. *Food Technol.*, Campaign 19(9), 1417–1420.

Light, N. D. and Bailey, A. I. (1979). Covalent cross-links in collagen. In: Parry, D. A. D. and Creamer, L. K. (Eds.), *Fibrous Proteins: Scientific, Industrial and Medical Aspects, Vol. 1*. London: Academic Press, 151–177.

Li-juan, C., Yan-ming, J. and Bi-xian, P. (1985). Determination of isoelectric point and its distribution of IAG gelatins by iso-electric focusing method. In:

Amman-Brass, H. and Pouradier, J. (Eds.), *Photographic gelatin, Proceedings of the Fourth IAG Conference*, Fribourg, 1983, 95–106.

Manzke, U., Schlaf, G., Poethke, R. and Felgenhauer, K. (1997). On the removal of nervous proteins from materials used for gelatin manufacturing during processing. *Drugs made in Germany*, 40(1), 32–36.

Marrs, W. M. and Wood, P. D. (1972). The gelatin and rupture properties of gelatin gels. In: Cox , R. J. (Ed.), *Photographic gelatin, Proceedings of the Royal Photographic Society Symposium*, Cambridge 1970. London: Academic Press, 63–80.

Marshall, A. S. and Petrie, S. E. B. (1979). *The Royal Photographic Society Symposium on Photographic Gelatin*, Oxford.

Mason, C. M. and Silcox, H. E. (1943), Water adsorption by animal glue. *Ind. Engng. Chem. Ind. Edition*, 35, 726–729.

Maxey, C. R. and Palmer, M. R. (1976). The isoelectric point distribution of gelatin. In: Cox, R. J. (Ed.), *Photographic gelatin II, Proceedings of the Royal Photographic Society Symposium*, Cambridge 1974. London: Academic Press, 27–36.

Melia, C. D. (1983). *Some Physical Properties of Gelatin Films on Relation to Hard Gelatin Capsules*. PhD Thesis, University of Nottingham.

Miles, C. A., Burjanadze, T. V. and Bailey, A. J. (1995). The kinetics of the thermal degradation of collagen in unrestrained rat tendon determined by DSC. *J. Mol. Biol.*, 245, 437–446.

Moll, F., Rosenkranz, H., and Himmelmann, W. (1976). The structure of gelatin cross-linked with formaldehyde. In: Cox, R. J. (Ed.), *Photographic gelatin II, Proceedings of the Royal Photographic Society Symposium*, Cambridge 1974. London: Academic Press, 197–213.

Nixon, J. R., Georgakopoulos, P. P. and Carless, J. E. (1966). The rigidity of gelatin–glycerin gels. *J. Pharm. Pharmacol.*, 18, 283–288.

Norris, T. O. and McGraw, J. (1964). Gelatin coatings and tensile strength of gelatin films. *J. Appl. Polym. Sci.*, 8, 2139–2145.

OIE (1999). International Animal Health Code – 1999 edition, Part 3, Section 3.2, Chapter 3.2.13, 1–10.

PAGI (Photographic and Gelatin Industries) (1992). Setting point test method. In: *Test Methods Of Gelatin For Photographic Film In Japan*, 7th edition, 4–5.

Pankhurst, K. (1947). 'Incipient shrinkage' of collagen and gelatin. *Nature*, 159, 538.

Piez, K. A. (1964). Non-identity of the three α-chains in codfish skin. *J. Biol. Chem.*, 239, 4315–4316.

Piez, K. A. and Gross, J. (1960). Amino acid composition of some fish collagens: the relationship between composition and structure. *J. Biol. Chem.*, 235, 995–998.

Pikkarainen, J. (1968). The molecular structures of vertebrate skin collagens: a comparative study. *Acta. Physiol. Scand. Suppl.*, 309, 72.

Pouradier, J. and Hodot, A. M. (1972). On the interactions between gelatin and glycerol. In: Cox, R. J. (Ed.), *Photographic gelatin, Proceedings of the Royal Photographic Society Symposium*, Cambridge 1970. London: Academic Press, 91–95.

Robinson, J. A. J., Kellaway, I. W. and Marriott, C. J. (1975a). The effect of ageing on the rheological properties of gelatin gels. *J. Pharm. Pharmacol.*, 27, 653–658.

Robinson, J. A. J., Kellaway, I. W. and Marriott, C. J. (1975b) The effect of blending on the rheological properties of gelatin solutions and gels. *J. Pharm. Pharmacol.*, 27, 818–824.

Saunders, P. R. and Ward, A. G. (1958). The physical properties of gelatin and its degradation products. In: Stainsby, G. (Ed.), *Recent Advances in Gelatin and Glue Research*. London: Pergamon Press, 197–203.

Sheppard, S. E., Hauck, R. C. and Dittmar, C. (1940). The structure of gelatin sols and gels. VI. The adsorption of water vapour and the electrical conductivity. *J. Phys. Chem.*, 44, 185–207.

Smith, P. I. (1929). *Glue and Gelatine*. London: Pitman Press.

SSC (1999). Scientific Report and Opinion on the Safety of Gelatine, adopted by the Scientific Steering Committee, updated version of 18–19 February 1999.

SSC (2000). Updated Opinion on the Safety with regard to TSE Risks of Gelatine Derived from Ruminant Bones or Hides, adopted by the Scientific Steering Committee, 12–13 September 2002.

SSC (2002a). Draft Report on the Qualitative Assessment of the Residual BSE Risk in Bovine-Derived Products: Gelatine, Tallow and Dicalcium Phosphate from Bone, Tallow from Fat Tissues, Tallow from Rendered Mixtures of Tissues (Version III).

SSC (2002b). Update of the Opinion of the Scientific Steering Committee on the Geographical Risk of

Bovine Spongiform Encephalopathy (GBR), SSC/11/01/2002/6.2.c1.

Stainsby, G. (1958). Gelatine and Glue Research Association Research Report Cl7.

Stainsby, G. (1977a). The physical chemistry of gelatin in solution. In: Ward, A. G. and Courts, A. (Eds.), *The Science and Technology of Gelatin*. London: Academic Press, 109–136.

Stainsby, G. (1977b). The gelatin gel and sol–gel transformation. In: Ward, A. G. and Courts, A. (Eds.), *The Science and Technology of Gelatin*. London: Academic Press, 179–207.

Steven, F. S. and Tristram, G. R. (1962). The denaturation of acetic acid-soluble calfskin collagen. Changes in optical rotation, viscosity and susceptibility towards enzymes during serial denaturation in solutions of urea. *Biochem. J.*, 85, 207–210.

Tabor, B. E. (1968). Cross-linking efficiency of gelatin hardeners. *J. Appl. Polym. Sci.*, 12, 1967–1979.

Taylor, S. K., Davidson, F. and Ovenall, D. W. (1978). Carbon-13 nuclear magnetic resonance studies on gelatin cross-linked with formaldehyde. *Photogr. Sci. Engng.*, 22, 134–138.

Tiemstra, P. J. (1968). Degradation of gelatin. *Food Technol.*, 22, 1151–1153.

Toda, Y. (1985). The relationship between the isoelectric point distribution and molecular weight distribution of gelatin. *The Royal Photographic Society Symposium on Photographic Gelatin*, Oxford.

Torrado Valeiras, J. I. (1966). Compatibildad de colorantes con gelatina. *An. Real. Acad. Farm.*, 32, 353–359.

United States Department of Agriculture (USDA) (1988). Federal register, Vol. 63, No. 3, 406–408, January 16 1998.

United States Pharmacopoeia/United States National Formulary (USP/NF) (2003) 26th/21th edition, 2003 and Supplement 1, January 2003, United States Pharmacopeia Commission, Washington.

Wainewright, F. W. (1977). Physical tests for gelatin and gelatin products. In: Ward, A. G. and Courts, A. (Eds.), *The Science and Technology of Gelatin*. London: Academic Press, 507–534.

Ward, A. G. (1953). British gelatine and glue research association: residential conference. *Nature*, 171, 1099–1101.

Ward, A. G. and Saunders, P. R. (1958). The rheology of gelatin. In: Eirich, F. R. (Ed.), *Rheology, Theory and Applications, Vol. 2*. London: Academic Press, 313–362.

WHO (1997). Report of a WHO Consultation on Medicinal and other Products in Relation to Human and Animal Transmissible Spongiform Encephalopathies, Geneva, Switzerland, 24–26 March 1997.

WO 99/33924 (1999). Gelatine Compositions, assigned to the Warner-Lambert Company, 8th July, 1999.

Yannas, I. V. (1972). Collagen and gelatin in the solid state. *J. Macromol. Sci., Rev. Macromol. Chem.*, 7(1), 49–104b.

3

Gelatin alternatives and additives

Brian E Jones

Capsule shells contain small quantities of additives that enable either the capsule to be formed more easily or to improve their performance in use. The materials, apart from gelatin (see Chapter 2) can be put into six main categories: gelatin alternatives, colouring agents, plasticizers, process and performance aids, preservatives and capsule shell coatings.

Gelatin alternatives

The initial reason for looking for gelatin alternatives in the 1830s was to find ways of overcoming Mothes's original patent. In the 19th century, capsules were made from a variety of materials such as decoctions of starch, gluten, animal membranes and a vegetable 'gelatin' prepared from carragheen moss. However, the only single material with the correct properties that has stood the test of time has been gelatin. Despite the lack of success in finding alternatives, all the major pharmacopoeias have permitted the use of materials other than gelatin for the manufacture of capsules for several years; the Europe Pharmacopoeia (PhEur.) states 'made of gelatin or other substance', the Japanese Pharmacopoeia states 'made of gelatin or a suitable material' and the US Pharmacopoeia states 'are usually formed from gelatin: however they also may be made of starch or other suitable material'. Thus pharmaceutical grade capsules can be made from any pharmacopoeial quality material provided that it has the necessary properties.

There has been a lot of activity during the last 150 years in the search for alternative materials, which is reflected in the number of patents granted. The first property that is required of a gelatin alternative is that it must be a good film former. The two-piece capsule has a shell wall that is only about 0.1 mm thick and this needs to be tough and flexible. A one-piece capsule has a much thicker wall and it must be suitably plasticized to form a film that is both pliable and easy to seal. Secondly, the capsule must rapidly dissolve in biological fluids at 37°C and, thirdly, it must exhibit a gelation property so that a capsule film can be cast or dipped, thus enabling existing capsule manufacturing machinery to be used. Gelling can be brought about either by a temperature change or by the removal of sufficient solvent to cause a significant increase in concentration. Most of the patents have had this objective in mind although there have been suggestions for a radical change in method such as using injection moulding (Eith *et al.*, 1987). Any new material must have good toxicological credentials because of the widespread use of capsules in medicines and nutraceuticals.

The cellulose ethers are the replacement materials most commonly mentioned in the literature that have been widely used in pharmaceuticals and foodstuffs. They can be used on standard capsule-making machines, but if used alone the process needs to be modified because of their temperature-related properties. The viscosity of their solutions, when heated from room temperature, first decreases to a minimum value then it increases rapidly to a gel point, which is only a few degrees above the point of minimum viscosity. To carry this out requires hot metal moulds to be dipped into cellulose solutions at ambient temperatures instead of the conditions used for gelatin where moulds at room temperature are dipped into hot solutions.

In 1950, H W Murphy of Eli Lilly & Company obtained an American patent for methylcellulose capsules and the process for their manufacture (Eli Lilly & Co, 1950). Two-piece capsules made to this patent were produced in both the USA and the UK in the early 1950s. The patent covered all cellulose ethers and in particular methylcellulose. The stacker type of manufacturing machines was used. The methylcellulose solution used had a gel point of 65°C. The solution in the dip pan was held at 18°C and the mould pins heated to >65°C were dipped into it, the film was formed and then the mould pins were passed through the standard drying kilns. The moisture content of the finished capsules was about 2–3%. They were used in significant numbers for a time. Unfortunately in use their *in vivo* disintegration performance was poor, so much so that the capsules sometimes passed through the body unaffected. When this became known their manufacture was discontinued.

Since that time several patents have been granted for the production of other modified celluloses with better *in vivo* performance characteristics. The Dow Chemical Co. has been particularly active in this field and was granted a series of patents. They used various celluloses and manufacturing methods: methylcellulose and hydroxypropylcellulose (Dow Chemical Company, 1957), hydroxyalkyl-alkylcellulose ethers including hydroxypropyl methylcellulose (HPMC) using solvent evaporation to form a film (Dow Chemical Company, 1969), using thermal gelling process at temperatures $\geqq 100°C$ (Dow Chemical Company, 1973) and then using moulds heated to 200°C using PTFE granules (Dow Chemical Company, 1974). They also patented the use of hydroxyalkylstarch for the manufacture of capsules (Dow Chemical Company, 1977).

In the late 1980s and early 1990s the problem of making capsules from cellulose ethers was solved in two different ways and both involved the use of HPMC. One involved using the viscosity/temperature relationship of the polymer and the other the production of a gelling system by the use of additives.

One solution was developed by the company G S Technologies Inc. (now R P Scherer West Inc USA) who introduced a vegetable capsule, the Vegicap, to satisfy the needs of the vegetarian sector of the nutraceuticals market in the USA. They applied similar methodology to the process as that used by Eli Lilly in the 1950s for the manufacture of methylcellulose capsules. They overcame the problems inherent in the process and were granted a series of patents (GS Technologies, 1997a, 1997b, 1997c, 1997d, 1997e, 1998a, 1998b). However, capsule shells of HPMC made to the same thickness as standard gelatin capsules did not have enough mechanical strength, so this was obtained by doubling the thickness. There was a problem in stripping the dried films from the mould pins because it was difficult to grip them tightly enough without damaging them. The solution was to use a stripper jaw that had dimples on the inner surface to improve the grip. Films were formed on the mould pins by dipping them into a solution of HPMC at a temperature of 70°C, which is above its thermal gelling point. When the pins are removed from the dipping pan and move to the drying tunnels on the machine, drying begins as soon as water evaporates and the temperature falls. If this were allowed to happen the HPMC film would revert to a liquid with a consequent distortion of the capsule shell wall. This problem is avoided by a system of induction heating for the mould pins to ensure that the correct temperature is maintained in the wet HPMC films until they are dry enough to maintain their shape.

The other solution was developed by Shionogi Qualicaps Co. in Japan, who in the late 1980s were trying to make capsules that did not have the problems associated with the moisture content of gelatin capsules, i.e. brittleness at low humidities and a water content that gives problems with labile actives (Ogura *et al.*, 1998). They developed an HPMC gelling system that had similar properties to gelatin and which was achieved by adding in small quantities of carrageenan to act as a gelling aid and potassium chloride to act as a gelation promoter (Japan Elanco, 1993, 1995, 1998; Shionogi Qualicaps, 1999, 2002; Nippon Elanco, 1996). These Quali-V capsules are produced on standard capsule manufacturing machines with the same dimensions and filling properties as gelatin capsules. They are produced in all sizes from 00 through to size 4.

Since this initial work, other companies have entered into this field. Warner Lambert obtained a patent for a capsule, Vcaps, using a different HPMC gelling system based on using gellan gum as a gelling aid and either ethylenediamine-tetraacetic acid (EDTA) or sodium citrate as a gelation promoter (Warner Lambert, 1998). They produce Vcaps in sizes 00 through to 4.

The products filled into capsules that can be most easily converted to HPMC capsules are those used for nutraceutical products. Current HPMC capsule usage can be seen from a search of the Internet. In the USA many companies have health and nutritional products that particularly appeal to the vegetarian market listed as filled into HPMC. In Europe the Internet is not so widely used for selling such products. There are databases of medicines available in Europe that have this information, for example BIAM, which is run as a service for those in the medical and associated professions by the Centre Recherche en Informatique of L'Ecole Supérieure des Mines de Paris. This lists medicines that are registered in France together with their composition and includes many HPMC hard capsules, which contain herbs and natural oils. An examination of recent patents shows that there is still a lot of activity in searching for alternative materials to gelatin, amongst them pullulan and modified starches, but as yet none of these capsules have been commercialised (Warner Lambert, 2001). For further information on soft capsules see Chapter 11.

Colouring agents

The chief reasons for colouring pharmaceutical products are for the aesthetic effect, ease of identification and the psychological effect on patients. Pharmaceuticals are coloured by the use of colourants and these materials need to have certain basic properties (Jones, 1993). They must be non-toxic, acceptable throughout the world on a long-term basis, have a high tinctorial value and be stable under the conditions of capsule manufacture and storage. Colouring agents may also be used to improve the product stability by applying their light-screening properties to protect photolabile substances filled into capsules.

Aesthetic effects

Products must have a constant standard appearance in order to be acceptable. The addition of a colouring agent effectively masks variations in the colour of the contents and raises the confidence of the user in its quality and stability.

Product identification

Coloration is the simplest and best method for quick identification of a product. It cannot be relied upon absolutely, but it does have practical advantages over methods such as embossing or printing. Colour plays an important role in good manufacturing practice (GMP) by enabling differentiation of products during the course of manufacture and distribution right up to the moment when the patient takes their medicine. Colour provides a key to safety for the geriatric patient on a multiple drug regimen. Colour plays a commercial role in the marketing of pharmaceutical products by helping to give them a brand or company image. Product appearance has some legal protection and this has caused certain problems in the area of generic products (Llewelyn, 1981).

Psychological effect on the patient

The psychological effect of colour on the users of medicines has been recognised since the earliest times. The Egyptians used to colour their draughts (Cooper, 1956) and it is also well known that certain colours elicit specific associated effects in many people (Swartz and Cooper, 1962). This is especially true of flavours (Peacock, 1949), those most commonly used being green for peppermint and red for strawberry. However, the effect of the colour of medicines on the patient has not really been successfully evaluated. Peacock (1949) described the coloured placebo effect that is particularly relevant for the psychiatrist treating a hypochondriac patient. Madsen (1957) undertook a series of clinical trials with medicines in a variety of coloured solutions in order to test their patient acceptability. Schapira *et al.* (1970) studied the effect of the colour of oxazepam tablets given to patients suffering from anxiety states. They used three colours, red, yellow and green and found that anxiety states responded best to green whereas depressive symptoms responded best to yellow. Hussain and Ahad (1970) commented on the results of this paper. They carried out a cross-over trial using chlordiazepoxide in patients suffering from anxiety states. The same dose was given either as a capsule or as a tablet preparation. They found that the consumption rate for the capsules was better than for the tablets, which they ascribed to the patients preferring the capsule form, making them thus more likely to take them as prescribed.

The specific effect of the colour of the capsule on patients has been studied by a group of Italian research workers. In their first study, which was on the effects of placebo treatments, they included capsule colour amongst their factors (Cattaneo *et al.*, 1970). They used blue and orange hard gelatin capsules and found an influence of colour that appeared to be gender-related. In their second trial, which was on the hypnotic effect of heptabarbitone, again they used orange and blue capsules (Lucchelli *et al.*, 1978) and concluded that blue capsules were better than orange for treating insomnia but that orange was a better colour for evaluating a sedative agent against a placebo. A similar experiment was performed by Bailie and Kesson (1981) using soft gelatin capsules of temazepam. They compared the effect of colour, yellow or green, patient's sex and preference on the hypnotic efficiency of the active and concluded that the men preferred yellow capsules but the women showed no clear preference for a colour.

The colour of a capsule can also influence the patient's perception of what a product will do for him or her. Jacobs and Nordan (1979) used a group of subjects to place six different coloured capsules into three classifications of drug effects. They found that red and yellow placebos were classified as stimulants and blue placebos were classified as depressants. Buckalew and Coffield (1982) used a panel of students to ascertain the

perceived characteristics of hard capsule colour and size and their relationship to tablets. Their results indicated that certain colours were associated with the treatment of specific conditions and that capsules were seen to be more potent than tablets. The experiences of Sandoz in Switzerland on the influence of appearance on patient compliance were reported (Anon, 1989). They commented that they had found a 'reasonably consistent pattern of preference, certain colours having potentially significant associations in people's minds'. These factors were used by them in conjunction with a commercial testing organisation, in deciding upon dosage from appearance and package design.

Light protection

The colour of the capsule governs both the amount and the wavelength of the light passing through the shell wall. The stability of photolabile substances can be improved by using capsules of an appropriate colour. Most photochemical breakdown occurs at specific wavelengths. It is possible to produce a capsule colour that has the maximum screening power for a particular substance using the appropriate colourants. The transmission of light through gelatin and HPMC film strips, whose thickness

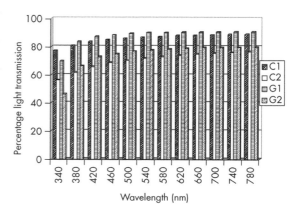

Figure 3.1 Transmission of light through gelatin (G) and HPMC (C) films corresponding to the formulation and thickness of the single wall of a hard capsule shell (Jones, 2000). C1 and G1 are the results from single films and C2 and G2 are results from two films placed on top of each other.

corresponded to that of the wall of a hard capsule, has been measured (see Figure 3.1) (Jones, 2000). This shows that light transmission through the polymer films is very similar. Most capsules contain titanium dioxide as an opacifying agent and this significantly reduces the amount of light that passes through the shell walls. The amount of light that penetrates through the wall of a hard capsule varies depending upon its position within the capsule. This is because in the region of the cap and body overlap there is twice the wall thickness compared to the other areas. The amount of light in the visible spectrum penetrating a hard capsule shell wall containing 2%w/w titanium dioxide is about 1.0% for a single film thickness and 0.2% for a double film thickness (Jones, 2000). The screening effect of capsule colours on product stability has been demonstrated using capsules filled with photolabile compounds, e.g. menadione (Prista *et al.*, 1970).

Nomenclature for colouring agents

The colouring agents used in commerce usually have trivial names and there are often several synonyms for each substance. The complex chemical nature of the molecules and the fact that many of the dyes are not pure compounds mitigates against the use of the chemical name. To overcome this problem, several numbering systems have been introduced to assign a recognition code to each dye. The most universal system is the Colour Index Number (CI No.) which is published jointly by the UK Society of Dyers and Colourists and the American Association of Textile Chemists and Colorists. In this system, all food, pharmaceutical and other dyes are grouped into chemical classes and then assigned a number on the basis of complexity and points of substitution. Similar numerical systems such as the EINECS (European Inventory of Existing Commercial Substances Index number) and the CAS (Chemical Abstract Service number) are in use to identify chemical entities. All of these systems enable the dye to be identified chemically but give no information about the specification that would make it suitable for the manufacture of pharmaceutical products.

There are three main systems that give information on the specifications for dyes for use in pharmaceuticals, devised by the European Union, the American Food and Drug Administration (FDA) and the Food and Agriculture Organisation of the United Nations and the World Health Organisation (WHO/FAO). The first two systems were designed for application in specific markets, the EU and the USA, whereas the latter one was designed as a system suitable for use worldwide. The EU system for pharmaceutical colourants is part of the legislation that covers all food additives. All approved additives are assigned an 'E' number. This signifies that they have been approved as safe for use for specific applications if they comply with the specification, which precisely defines the material that can be used (Official Journal, 1995). The colourants that can be used in pharmaceuticals are described in a corresponding directive (Official Journal, 1994). The responsibility to ensure that the colourants comply with the specifications rests with the user, i.e., the pharmaceutical manufacturer. The American FDA system divides food and pharmaceutical colourants into a variety of classes depending on the uses for which they have been approved: F D & C dyes can be used in foods, drugs and cosmetics, D & C dyes are for drugs and cosmetics only and Ext. D & C dyes are for drugs and cosmetics used externally. The specification for each colourant is published in the Federal Register. Every lot of colourant manufactured for use in the USA must be certified before use. In order to do this, the colourant manufacturers have to submit samples to the FDA for analysis and if approved, a document certifying that it is fit for its intended use is issued. Pharmaceutical pigments are not certified, unlike soluble dyes. Specifications are published in the Federal Register and, like the EU system, the responsibility for assuring their quality is the responsibility of the user. Tables 3.1, 3.2 and 3.3 list the colouring agents that have been reported as being used for the coloration of capsules. The FAO/WHO system (1984) was designed as a model example that could be adopted by countries and help in the harmonisation of colourant specifications.

Table 3.1 Pharmaceutical colourants, water-soluble dyes

Common name	Chemical Index number	EU number	USA name	Chemical class
Reds				
Allura Red	16035	E129	F D & C Red No 40	Azo
Amaranth	16185	E123		Azo
Azorubine	14720	E122		Azo
Ponceau 4R	16255	E124		Azo
Erythrosine	45430	E127	F D & C Red No 3	Xanthene
Phloxine B	45410		D & C Red No 28	Xanthene
Orange				
Sunset Yellow	15985	E110	F D & C Yellow No 6	Azo
Yellows				
Tartrazine	19140	E102	F D & C Yellow No 5	Azo
Quinoline Yellow	47005	E104	D & C Yellow No 10	Quino-Phthalene
Blues				
Indigo Carmine	73105	E132	FD & C Blue No 2	Indigoid
Brilliant Blue	42090	E133	FD & C Blue No 1	Triphenylmethane
Patent Blue V	42051	E131		Triphenylmethane

Table 3.2 Pharmaceutical pigments

Name	Colour Index number	EU number	Formula
Titanium Dioxide	77891	E171	TiO_2
Black Iron Oxide	77499	E172	$FeO.Fe_2O_3$
Red Iron Oxide	77491	E172	Fe_2O_3
Yellow Iron Oxide	77492	E172	$FeO(OH).nH_2O$

Table 3.3 Natural colourants used in pharmaceuticals

Name	Hue	Colour Index number	EU number
β-carotene	Orange	40800	E160a
Canthaxanthin	Orange	40850	E161g
Chlorophyll	Green	75810	E140, E141
Cochineal	Red	75740	E120

Types of colouring agent

FAO/WHO (1984) have classified food colours into a series of chemical groups: anthocyanin, anthraquinone, azo (mono- and bis-), betalin, carotenoid, flavone, indigoid, inorganic pigment, iso-alloxazine, phorbins, quinophthalone, triaryl-methane and xanthene. The colourants mainly used for colouring capsules are synthetic water-soluble dyes (azo, indigoid, quinophthalone, tri-arylmethane and xanthene), pigments (especially the opacifying agent titanium dioxide) and certain dyes of natural origin (carotenoids).

Synthetic dyes

The synthetic dyes are used primarily in their water-soluble form, which are as their sodium, calcium, potassium or ammonium salts. Sometimes they are used in the form of insoluble lakes, which are produced by precipitating the water-soluble forms on to an alumina or titanium dioxide base. They are drawn from several chemical classes. The principal one used to be the azo group, which until the 1980s accounted for the majority of food and pharmaceutical dyes used. Since then these dyes have fallen out of favour, mainly owing to consumer pressures rather than for toxicological reasons.

Azo dyes

These are characterised by having an azo linkage, i.e. a nitrogen-to-nitrogen double bond ($-N=N-$). Chemically, their molecules contain benzene and naphthalene rings joined by azo linkages and containing sulfonic acid groups. There are many members that are chemical isomers of other colourants in the same class. They are further subdivided into the monoazo and the bisazo groups, having one or two azo linkages in their molecules, respectively. Dyes in this group are the monoazo dyes Amaranth (CI No. 16185), Sunset Yellow FCF (CI No. 15985) and Tartrazine (CI No. 19140) and the bisazo dye Brilliant Black PN (CI No. 28440).

Indigoid dyes

This group of dyes is based on indigo, which has two isatin molecules joined by a double bond. Most members of the group are produced synthetically, but indigo itself occurs in nature. Pharmaceutically, the most commonly used dye in this class is Indigo Carmine (CI No. 73015).

Quinophthalone dyes

These dyes are based on the naphthalene-quinone structure. Quinoline Yellow (CI No. 47005) is the member of this group used pharmaceutically. Quinoline Yellow consists of a mixture of salts, principally the mono- and disulfonate. There is a significant difference between the EU and the USA in the definition of this colourant: the EU material contains not less than 50% of the disulfonate and not more then 50% of the monosulfonate, whereas the USA material contains not

less than 75% of the monosulfonate and no more than 15% of the disulfonate (Jones, 1993). This means that it is not possible to obtain material that complies with both specifications.

Triarylmethane dyes

The members of this group are based on a central triphenylmethane structure and are substituted amine derivatives, with or without sulfonic acid groups. They can be either cationic, as is Methyl Violet, a hydrochloride derivative, or anionic, like Brilliant Blue FCF, which is the disodium salt of sulfonic acid. The dyes in this class that are used pharmaceutically are Brilliant Blue FCF (CI No. 42090), Green S (CI No. 44090) and Patent Blue V (CI No. 42051).

Xanthene dyes

The members of this group are based chemically on a xanthene nucleus. Most of the pharmaceutical members are derived from fluorescein, being either brominated or iodinated derivatives, e.g. eosin, which is tetrabromofluorescein and erythrosine which is tetraiodofluorescein. Several members of this class are used topically, but only Erythrosine (CI No. 45430) and D & C Red No. 28 (Phloxine, which is a brominated derivative) are used orally.

Pigments

Oxides of iron

Iron oxide can be obtained in colours ranging from black to yellow through brown and red, the difference between them being caused by their state of oxidation. The three oxides most commonly used pharmaceutically are Black Iron Oxide [$FeO.Fe_2O_3$] (CI No. 77499), Red Iron Oxide [Fe_2O_3] (CI No. 77491) and Yellow Iron Oxide [$FeO(OH).nH_2O$] (CI No. 77492). These pigments are widely used in pharmaceuticals. However there is a limitation to their usage in the USA and countries that adopt its regulations. The amount of iron oxide that can be used as a colourant in a pharmaceutical is limited to a daily intake of not more than 5 mg of elemental iron. The proportion of elemental iron in each oxide is black $\leq 68\%$, red $\leq 68\%$ and yellow $\leq 60\%$ (Official Journal, 1995). This limit was introduced during the 1960s and has no physiological basis because iron oxides are only very slightly soluble in gastric juice and the iron that they contain cannot be absorbed (Hess and Shrank, 1979).

Titanium dioxide

This pigment (CI No. 77891) has been used extensively as an opacifying agent in the manufacture of capsules. It occurs in two crystalline forms, anatase and rutile. The anatase form is the one used in food and pharmaceuticals.

Natural colouring agents

The use of naturally occurring colouring agents in capsule manufacture is restricted because of their poor technical properties. Many of them are light sensitive, have poor water solubility, have a low tinctorial value, i.e. require a high concentration to make a saturated hue, and some are unstable in the presence of gelatin. Several types of natural dye have been successfully used to colour capsules. The carotenoids are widely distributed in nature, both in the plant and animal kingdoms, e.g. in carrots, tomatoes and lobsters. Those used in foods and pharmaceuticals are β-carotene (CI No. 40800) and canthaxanthin (CI No. 40850). The phorbins are the chlorophyllins obtained by solvent extraction from natural strains of edible plant material. These are used as the alkali metal salts obtained by saponification of the extracts, E140 (ii), or are the copper complexes, E141 (i), or the alkali salts of these complexes, E141 (ii). An anthroquinine, cochineal, the colouring principal of which is carminic acid is obtained by aqueous or ethanolic extraction from the dried bodies of the female insect *Dactylopius coccus Costa*.

Standards for colouring agents

The colourants which are used in pharmaceuticals have to comply with such standards of purity as are required to protect the consumer and to enable the substance to be clearly defined by regulatory authorities (FAO/WHO 1965, 1984; Official Journal, 1995). In most countries, the colourants used in capsules are those permitted for use in food. Usually a colourant is a mixture and the specifications include limits on the quantities of substances other than the main component(s), which are allowed to be present. Some substances, such as simple organic salts, may be present but are not deleterious from a functional or safety standpoint, whereas others, such as non-sulfonated amines, which may be present in the initial raw materials, must be eliminated as far as is practically possible. The following are the tests that are typically applied:

- Assay. This defines the total amount of colouring matter that may be present. It is usually determined by titration with titanous chloride, although in certain cases other methods, such as spectrophotometry, are used.
- Loss on drying. This limits the amount of water present.
- Chlorides and sulfates. These are calculated as sodium salts and are determined by standard analytical techniques. They are present owing to the use in the 'salting-out' of the colourant at the end of the manufacturing process. The limit for this is usually combined with that for loss on drying and the total varies, according to the substance, from 15 to 30%.
- Water-insoluble matter. This indicates the quantity of extraneous materials, e.g. filter aids, which can be present and is normally not more than 2%.
- Ether-extractable matter. This indicates the quantity of organic impurities, e.g. non-sulfonated amines, which may be present and is normally more than 0.2%.
- Arsenic and heavy metals. These are subject to the usual type of limit tests for pharmaceutical products and are expressed in mg kg^{-1}.
- Subsidiary colouring matters. Isomers and other related compounds are present in the raw materials, or are produced during the chemical synthesis. Those that are coloured are termed 'subsidiary colouring matters' and the limits vary from 1 to 10% depending on the colourant being examined.
- Organic compounds other than colouring matters. These are isomers or other related compounds which are not coloured. Limits may be given for individual specified impurities or for a combination of several impurities and vary from 0.01 to 5% depending on the nature of the impurity and the colourant being examined.

Identification of colouring agents

The positive chemical identification of dyes is often difficult because of their lack of absolute chemical purity, but it can be done by comparative techniques such as chromatography or spectrophotometry. To identify the colourants that have been used to colour capsules, the first step is to extract them, free of other interfering substances. Two methods are available, the choice being dependent on the quantity of colourant in the product (Jones, 1973). For capsules with a high level of soluble colourant a simple extraction using an ethanol/water 3:1 mixture is used. The extract is evaporated to dryness and then redissolved in the minimum quantity of distilled water. For capsules with a low level of colourant a larger sample of capsules is required and the dyes can be absorbed onto polyamide powder or an ion-exchange filter paper, e.g. Whatman AE81, in order to concentrate them.

Chromatography

The simplest and quickest method of identifying the extracted dye is by thin-layer chromatography. Chromatographic plates precoated with cellulose such as Cellulose MN 300 are the most suitable. The following solutions are applied to the plate: (a) a solution of the capsule extract, (b) reference solution(s) containing the dye(s) thought to be present and (c) a combined spot of the solution of capsule extract and the reference dye(s) solution to make allowance for any co-extracted material. The chromatogram should be run in two solvent systems which are chosen to give, if possible, different R_f values for the dyes under consideration. A satisfactory pair of solvents is 2% trisodium citrate in 5% ammonia solution and a mixture of isopropyl alcohol: strong ammonia solution (4:1). This system will distinguish between all the commonly used capsule colourants, except for the triphenylmethane dyes which have very similar R_f values and tend to produce long streaky spots of the same hue on the chromatogram. To distinguish between these, a plate coated with an ion-exchange cellulose such as diethylaminoethyl-cellulose (Cellulose MN 300 DEAE) (Turner and Jones, 1971) can be used. This is then run in the

ammoniacal sodium citrate solvent and will give a significant differentiation between the dyes.

Legislation for colouring agents

The colourants that can be used to colour pharmaceutical products are governed in nearly all countries by legislation, which has been drawn up on the basis of toxicity (Jones, 1993). However, the list of permitted colourants varies from country to country and is usually subject to a regular review system. The trend is for the list of permitted colourants to get shorter with time.

Plasticizers

The difference between hard and soft gelatin capsules is that soft capsules contain appreciable quantities of a plasticizer other than water. The function of the plasticizer in the capsule wall is to reduce the rigidity of the gelatin and make it pliable. For soft gelatin capsules this is particularly important because the film after filling must take up the form of the mould. For hard gelatin capsules, water is the plasticizer and this is clearly demonstrated by the fact that as their moisture content falls they become more brittle.

A large variety of chemical materials has been suggested for this application in soft gelatin capsules. However, glycerol was the first true plasticizer used in capsules that is still widely used today (Taetz, 1875). The other materials that have been used are some of the polyhydric alcohols, natural gums and sugars. Their exact function in the gelatin film does not appear to have been elucidated. Several of them are used in other applications as humectants and it is possible that they function by binding water molecules within the gelatin structure, thereby reducing rigidity.

Some materials that have been used as plasticizers in soft gelatin capsule manufacture are glycerol, sorbitol, propylene glycol, sucrose and acacia. The proportion of plasticizer needed varies depending on the type of soft gelatin capsule being made, but is usually in the region of 20–40% by weight. Certain other materials have been claimed to enhance the effect of the main plasticizer when added in concentrations of 2–6%. These include glycine, mannitol, acetamide, formamide and lactamide. See Chapter 11 for further information

Process and performance aids

Hard and soft capsules are prepared from coloured polymer solutions, with or without preservatives and plasticizers. To this mix, small quantities of other substances can be added to aid in the manufacturing process (process aids) or to aid in the subsequent performance of the capsule (performance aids). Process aids are materials that assist in the manufacturing process, e.g. surfactants, which enable the polymer solution to take up the shape of the moulds better. Performance aids are materials that help to improve the patient acceptability of the product, such as flavouring agents or to improve hard capsule performance on filling machines.

Sodium lauryl sulfate is a surfactant used in the production of hard gelatin capsules. Its use is specifically mentioned in the US Pharmacopoeia, which allows a suitable concentration to be present. It is added to the gelatin solution during the preparation stage. The stainless steel mould pins are lubricated prior to dipping into the gelatin solution and, because of this, sodium lauryl sulfate in solution is added to reduce the surface tension of the mix and cause the mould pins to be wetted more uniformly. An indication of a lack of sodium lauryl sulfate is the appearance of thin areas on the capsule walls where the surface of the mould pins has not been sufficiently wetted.

Capsules are often filled with materials such as antibiotics, which have an objectionable smell and taste. Capsules containing these medicaments can be made more attractive to the patient by flavour masking. Soft gelatin capsules have been produced with walls containing either flavouring agents such as vanillin, or volatile oils such as peppermint and menthol. Similarly hard capsules have been produced, not by adding the materials to the solution during the manufacturing process, but by adding them to the capsule surfaces after filling or by applying them to the wadding used in packaging.

Hard capsules on high-speed filling machines have to pass rapidly through a series of tubes and holes in bushings. If the flow is delayed or interrupted, the machine output is significantly reduced. Historically, capsule makers have dusted the empty shells after manufacture with very small amounts of materials to improve their flow such as sodium lauryl sulfate or edible waxes such as carnauba. Companies filling hard capsules in the past frequently dusted capsules with talc to increase output by improving the flow and feeding through their machines. Another way in which to reduce the surface friction of gelatin capsules is to add materials into the shell wall during manufacture. This has the effect of modifying the surface structure of the film thereby reducing its frictional properties. Silicon dioxide has been used for this purpose and it is listed as a component of the capsule shell in some of the registered products in the European Medicines Evaluation Agency (EMEA) 'list of authorised products'.

Preservatives

Gelatin is a good medium for bacterial and fungal growth, especially if sufficient moisture is available. During the capsule-manufacturing process the gelatin is in solution and is kept warm to prevent gelling, which gives almost ideal conditions for bacterial growth. The gelatin is not initially sterile but does have as low a bacterial count as is practicable to attain. Bacterial growth can alter the viscosity of the gelatin as well as contaminating it.

Formerly the simplest way to control the microbiological content of capsules during the manufacturing process was considered to be by the use of preservatives. These were usually added during the preparation of the gelatin solution. A few of these were bactericidal, but the majority were bacteriostatic in action. The substances that can be used are limited by two considerations. First, pharmaceutical legislation restricts the substances that can be added. Second, the preservative can affect the capsule. For example, beta-naphthol was at one time permitted and was used during the 1950s, but it caused the capsules to turn brown on storage. Sulphur dioxide is the preservative that was most widely used in the past. Its use was first mentioned in 1843 for the manufacture of membrane capsules and it was widely permitted for use in foodstuffs and pharmaceuticals. For capsule manufacturing it was usually added in the form of a solution of sodium sulfite or sodium metabisulfite. The quantity used is so that the concentration in the final capsule is less than 1000 ppm, calculated as SO_2. It is volatile and is lost during the manufacturing process. It has one major drawback in that, even at low concentrations, it affects many of the commonly used colourants. As little as 60 ppm will cause colour loss with Amaranth, Sunset Yellow FCF and Indigo Carmine. Sulphur dioxide is a reducing agent, but its bleaching action is not simple. It is thought that it reacts with azo dyes by adding 1:4 across the azo-linkage, in the presence of gelatin, to produce a colourless molecule (Wyatt, 1966). Its use has declined significantly since the 1970s in Europe and North America when there were well publicised problems of sensitisation reactions associated with its use as an anti-browning agent on lettuce in fast-food restaurants. The Pharmacopoeia of the People's Republic of China (2000) includes a limit test for sulfite in hard gelatin capsules.

The other main preservative type used has been the esters of *p*-hydroxybenzoic acid. They are considered to be bacteriostatic rather than bactericidal in action and are employed in concentrations up to 0.2% w/w in the finished capsule. Each ester is effective against a different spectrum of organisms and as a result they are usually used in combination. The most commonly used combination is methylhydroxybenzoate (methylparaben) with propylhydroxybenzoate (propylparaben) in a 4:1 methyl to propyl ratio. A number of organic acids, such as benzoic, propionic and sorbic, and their sodium salts have been used in capsule manufacture, chiefly for their activity against moulds and yeasts, which particularly affect soft gelatin capsules. The compounds have general bacteriostatic properties and are used at concentrations up to 1% w/w. Their activity is influenced by pH, their optimum effect being obtained at pH 4.5–5.5.

The philosophy of modern GMP does not encourage the use of preservatives because they could cover up bad practices. The GMP way is to reduce the possibility of contamination during the manufacturing process. When the gelatin solutions are being prepared they should be heated to a temperature, >50°C, at which bacterial growth will decline significantly; if the solution is maintained at this temperature, is circulated constantly and no local cool spots are allowed to exist, then bacterial growth will be suppressed. If the bacterial level can be held down, through to the finished capsules, the normal moisture levels of both hard and soft capsules mean that the water activity will be sufficiently low to prevent the growth of any organisms during storage (Anon, 1980).

Sterilising agents

It was thought to be useful to sterilise capsules at the end of the manufacturing process. This can be done by treatment with ethylene oxide, which is widely used to sterilise medical equipment, particularly dressings and thermolabile materials. It is used in combination with inert gases because it forms an explosive mixture with air. The most frequently employed mixture is 10% ethylene oxide and 90% halogenated hydrocarbons. The sterilising efficiency depends upon the exposure conditions, namely time, temperature, relative humidity and concentration. For the treatment of capsules, only normal ambient conditions can be used, otherwise the gelatin will be affected. As a result, only count reduction can be brought about, not a complete sterilisation. The process is applied to hard gelatin capsules after their final packaging when the appropriate dose is injected into the container, which is then sealed. The dose is so that the residual levels are less than 40 ppm in the capsules (Gold Sheet, 1978). Ethylene oxide breaks down to produce ethylene glycol and ethylene chlorhydrin and even stricter limits are placed on the residual quantities of these compounds, namely 5 ppm and 3 ppm, respectively. This system was widely used during the 1970s and 1980s but it is no longer used in Europe and the USA. It is used in the People's Republic of China and there is a limit for 'chlorethanol' (ethylene chlorhydrin) in their Pharmacopoeia (2000).

Like ethylene oxide treatment, γ-irradiation is widely used to sterilise surgical dressings and delicate equipment. It has been suggested for the sterilisation of capsules, but unfortunately radiation may change both the colour and the performance of capsules. The sterilising dose of γ-radiation is 3 megarads (30 Gy) and at this level treated capsules change colour (Robson and Allan, 1972). The gelatin itself changes from a pale translucent yellow-brown to a darker brown, almost as if charred and several of the more commonly used dyes fade during the treatment. At exposures of 10 megarads (100 Gy), the disintegration times of hard gelatin capsules do not increase but those of soft ones do (Hüttenrauch, 1971). The plasticizer content of soft gelatin capsules probably makes them more susceptible to radiation damage.

All the patents for capsules produced from synthetic polymers claim that such capsules have improved stability, being less susceptible to bacterial growth and moisture. Shionogi Qualicaps have a claim in a patent for manufacturing HPMC capsules that includes an ultraviolet germicidal irradiation of the capsule films before they are stripped from the mould pins to improve their whiteness (European Patent, 1999).

Capsule shell coatings and treatments

Most capsules are formulated to release their contents as soon as they dissolve in the gastric juices. However, some active ingredients need to be released in a different way and capsules are coated with various polymers or the capsule shell is treated to modify its solubility properties. An enteric capsule is one that resists the action of the gastric juices but dissolves under the less acid conditions in the duodenum and intestines (Schroeter, 1965). There are several reasons for using such coatings: drugs may be unstable at the pH of gastric juice, or may irritate the gastric mucosa; they may interfere with gastric metabolism, or the site of their absorption or action may be in the duodenum or intestines. Enteric

capsules have been known since the end of the 19th century but their development has not been as prominent as that of enteric-coated tablets, largely owing to the difficulties of rendering the capsule completely resistant to gastric juice. Early attempts to make enteric capsules have been described in Chapter 1. There are three ways in which standard capsules can be used to make enteric products: the formulation of the shell can be modified to change its solubility characteristics, the shell can be covered with a coating that has the correct solubility characteristics or the capsule can be filled with a product that is formulated to give an enteric release, e.g. enteric-coated pellets.

The solubility of gelatin capsules can be changed significantly by treating them with an aqueous solution of formaldehyde. This causes cross-linkages within the gelatin to be formed, which reduce its solubility making the capsules take longer to dissolve. The release of product from these capsules is controlled by time rather than by a change in intestinal pH like a true enteric coat. Thus the release-controlling step is the gastric residence time of the product. The literature indicates that the reaction of formaldehyde with gelatin is a cross-linking reaction that once initiated is difficult to stop. This reaction was used to produce 'stressed' hard and soft capsules as part of the Capsule Group (FDA/Industry/Academic working party) investigation into changes in the dissolution rate of gelatin capsules after accelerated storage in the 1990s (Cole et al., 1997). However, more recently some workers have claimed to have perfected a system using formaldehyde to produce gastroresistant capsules whose properties do not change on storage (Pina et al., 1996, 1997; Pina and Sousa, 2002). They soaked capsules which had been filled and banded in ethanolic formaldehyde solutions, dried them at 37°C for 30 min, washed them in 90% ethanol for 30 min and then re-dried them (Pina et al., 1996). They claimed that the drying steps completed the reaction between the formaldehyde and the gelatin. They carried a series of dissolution studies using actives with a range of solubilities and found that it was necessary to alter the amount of formaldehyde in relation to their solubility (Pina et al., 1997). They also found no change in disintegration times in simulated intestinal fluid after storage in 'well stoppered' bottles at ambient conditions for 6 months (Pina *et al.*, 1997; Pine and Sousa, 2002).

The fact that capsules can be made insoluble with formaldehyde treatment has been used successfully to develop a position-release gelatin capsule. Biorex Laboratories (1967) obtained a patent for the manufacture of a product for the treatment of duodenal ulcers using this principle. The capsule contained a mixture of the active ingredient, carbenoxalone sodium, an effervescent mixture and sugar filled into a size 00 capsule. After filling, the capsules were pulled slightly apart to elongate them and then were sealed by banding. They were dipped in a formaldehyde solution and then dried. This made the capsules completely insoluble. After a capsule had been swallowed, the sugar drew water into the shell and gas was produced by effervescence. The increased internal pressure resulted in the capsule inflating because the walls were softened after being permeated by the liquid. Once inflated, the capsule was too large to pass through the pyloric sphincter where it became stuck. When the sphincter contracted during the normal stomach activity, the capsule was burst expelling its contents into the duodenum, the required site of action.

Alternatively, capsules can be coated with a solution of polymer that has the required solubility characteristics. The first polymer to be produced that met these needs was cellulose acetate phthalate (CAP). This was developed by Eastman Kodak in 1940 (Malm *et al.*, 1951). The first patents to use CAP with capsules were based on using mixtures with gelatin to form hard shells (Parke Davis & Company, 1949, 1951). However, CAP/gelatin mixtures produced capsules that were unstable on storage because of a slight decomposition of the CAP, which liberated acetic acid and they became less soluble in the intestines. They were also very brittle, because of stresses in the wall caused by incompatibility between the polymers. All enteric polymers by their chemical structure are incipiently brittle and thus despite many patents being granted none using gelatin/polymer mixtures have proved to be viable to date.

Capsules can be coated by any of the classical pharmaceutical methods (Jones, 1970), though the method of choice is the fluidised-bed air-suspension technique. Methods using tablet-coating pans do not work well with capsules because of their low bulk density and awkward shape, which means that they do not roll as evenly as tablets in the pan, and so the coating tends to be uneven. The air-suspension technique allows coating material to be readily applied by spray over all the surface of the capsules. The polymers that can be applied in this way are the standard enteric film coats, CAP, polyvinyl acetate phthalate, methacrylic acid polymers and HPMC phthalate. They were originally applied to the capsules from solution in volatile organic solvents. Plasticizers are included in the formulations in order to improve the mechanical strength and flexibility of the coat and to decrease the transmission of water vapour through it. The most commonly used plasticizers are diethylphthalate, dibutylphthalate, propylene glycol and triacetin.

The problems in applying coatings to the surface of capsules are capsule separation and appearance. In all the coating systems, the capsules are subjected to vigorous movement and can come apart unless self-locking. One of the important features of a gelatin capsule is its high gloss finish. Anything deposited on its surface will detract from its appearance unless it is applied carefully and even the best enteric capsules have a somewhat matt appearance.

The increased awareness of pollution has meant that film coating of tablets and capsules is now principally performed using aqueous systems. Work has been published for both soft and hard capsules. The same equipment is issued as was used for organic solvents systems. The factors that have to be taken into account depend on the individual properties of the two types of capsules.

Several workers have published studies using an aqueous system to coat hard gelatin capsules. Aqueous dispersions of CAP with diethylphthalate as a plasticizer were applied in a fluidised-bed coater (Charreté and Plaizier-Vercammen, 1992). Capsules were filled with either aspirin or sodium salicylate. A three-hour resistance to the acid part of the enteric product test was obtained with aspirin. For sodium salicylate it was found that a subcoat of Eudragit RS 100 was needed. A comparison was made of the coating performance of ammoniated solutions of CAP and cellulose acetate trimellitate (CAT) using several plasticizers (Plaizier-Vercammen and Steppé, 1992). The capsules were filled and the coating was applied in a fluid-bed coater. The amount deposited was related to the acid resistance of the capsule and better results were obtained with CAP then CAT.

Felton *et al.* (1995, 1996) carried out two studies with soft gelatin capsules using an aqueous latex of methacyrlic resin copolymer, Eudragit L30 D-55. In the first study they found that both the type of fill and the plasticizer influenced the disintegration of the capsules (Felton *et al.*, 1995). In the second study they investigated the mechanical properties and the film adhesion properties of the capsules (Felton *et al.*, 1996). The physical analysis was carried out using a Chatillon force gauge attached to a motorised stand, which enabled force–displacement curves to be obtained. This could measure both the break in the film coating and the break in the capsule shell. Both these studies showed that the best plasticizer for the enteric film was triethyl citrate and that the fill formulation also affected the mechanical properties of the capsules.

Enteric-coated hard gelatin capsules are frequently used in early trials particularly for proof of concept work. Fugono *et al.* (2002), in a study examining the use of vanadyl sulfate in the treatment of type 2 diabetes, used special minicapsules, which are designed for administration to rats. They compared the administration of the same quantity of active given as either a solution, a mini-capsule or a mini-capsule enteric coated with HPMC phthalate. They found that the bioavailability of the vanadyl sulfate from the enteric-coated mini-capsules was almost twice that from either a solution or a standard mini-capsule. Hosny *et al.* (2002) carried out a similar study using dogs. They compared a variety of formulations of insulin. They made granules from a mixture of insulin and sodium salicylate, which was added as an absorption promoter, by either wet granulation or by incorporating it into either polyethylene glycol 4000 or Witepsol W35. The resulting granules were filled into capsules and spray coated with a solution of Eudragit S100 in acetone. The capsules were administered to beagle dogs who had been rendered diabetic by food deprivation and the administration of a combination of drugs. The best results were obtained with the Witepsol W35 formulation. An *in vivo* trial in humans was carried out with hand-dipped capsules for the delivery of 4-aminosalicylic acid to the colon (Tuleu *et al.*, 2002). Hard HPMC capsules were coated with a mixture of amylose and ethylcellulose. The passage of the capsules through the intestines was tracked using gamma scintigraphy. Uncoated HPMC capsules released their contents in the stomach within 10 min after swallowing. The coated capsules did not start to release their contents until they reached the colon.

Several simple devices have been reported in the literature for the enteric coating of small numbers of capsules, i.e. <200. Ekberg and Källstrand (1972) first introduced the idea of a coating tower that consisted of a plastic container with a base in which were two inlets for compressed air and a tube through which a coating solution could be introduced. The jets of compressed air caused the capsules to swirl around the walls and rise up the sides. When they lost their momentum they fell back into the centre where they were wetted with coating solution. The air jets then carried them to the side walls where they were carried round again. By this means a coating was gradually deposited and dried on the capsules. Further workers suggested modifications to the apparatus and it evolved into a very sturdy device based on a 750 ml tall-form stainless steel beaker (Forbes and Jones, 1974; Evans *et al.*, 1979). Remon *et al.* (1983) proposed an apparatus for coating even smaller numbers of capsules (<20). This consisted of a rotating disc around the edge of which were clips to hold capsules. The assembly was rotated so the edge of the disc passed through a trough containing the coating solution and the capsules were dipped up to five times to obtain a sufficient thickness of coating. ter Horst *et al.* (1989) proposed an apparatus made of glass, which they recommended for use in either a community or hospital pharmacy. This could coat between 30 and 100 capsules at a time and they gave a detailed protocol for its use.

References

Anon (1980). Equilibrium moisture content and its significance for microbiological stability. In: *Gelfax, a series of useful facts on all aspects of gelatin quality.* No. 20, Croda Gelatin Ltd, Widnes, UK, pp 3.

Anon (1989). Can the appearance of medication improve compliance. *Pharm. J.*, 242 (Jan 28), 98.

Bailie, G. R. and Kesson, C. M. (1981). The effect of capsule color on hypnotic efficiency. *Drug Intell. Clin. Pharm.*, 15, 492–493.

Biorex Laboratories Ltd. (1967). *Improvements relating in or to dosage unit forms for the administration of medicaments and diagnostic agents.* British Patent No. 1 093 286.

Buckalew, L. W. and Coffield, K. E. (1982). An investigation of drug expectancy as a function of capsule color and size and preparation form. *J. Clin. Psychopharmacol.*, 2, 245–248.

Cattaneo, A. D., Lucchelli, P. E. and Fillippucci, G. (1970). Sedative effects of a placebo treatment. *Eur. J. Clin. Pharmacol.*, 3, 43–45.

Cadé, D., Madit, N. and Cole, E.T. (1994). Development of a test procedure to consistently cross-link hard gelatin capsules with formaldehyde. *Pharm. Res.*, 11(10) Suppl., p. S-147 PT 6067.

Charreté, I. and Plaizier-Vercammen, J. A. (1992). Evaluation of a water dispersion of Aquateric for its enteric coating properties of hard gelatine capsules manufactured with a fluidized bed technique. *Pharm. Acta. Helv.*, 67, 227–230.

Cooper, I. J. (1956). The use of colorants and drugs in drugs. *J. Am. Pharm. Assoc., Pract. Pharm. Edn.*, 17, 640–643.

Dow Chemical Co. (1957). *Thermoplastic compositions of water soluble cellulose ethers.* US Patent No. 2 810 659.

Dow Chemical Co. (1969). *Preparation of medicinal capsule shells from hydroxyalkyl-alkylcellulose ethers.* British Patent No. 1 144 225.

Dow Chemical Co. (1973). *Dip coating process for preparing cellulose ether film products.* British Patent No. 1 310 697.

Dow Chemical Co. (1974). *Apparatus for heating capsule forming pins.* US Patent No. 3 842 242.

Dow Chemical Co. (1977). *Capsule shell.* US Patent No. 4 026 986.

Eith, L., Stepto, R. T. F., Wittwer, F. and Tomka, I. (1987). Injection-moulded drug-delivery systems. *Manuf. Chem.*, 58(1), 21, 23, 25.

Ekberg, L. and Källstrand, G. (1972). Magensaftresistent dragéring av operkulatkapsular i recepturskala. *Svensk. Farm. Tidskr.*, 76, 375–378.

Eli Lilly & Co. (1950). *Methyl cellulose and the process of manufacture.* US Patent No. 2 526 683.

Evans, B. K., Fenton-May, V. G. and Lee, M. G. (1979). Enteric-coated capsules; an oral preparation for sodium diethyldithiocarbamate. *J. Clin. Pharmacol.*, 4, 173–177.

FAO/WHO (1965). Joint Expert Committee on Food Additives, Eighth Report, FAO Nutrition Meetings Report Series No. 38 (1965); Technical Report Series WHO No. 309, 1965 and FAO Nutrition Meetings Report Series No. 38B. 1966; WHO/Food Add./66.25.

FAO/WHO (1984). Specifications for identity and purity of food colours. Joint Expert Committee On Food Additives, FAO Food and Nutrition Paper No.31/1, Rome, pp 200.

Felton, L. A., Haase. M. M., Shah, N. H., Zhang, G., Infeld, M. H., Malick, A. W. and McGinty, J. W. (1995). Physical and enteric properties of soft gelatin capsules coated with Eudragit L 30 D-55. *Int. J. Pharm.*, 113, 17–24.

Felton, L. A., Shah, N. H., Zhang, G., Infeld, M. H., Malick, A. W. and McGinty, J. W. (1996). Physical-mechanical properties of film-coated soft gelatin capsules. *Int. J. Pharm.*, 127, 203–211.

Forbes, D. R. and Jones, B. E. (1974). Hard gelatin capsules in hospital pharmacy. *J. Hosp. Pharm.*, 32, 209–214.

Fugono, J., Yasui, H. and Sakuri, H. (2002). Enteric-coating capsulation of insulinomimetic vanadyl sulfate enhances bioavailability of vanadyl species in rats. *J. Pharm. Pharmacol.*, 54, 611–615.

Gold Sheet (1978). Editorial, 12 (No.7), FDC Reports, Chevy Chase, Maryland, USA.

GS Technologies Inc (1997a). *Method and apparatus for manufacturing pharmaceutical cellulose capsules – drying the capsules.* European Patent 0 781 540 A2.

GS Technologies Inc (1997b). *Method and apparatus for manufacturing pharmaceutical cellulose capsules – sizing the capsule.* European Patent 0 781 541 A2.

GS Technologies Inc (1997c). *Method and apparatus for manufacturing pharmaceutical cellulose capsules – fully gelatinizing.* European Patent 0 781 542 A2.

GS Technologies Inc (1997d). *Method and apparatus for manufacturing pharmaceutical cellulose capsules – removing capsule from pin*. European Patent 0 784 696 A2.

GS Technologies Inc (1997e). *Method of manufacture of pharmaceutical cellulose capsules*. US Patent 5 698 155.

GS Technologies Inc (1998a). *Apparatus for the manufacture of pharmaceutical cellulose capsules*. US Patent 5 750 157.

GS Technologies Inc (1998b). *Method for the manufacture of pharmaceutical cellulose capsules*. US Patent 5 756 036.

Hess, H. and Schrank, J. (1979). Colouration of pharmaceuticals – possibilities and technical problems. *Acta Pharm. Technol.*, 8 Suppl., 77–87.

Hosny, E. A., Al-Shora, H. I. and Elmazar, M. M. A. (2002). Oral delivery of insulin from enteric-coated capsules containing sodium salicylate; effect on relative hypoglycemia of diabetic dogs. *Int. J. Pharm.*, 237, 71–76.

Hussain, M. Z. and Ahad, A. S. (1970). Tablet colour in anxiety states. *Br. Med. J.*, iii, 466.

Hüttenrauch, R. (1971). Härten von gelatinehaltigen Arzneiformen, insbesondere Kapseln, durch energiereiche Strahlung. *Pharmazie*, 26, 506–507.

Jacobs, K. W. and Nordan, F. M. (1979). Classification of placebo drugs: effect of color. *Percept. Mot. Skills*, 49, 367–372.

Japan Elanco Company (1993). *Hard capsule for pharmaceutical drugs and method for producing same*. US Patent 5 264 223.

Japan Elanco Company (1995). *Hard capsule for pharmaceutical drugs and method for producing the same*. US Patent 5 431 917.

Japan Elanco Company (1998). *Capsule shell*. US Patent 5 756 123.

Jones, B. E. (1970). Production of enteric coated capsules. *Manuf. Chem.*, 41(5), 53–54, 57.

Jones, B. E. (1973). *Hard Gelatin Capsule Colourants, A Qualitative And Quantitative Study*. MPharm Thesis, University of Wales.

Jones, B. E. (1993). Colours for pharmaceutical products. *Pharm. Tech. Int.*, 5(4), 14–20.

Jones, B. E. (2000). Personal results.

Llewelyn, D. (1981). Legal protection for the coloured get-up of ethical pharmaceuticals. *International Review of Industrial Property and Copyright Law IIC*, No. 2, pp 185–197, Max-Planck-Institute for Foreign and International Patent, Copyright and Competition Law, in association with Verlag Chemie GmbH, Weinheim, Germany.

Lucchelli, P. E., Cattaneo, A. D. and Zattoni, J. (1978). Effect of capsule color and order of administration of hypnotic treatments. *Eur. J. Clin. Pharmacol.*, 13, 153–155.

Madsen, E. (1957). Farrekorrigenser til medicinsk brug. Inledende forsøg. Testrung af Dansk Dørn i aldem 7–12 Ar. *Dansk Tidsskrift for Farmaci*, 31, 29–51.

Malm, C. J. Emerson, J. and Hiatt, G. D. (1951). Cellulose acetate phthalate as an enteric coating material. *J. Am. Pharm. Assoc., Scient. Edn*, 40, 520–525.

Nippon Elanco (1996). *Capsule shell composition and their use*. European Patent Application 0 714 659 B1.

Official Journal of the European Communities (1994). L237, 13, pp 13–29. European Parliament and Council Directive 94/36/EC of 30 June 1994 on colours for use in foodstuffs.

Official Journal of the European Communities (1995). L226, 38, pp 1–45. Commission Directive 95/45/EC of 26 July 1995 laying down specific criteria concerning colours for food use.

Ogura, T., Furuya, Y. and Matsuura, S. (1998). HPMC capsules – An alternative to gelatin. *Pharm. Technol. Eur.*, 10(11), 32, 34, 36, 40, 42.

Parke, Davis & Co. (1949). *Enteric capsule*. US Patent No. 2 491 475.

Parke, Davis & Co. (1951). *Process and apparatus for manufacturing capsules*. US Patent No. 2 575 789.

Peacock, W. H. (1949). *The Application Properties of Certified 'Coal Tar' Colourants*, Calco Technical Bulletin, No. 715, New Jersey, USA.

Pharmacopoeia of the People's Republic of China (2000). English edition, The State Pharmacopoeia Commission of P. R. China, Chemical Industry Press, Beijing.

Pina, M. E. and Sousa, A. T. (2002). Application of hydroalcoholic solutions of formaldehyde in preparation of acetylsalicylic acid gastro-resistant capsules. *Drug Dev. Ind. Pharm.*, 28, 443–449.

Pina, M. E., Sousa, A. T. and Brojo, A. P. (1996). Enteric coating of hard gelatin capsules. Part 1. Application of hydroalcoholic solutions of formaldehyde in preparation of gastro-resistant capsules. *Int. J. Pharm.*, 133, 139–148.

Pina, M. E., Sousa, A. T. and Brojo, A. P. (1997). Enteric coating of hard gelatin capsules. Part 2. Bioavailability of formaldehyde treated capsules. *Int. J. Pharm.*, 148, 73–84.

Plaizier-Vercammen, J. A. and Steppé, K. (1992). Evaluation of enteric coated capsules coated with ammoniated water solutions of cellulose acetate phthalate and cellulose acetate trimellitate. *Pharm. Ind.*, 54, 1050–1052.

Prista, L. N., Morgado, R. R. and Pinho, A. A. (1970). Ensios com cápsulas de gelatina. II – Efeito das radioções ultravioleta. *An. Fac. Farm. Porto*, 30, 35–46.

Remon, J. P., Gyselnick, P., Van Severen, R. and Braeckman, P. (1983). A new small scale apparatus for enteric coating of hard gelatine capsules. *Acta Pharm. Tech.*, 29, 25–27.

Robson, C. and Allen, G. A. (1972). Personal communication, Eli Lilly & Co. Ltd.

Schapira, K., McClelland, H. A., Griffiths, N. R. and Newell, D. J. (1970). Study on the effect of tablet colour in the treatment of anxiety states. *Br. Med. J.*, ii, 446–449.

Schroeter, L. C. (1965). In: E. W. Martin *et al.* (Ed.), *Remington's Pharmaceutical Sciences*, 13th edition. Easton, Pennsylvania: Mack Publishing, p. 601.

Shionogi Qualicaps Company (1999). *Hard capsules*. European Patent 0 592 130 B1.

Shionogi Qualicaps Company (2002). *Process for producing hard capsules*. US Patent 6 413 463.

Swartz, C. J. and Cooper, J. (1962). Colorants for pharmaceuticals. *J. Pharm. Sci.*, 51, 89–99.

Taetz, R. (1875). Un genre de capsule élestique destinée à faciliter la déglutition des medicaments. French Patent No. 106 325.

ter Horst, H. J., Kloeg, P. H. A. M. and van Drunen, J. R. (1989). De 'capsule coater'. Een apparaat voor de bereiding van maagsapresistente capsules in de apotheek., *Pharm. Weekbl.*, 129(29) 530–538.

Tuleu, C., Basit, A. W., Waddington, W. A. Ell, P. J. and Newton, J. M. (2002). Colonic delivery of 4-aminosalicylic acid using amylose-ethylcellulose-coated hydroxypropyl methylcellulose capsules. *Aliment. Pharmacol. Ther.*, 16, 1771–1779.

Turner, T. D. and Jones, B. E. (1971). The identification of blue tri-phenylmethane food dyes by thin layer chromatography. *J. Pharm. Pharmacol.*, 23, 806–807.

Warner Lambert Company (1998). *Polymer film composition for capsules*. International patent application (PCT) WO 98/27 151.

Warner Lambert Company (2000). *Polymer film composition for capsules*. European Patent 1 057 862 A2.

Warner Lambert Company (2001). *Pullulan film compositions*. International patent application (PCT) WO 01/07 507.

Wyatt, M. A. (1966). Personal communication.

4

Manufacture and properties of two-piece hard capsules

Brian E Jones

Introduction

The process for the manufacture of hard two-piece capsules has remained essentially unchanged since the method proposed by Lehuby in his original patent in 1846. It has since been refined and automated by successive generations of chemists, engineers and pharmacists. However, like many other European inventions, it was first developed and then exploited in the United States of America and was only later reintroduced back into Europe. The industry in its present form has evolved from American-based companies and the two that have contributed most to this development are Eli Lilly & Company of Indianapolis (now Shionogi Qualicaps Company Ltd) and Parke, Davis & Company of Detroit (now the Capsugel Division of Pfizer Inc.) (see Chapter 1). Since the first edition of this book there has been a significant increase in the number of manufacturers around the world, particularly in China and India. Capsules are being made in three of the continents, Europe, Asia and America. In most of these places conventional modern machinery is used, but in others, a small number of capsules are still being made on a cottage industry basis using hand-held moulds as described by Lehuby.

The original patent described the use of several materials including carrageenan and acacia. However, the material of choice soon became gelatin because of its technical properties particularly its conversion from a liquid to a solid, a sol to a gel, at a temperature close to ambient conditions. Over the years there have been many attempts to use alternative polymers and since the first edition of this book, capsules are now being made successfully from non-animal origin polymers with the principal one being hydroxy-propyl methylcellulose (HPMC). Capsules can be made from this latter polymer by two completely different routes (see Chapter 3). Both methods use the established standard manufacturing machines, although the Vegicap system requires extensive modifications to be made to enable the mould pins to be heated prior to and immediately after dipping and to the automatic stripping section to allow the dried films to be removed from the pins. The text below describes the manufacture of hard gelatin capsules and the principles used can be applied to capsules made from other polymers that are manufactured on this type of machine.

The manufacturing process

The working of a modern high-speed capsule-manufacturing machine and the essential operations can be summarised as follows. A stock solution is prepared, containing the gelatin itself together with colourants and various process additives. Capsules are made by dipping stainless steel mould pins into this solution and a film, which is subsequently dried, is formed on their surfaces. The dried capsule parts are removed from the moulds, cut to the correct lengths and the caps and bodies are assembled. The assembled capsules are sorted, printed and packaged (Norris, 1959, 1961; Martyn, 1974; Jones, 1982; Hostetler, 1986). During the time that hard capsules have been manufactured there have been significant improvements in the output of capsules per machine per day. This has been achieved through a better understanding of the process and the raw materials. In more recent times further gains have been made by the use of computers, which have allowed close monitoring of the processes, enabling them to be optimised (Martyn, 1974; Hannon and Markowski, 1996).

The gelatin solution

The materials used in the preparation of hard gelatin capsules are the gelatin itself, colourants, which may be natural or synthetic water-soluble dyes or pigments, preservatives and surfactants. The gelatins used have been described in Chapter 2. The manufacturer purchases these against a strict specification and it is a common practice to draw material from more than one supplier and blend different lots together to give the optimum combination of properties.

A stock solution of gelatin is first prepared by dissolving the gelatin in hot (60–70°C) demineralised water in jacketed stainless steel pressure tanks. This concentrated solution contains from 30–40% w/w of gelatin. This solution is very viscous and contains bubbles of entrapped air, which can be removed by applying a vacuum to the vessel in which the solution is prepared. The batch size of the mix is governed by the rate of use. Gelatin in a hot solution hydrolyses and its Bloom strength and viscosity gradually decrease. Thus the manufacturer must balance economy in work against this decline in physical properties. In practice, comparatively small batches are prepared to ensure a quick turnover of material.

When the gelatin solution has matured, aliquots are withdrawn from the manufacturing vessel and are prepared for the requirements of each individual machine. First, the colourants are added. These are in the form of solutions of dyes or suspensions of pigments. At the same time, other materials may be incorporated into the mix, such as a surface tension modifier, e.g. sodium lauryl sulfate solution. The viscosity is measured and then adjusted by the addition of hot demineralised water to a target value, which is dependent upon the size of the capsule and whether it is for the cap or the body. The thickness of the resulting capsule shell wall is governed by the viscosity of the initial gelatin solution. The viscosity can be measured directly by an electromechanical rotational viscometer, or by a standard glass U-tube, or inferentially by measuring the density of the solution with a Baumé hydrometer (Norris, 1961).

Solutions of more than one viscosity are supplied to each machine, the initial solution having a higher viscosity than subsequent deliveries. The reason for this is that the first solution is at the correct viscosity to make the required capsule thickness. The temperature of the solution at dipping is from 45–55°C and as a result a significant amount of water is continually being lost by evaporation. The second and subsequent gelatin solutions prepared for delivery to the machine are made at a lower viscosity than the first one to compensate for this loss.

Capsule formation

The moulds on which capsules are formed are called pins and groups of these are set in line on metal bars; the whole assembly is called a pin bar. Hubel made his original moulds from gauged iron rod, which resulted in capsules that became discoloured on storage. To prevent this, other metals were used, initially phosphor bronze and then stainless steel, which is the material of choice today. The moulds for both cap and body have the same general form, the body being the longer of the two. Their form is so that the capsule shell consists of two parts which are slightly but regularly tapered towards their closed ends and the diameters are so arranged that the inner diameter of the cap near its closed end is slightly less than the outer diameter of the body at its open end (Eli Lilly & Company, 1966; Parke, Davis & Company, 1968). The taper, which is of the order of $0.1–0.3$ mm cm^{-1} in length, is to ensure that the capsule parts can be removed from the mould pins after forming and drying. If the sides were parallel, a vacuum would be created that could cause a collapse of the capsule wall when its removal was attempted.

The gelatin solution is taken to the machine and placed in a jacketed storage hopper. From there it is fed into a jacketed and stirred container called a 'dip pan' or 'dip pot' where the capsule shells are formed. The amount of gelatin solution in the dip pan is controlled by a level-sensing device, which regulates the amount fed from the storage hopper. The dip pan jacket is provided with heating controls so that the temperature of the gelatin solution can be accurately maintained. The level of the gelatin solution in the area of the dip pan, where the capsules are formed, is maintained constant by means of a rectangular box positioned inside it. The solution is pumped up through the centre of the box so that it overflows the edge thus ensuring a constant level inside it even if the amount in the dip pan varies from time to time. The dipping section of the machine is shown in Figure 4.1.

The pin bars are gently lowered into the gelatin solution, then slowly withdrawn. Gelatin is picked up on the mould pins owing to gelation and the quantity retained is governed by the viscosity of the solution. The viscosity also governs the way in which the gelatin solution runs down and off the mould pins. Immediately after withdrawal, there is an accumulation of gelatin on the tops of the pins formed as the pin breaks free from the surface of the solution. To spread this gelatin evenly over the surface of the mould pins, the pin bars are rotated about a horizontal axis as they are transferred from the lower level to the higher level of the machine. As they rise, they pass through a stream of cool air, which helps to set the gelatin solution and fix the film on the mould. There is no further movement in the gelatin film after the pin bars reach the upper level of the machine.

Drying

The pin bars are then passed through a series of drying kilns. In these, large volumes of controlled humidity air are blown directly over the pins to dry the film. The air is heated to a few degrees

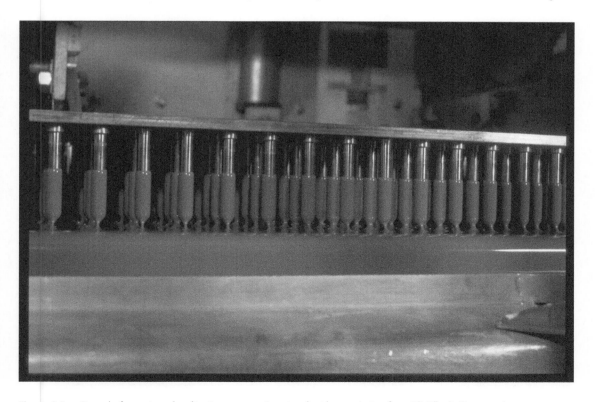

Figure 4.1 Capsule formation: the dipping process (reprinted with permission from Eli Lilly & Company).

above ambient conditions (22–28°C). Higher temperatures cannot be employed because the gelatin gel could revert to a sol and the moulds have to return to the correct operating temperature for redipping, otherwise the viscosity control would be upset. At the rear of the machine the pin bars are transferred back to the lower level of the machine and are returned to the front end through further drying kilns. The drying conditions are adjusted so that the drying rate is low to begin with, slowly reaches a maximum at the rear of the machine and then slows down again as it approaches the front. Excessive drying rates will cause splits to occur in the gelatin films. When the capsule parts emerge from the last kiln they contain about 15–18% w/w of water. This is higher than that required in the finished capsule, 13–16%, but it must be high at this stage, otherwise problems could arise in removing the cast films from the moulds. A balance has to be struck between drying the film until it is strong enough

to handle and overdrying with consequent brittleness. An overall view of the machine is given in Figure 4.2.

Capsule removal and assembly

The gelatin films formed on the pin moulds are longer than required for the finished capsule. This is because the upper edges of the films are of variable thickness, 'feathery', owing to the dipping process so that they need to be trimmed to give a good clean edge at the length required. The capsule shells are removed from the pins by sets of soft metal jaws, which are placed around each pin on a bar. The jaws are closed and pulled back along the pin and, in the process, transfer the shell into a metal holder or collet. The holder is rotated against a sharpened and specially hardened knife that is adjusted to cut the capsule part to the specified length. Gelatin is a tough material

Figure 4.2   Capsule-manufacturing machine (reprinted with permission from Eli Lilly & Company).

and the knives used are made from materials that can retain their sharp edges for long periods. Currently the most successful material that has been used is a ceramic, the same as is used for the tiles on the space shuttle. The trimmings are removed by suction and usually recycled, gelatin being an expensive material. The capsule parts are then transferred to a central joining block where the two halves are fitted together and ejected from the machine. The completed capsules are carried by means of a revolving belt from the machine into a receiver. The output of a manufacturing machine is more than 1 million capsules per 24 hr.

The pin bars are then passed through another section where they are cleaned with felt pads held in metal holders. At the same time a small quantity of lubricant is applied to them. The lubricant or release aid, colloquially referred to as a grease, is necessary to enable the gelatin film to be stripped from the pins. The function of this lubricant is first to prevent the gelatin film from adhering too strongly to the mould pin and second to enable the gelatin film to slide easily over the surface of the metal mould. The materials used are a specific formulation of mixtures of pharmaceutical grade materials, each particular to the individual capsule manufacturer. The pin bars are then assembled in groups of five or more and are ready to enter into the cycle once more.

During the machine cycle the operator takes samples of capsules and checks the lengths of the cap and body, the thickness of the capsule wall and the end wall or dome of the capsule. The thickness of the wall is usually expressed as an average or 'double wall thickness' and is measured by flattening the capsule between the anvils of a measuring gauge. The information obtained from the sample is used to make adjustments to the machine, which can be made either manually on the basis of an operator decision, or can be controlled automatically by a computer (Martyn, 1974; Hannon and Markowski, 1996).

Capsule sorting

Capsules must satisfy dimensional requirements as well as being fault free. Examples of the dimensional specifications for two of the main suppliers on the European market are shown in Tables 4.1 and 4.2.

The lengths of the cap and the body can be measured by using standard gauges, either mechanical or digital. These values are related to each other, in that when the capsule is closed to the correct length after filling (the closed joined length) the cap length is exactly half of the total length. The length of the body is chosen so that when the capsule is closed to the correct closed joined length it extends into the cap as far as the base of its dome. In this position the locking features on the inside of the cap and on the outside of the body are correctly engaged. The unclosed joined length is the length of the empty capsule as it leaves the manufacturers. It is important because, if it varies outside the allowable limits, the capsules will fall apart if they are too long and will fail to be separated on a filling machine if too short because the locking features may be partially engaged. Capsules are made on moulds, which are tapered and as a result are slightly bell-shaped. The diameters are important at two positions in the filling process: the body diameter at the open end because this influences how well it can re-enter the cap during high-speed reclosing and the cap diameter at the closed end, which influences how well the capsule halves stay together after closing.

The hard gelatin capsule is manufactured to tight engineering tolerances despite the fact that it is made from a polymer of natural origin by a dipping process. The nature of the method and the process causes a small percentage of defective capsules to be produced. Immediately after manufacture the empty capsules are sorted either manually, mechanically, or electronically. On ejection from the manufacturing machines capsules are fed on to the surface of honeycomb-like plates. The diameter of the holes on these plates is such that only good capsules can pass through them. Damaged capsules get stuck in the holes and are removed. For manual inspection empty capsules are placed in a hopper and are fed by means of a vibrating plate or moving band past one or more inspectors. Their job is to remove any defective capsules that they see from the bulk. The faults which they look for are defects caused either by poor initial formation of the capsule film, such as thin areas or bubbles, or those produced as the

Table 4.1   Dimensions of Coni-snap capsules made by Pfizer Inc. Capsugel Division, measured at a moisture content between 13% and 16% w/w (Capsugel, 1999, 2000)

Capsule size	Cap length (mm)	Body length (mm)	Cap diameter (mm)	Body diameter (mm)	Closed joined length (mm)
000	12.95	22.20	9.91	9.55	26.1
00el	12.95	22.20	8.53	8.18	25.3
00	11.74	20.22	8.53	8.18	23.3
0el	11.68	20.19	7.65	7.34	23.1
0	10.72	18.44	7.64	7.34	21.7
1	9.78	16.61	6.91	6.63	19.4
2	8.94	15.27	6.35	6.07	18.0
3	8.08	13.59	5.82	5.57	15.9
4	7.21	12.19	5.32	5.05	14.3
5	6.20	9.30	4.91	4.68	11.1
Tolerance	± 0.46	± 0.46	± 0.46	± 0.46	± 0.3

capsules pass through the automatic section of the machine, such as poorly cut edges, splits or holes caused by the stripper jaws. Further information on defects is given in Chapter 13.

In the 1970s Eli Lilly & Company were the first to patent a system that used an optical electronically controlled device to inspect and sort capsules. In this machine the capsules are conveyed on a chain with buckets from a hopper to a revolving disc. This orientates the capsules and spins them at high speed (10 000 rpm) in a beam of light. Defects in the capsule surface will deflect the beam; this deflection is detected by sensors that activate a rejection mechanism to remove the faulty capsules. More recently image-analysing equipment attached to digital cameras has been used to inspect the surface of capsules to remove faults.

Capsule printing

Hard gelatin capsules are usually printed with a variety of information such as the product name, the approved chemical name, product strength or the company name and logo or symbol. This allows rapid identification of the contents of the capsule, which is particularly important in cases of poisoning and it helps to promote the identity of the manufacturer. The process that is used most often is an 'offset' method using an edible pharmaceutical grade ink on an automatic

machine. The process is normally carried out by the manufacturer of the empty shells, although a very small number of companies do print capsules after they have been filled.

Printing machines

Printing machines all have the same basic parts that are essential for an 'offset' process and differ from each other only in the way in which the capsules are handled. The legend to be printed is engraved on a highly polished metal cylinder, made either of stainless steel or of a softer metal such as brass, and which has been chromium-plated. This is called the 'rotogravure cylinder'. On the machine it revolves in a reservoir of printing ink. The ink fills the engraving and as the cylinder revolves it comes into contact with the sharp edge of a metal strip called the 'doctor blade', which removes the ink from its surface leaving only that which fills the engraving. The ink is transferred to a rubber-offset roll geared to run in contact with the rotogravure cylinder. The capsules pass under this rubber roll and the ink is transferred to them. On some machines another rubber roller is used at this position, 'the pressure roll', to ensure that the capsules press against the offset roll to pick up the ink.

There are only a few specialist manufacturers of such machines. Between them they provide models ranging in capacity from 20 000 to one million capsules per hour. R. W. Hartnett

Table 4.2 Dimensions of Posilok capsules made by Shionogi Qualicaps, measured at a moisture content between 13% and 16% w/w (Shionogi Qualicaps, 2002)

Capsule size	Cap length (mm)	Body length (mm)	Cap diameter (mm)	Body diameter (mm)	Closed joined length (mm)
00	11.7	20.4	8.56	8.21	23.6
0E	12.0	20.9	7.66	7.36	24.0
0	10.9	18.6	7.66	7.36	21.8
1E	10.6	18.4	6.95	6.65	21.2
1	9.7	16.7	6.95	6.65	19.5
2	8.9	15.3	6.38	6.08	17.8
3	7.9	13.5	5.85	5.60	15.8
4	7.2	12.4	5.34	5.07	12.4
Tolerance	± 0.3	± 0.3	± 0.3	± 0.3	± 0.3

Company produces a small-output machine, the Delta, in which the capsules are carried in buckets from the hopper to the printing head. This machine can be used to print legends on both sides of the capsule simultaneously. There is a 'B' series of machines with capacities up to one million per hour, on which the capsules are carried from the hopper to the printing roll in a multiple-slotted track in 16 rows. Markem Machines Ltd produces two types of machine.

The smaller and older versions use a metal disc with a slotted circumference to carry the capsules past the printing head. The larger and newer machines use a revolving cylinder with slots cut in it which are filled by centrifugal force, enabling higher outputs, up to 150 000 capsules per hour, to be achieved. This latter machine can also print around the capsule by using a soft rubber offset roll.

One hard capsule manufacturer, Shionogi

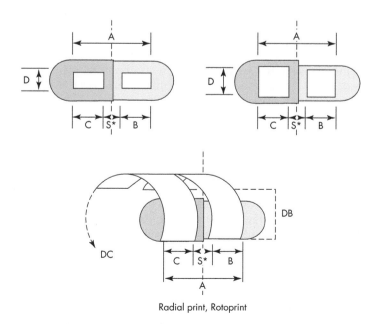

Radial print, Rotoprint

Figure 4.3 Key to printing dimensions on capsules, specified in Table 4.3.

Qualicaps, builds printing machines for its own use. These were designed to meet the quality needs of the Japanese market. The most significant difference between their design and the other manufacturers is that there is a longer conveyor belt after the printing roller than the other machines. This allows more time for the ink to dry on the capsule shells before they come into contact with one another in the collecting hopper and more space for the positioning of automatic inspection devices.

Some modern printing machines are able to rectify capsules before printing if so required. Thus it is possible to print specific information solely on either the cap or the body of capsules and in addition certain machines are able to print messages in two different ink colours. Additionally when the capsules are printed radially it is possible to print bands of colour on the capsules to make a more complicated get-up.

Printing inks

The inks that are used all have the same basic formulation. Insoluble colourants are dispersed in a volatile solvent system that contains a film-forming polymer. The ingredients must be of oral pharmaceutical grade and comply with the legislation in force in the country in which the capsule products are used.

Two basic types of colourant are used: pigments and lake dyes. The pigments are insoluble substances and the most commonly used ones are carbon, iron oxides (red, yellow and black) and titanium dioxide. Lakes are dyes that have been adsorbed on to a substrate of alumina or titanium dioxide. They are made from the same dyes used to colour capsules. The quantity of colourant used is in the range 20–40% w/v of the ink. A large range of colours of ink could be produced using the available colourants, but in practice this is not done and the most common inks are the basic colours black, blue, grey, red and white. The reason for this is that the ink is being applied to the coloured surface of a capsule and the printing characters are very small, so that fine distinctions of colour are very difficult to see.

A film former is the most important constituent of the ink, the one most widely used being shellac. Other substances, such as modified celluloses, have been used, but shellac works well and has the added advantage that it is universally acceptable in medicines. It is a resinous substance produced by a scale insect, *Laccifer lacca Kerr* (Coccidae), which exudes the material onto the branches of the trees on which it lives. It is collected manually, sorted and purified before use. It is readily soluble in organic solvents and in concentrations of 30–40% w/v gives a solution of high viscosity capable of suspending the large quantities of colourants used. When the solvents evaporate it is deposited as a tough shiny film which adheres to the gelatin surface. One drawback in the use of shellac is that it is poorly soluble in water, which makes it difficult to clean equipment after use. However, one ink manufacturer, Colorcon in the 1970s patented a process for chemical modification of the shellac to make it water-miscible, thus considerably reducing the handling problems of the capsule printer.

Various substances are added to the ink to improve its performance on the printing machines. These may be either surfactants or suspending agents. Surfactants improve the spreading characteristics of the ink by lowering its surface tension. They ensure good pick-up of the ink by the engraving on the rotogravure roll and improve its transfer to the rubber-offset roll and hence to the capsule surface. Those used have included dimeticone and lecithin. Suspending agents improve the stability of ink on storage and ensure that the suspension maintains its homogeneity on the printing machines. Modified celluloses, such as HPMC, have been used for this purpose.

Solvents are chosen according to their volatility, as this is the most critical parameter in the performance of an ink. It must dry as soon as possible after it is transferred onto the capsules, because they come into contact with each other almost immediately when ejected from the machine. The ink, however, must not dry too fast, otherwise it would adhere to the rotogravure cylinder and the rubber offset roll and would fail to be transferred to the capsule. The solvents are mainly mixtures of lower alcohols and are blended to give the correct evaporation rate for each particular application.

The printing ink on the machine is contained in an open reservoir in which the rotogravure cylinder rotates and there is a significant loss of solvent, which needs to be replaced. An excessive loss of solvent causes the ink viscosity to increase, which worsens the ink pick-up by the engravings. This problem can be overcome by shrouding the reservoir and by providing a continuous solvent drip feed to make good the loss. An extra portion of solvent present in the ink formulation can be added to the reservoir to modify the evaporation rate and correct any printing faults that may occur. If the ink is drying too slowly, ink transfer will occur from one capsule to another as they leave the machine. Increasing the proportion of isopropanol will correct this. If the ink is drying too fast, there will be poor ink transfer to the capsules and addition of butan-2-ol or 2-ethoxy-ethanol is needed. If there is poor ink pick-up by the engraving, it can be overcome by the addition of ethanol to the reservoir.

The solvents that are used in ink manufacture are covered by a Guidance based on a harmonised regulation from the International Conference on Harmonisation (ICH) initiative (EMEA, 1997; USDHH, 1997). This guidance recommends the acceptable amounts of residual solvents in pharmaceuticals in terms of patient safety. Solvents are divided up into three categories depending upon toxicological data: class 1, solvents to be avoided, known to cause unacceptable toxicities; class 2, solvents to be limited, associated with less severe toxic reactions; and class 3, solvents with low toxic potential. The objective is to encourage the use of class 3 solvents, but it recognises the fact that solvents in class 1 and 2 may have to be used in the manufacturing process for technological reasons and it encourages their replacement by less toxic solvents. The solvents in each class are assigned a permitted daily exposure (PDE) limit and methods of calculating the daily exposure to a pharmaceutical product are given. Solvents used in ink manufacture are from class 3, except for 2-ethoxyethanol, which is used in low concentrations in a number of special inks. The usage levels of ink on capsules or tablet products mean that the levels of residual solvents are extremely low. The volume of solvent that is present in an ink application to a single capsule can be calculated from the volume of the engraving on the rotogravure cylinder and the quantitative composition of the ink. These calculations show that the amount of solvent per capsules applied, i.e. assuming none is lost by evaporation, is significantly lower than the PDE limits. The volume of ink applied per capsule is normally less than 2 µL and in practice most of the solvent is lost by evaporation, which is verified by the fact that the ink is dry to the touch within seconds of the ink being applied.

Design of capsule print

Most typical printing involves a two-part inscription, one part on each half, laid along the long axis of the capsule, axial printing. Alternatively, the inscription can be orientated perpendicularly to the long axis of the capsule, radial printing. In practice, axial printing is still the most commonly used. The surface area available for axial printing is limited by two factors, the position of the cap/body junction and the curvature of the capsule wall, because the empty shell may not be positioned centrally in its holder as it passes under the transfer roll. Thus the size of the inscription that can be applied is reduced so that it does not cross the junction. The typical dimensions of the areas that can be used are given in Table 4.3.

In standard axial printing, an accuate surface subtending an angle of 42° at the axis of the capsule is all that can be used. Shionogi Qualicaps were the first company to devise a method of increasing this angle to 84°, which represents the maximum area of the capsule surface that the eye can see at one glance. This was achieved by the use of a softer rubber transfer roll. In radial printing, a much larger area subtending an angle of almost 360° can be printed but this can only be read by turning the capsule around.

The most important criterion in designing the print on a capsule is legibility. The engraving must be made so that it makes the best use of the available capsule surface and assists problem-free running on the printing machines. To illustrate the points that are taken into account, consider the printing of the word 'tetracycline' axially on a capsule. There are four possible arrangements (see Figure 4.4) (Jones, 1974). The first example shows the word in upper- and lower-case letters so that it completely fills the maximum length available on each half of the capsule. This fails to use the height of the printing area effectively. The shorter the legend the better the appearance, so that long words such as tetracycline are better hyphenated and reproduced on two lines using large print. The second and third examples illustrate this point. In the second example, upper- and lower-case letters are used and in the third example, capital letters only, which even more fully occupies the maximum printable area. If the word is suitable for division into two, one half can be printed on each half of the capsule, using the largest possible print size. This is shown in the fourth example.

It is necessary for the engraving to deposit discrete quantities of ink onto the capsule for maximum clarity. Two types of engraving are used on the rotogravure cylinder. For fine lines, such as in the first example, an 'open etch' engraving is used, which is a simple canal that picks up the ink. For thicker lines, such as in the fourth example, a screened engraving is used, which consists of a grid of holes separated by lands. This picks up less ink and as a result reduces the possibility of ink smudging. Adequate spacing between letters is also important to reduce smudging, particularly in the case of adjacent vertical lines such a between the 'I' and 'N' in the fourth example and where small spaces are surrounded by ink, as in the letter 'R'.

Table 4.3   Dimensions for the available printing area of capsules (Shionogi Qualicaps, 2002) (see Figure 4.2)

Capsule size	Dimension (mm)				
	A	D	B	C	S
Standard axial print					
00	17.5	3.10	6.20	6.20	5.10
0E	17.8	2.80	6.35	6.35	5.10
0	15.9	2.80	5.80	5.80	4.30
1E	15.0	2.70	5.35	5.35	4.30
1	14.2	2.55	4.95	4.95	4.30
2	12.8	2.30	4.25	4.25	4.30
3	11.4	2.10	3.55	3.55	4.30
4	10.8	1.90	3.25	3.25	4.30
5	7.3	1.40	1.90	1.90	4.30
Double-width axial print					
00	17.5	5.70	6.20	6.20	5.10
0E	17.8	5.10	6.35	6.35	5.10
0	15.9	5.10	5.80	5.80	4.30
1E	15.0	4.60	5.35	5.35	4.30
1	14.2	4.60	4.95	4.95	4.30
2	12.8	4.20	4.25	4.25	4.30
3	11.4	3.90	3.55	3.55	4.30
4	10.8	3.55	3.25	3.25	4.30
Radial print					
	A	B & C	S	DB	DC
0E	15.8	4.10	3.60	23.0	23.9
0	14.0	5.20	3.60	23.0	23.9
1E	13.2	4.80	3.60	20.7	21.6
1	11.3	3.90	3.50	20.7	21.6
2	11.4	4.00	3.40	19.0	19.9
3	9.1	2.90	3.30	17.4	18.2
4	7.8	2.30	3.20	15.8	16.7
5	5.5	1.70	2.50	13.6	14.4

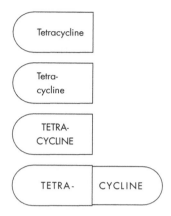

Figure 4.4   Four arrangements for axial printing on capsules.

Packing and storage

Capsules that have passed through sorting, printing and quality control are then packaged for transport and shipment. Up until the 1970s, the majority of empty capsules were packed in polythene-lined fibre drums, which had metal or plastic lids. Developments in packing materials, in the design of handling systems and the economics of freight have meant that drums have been replaced in most cases by cardboard boxes. These have the advantage that they can be stored flat when empty, thereby occupying much less storage space than drums and can be packed more efficiently into large transport containers. The capsules are not placed directly in the boxes but are put inside a liner. The liner may be either a heat-sealed paper sack with saran lining or a polythene bag sealed with a tie, which is adequate in temperate climates where proper warehouse facilities are available. However, multilaminate aluminium foil sacks, which are heat sealed, are required in severe climatic conditions or where the warehousing facilities cannot be maintained at uniform temperatures.

Capsules can be stored in the unopened packs as supplied by the manufacturers for long periods without any undue deleterious effects provided a few simple rules are observed. The storage area preferably should have a temperature between 15 and 30°C and a relative humidity (RH) less than 70%. If this is maintained, the only causes of

capsule damage could be sudden heat changes that result in localised moisture transfer effects within the container. Boxes of empty capsules are comparatively light and because of this they are often stored on the top levels of the racking in a warehouse. In this location there are often windows set into the roof and sunlight could shine on part of a carton causing a high local temperature or other heat sources such as overhead space heaters that could cause a similar problem. These conditions will cause poor separation of capsules on filling owing to changes induced in capsule dimensions by moisture transfer or at worst will cause clumping of capsules inside an unopened carton owing to moisture being concentrated at the spot where the highest temperature is reached. Even if the temperature in the warehouse is well controlled, care must be taken if the temperature in the production filling area is significantly different. For example, in winter months the temperature in the warehouse could be lower than in the production area. If this is the case, time should be allowed for the boxes of capsules to come to equilibrium with their new surroundings before they are opened. This will allow time for the humidity inside the cartons to change and come to an equilibrium, thus avoiding any dimensional changes in the capsule shells.

Quality control

During all stages of the manufacturing operation the requirements of good manufacturing practices (GMPs) are observed and many quality control operations are carried out. An outline of the tests carried out in a typical manufacturing operation is given below. Further information on quality control is given in Chapter 13.

Raw materials

The gelatin is assayed for physical properties: Bloom strength, viscosity and viscosity loss, chemical properties associated with purity, limits for heavy metals, arsenic, ash and microbiological properties such as the total count of organisms present and the absence of coliform and certain test pathogenic organisms. The colourants are assayed for chemical properties: purity and limits for heavy metals and arsenic, colour properties, dye content, subsidiary dye content and colour value. The process aids are assayed for chemical purity and the water for chemical and microbiological purity.

Machine output

The output of the manufacturing machines is continuously monitored for dimensional correctness during the production of each lot. The colour of the capsules produced is checked visually against a standard capsule or a film strip and, if necessary, adjustments are made to the gelatin solution on the machine by adding standardised dye solutions. The colourant composition of the capsules is monitored using thin-layer chromatography to ensure that the correct ones are present. Further colour checks can be made against an absolute numerical standard by measuring the colour of the capsule using a reflectometer, e.g. Hunter Laboratories LabScan XE.

The moisture content of the capsules is monitored throughout the process and the data are used to adjust the drying kilns on the machine. The traditional laboratory method of loss on drying in an oven takes too long so several other methods have been used that give a result in minutes rather than hours. They do not have the absolute precision of the oven technique but the results are sufficiently accurate to make machine adjustments. Infrared balances that dry the sample rapidly have been used but if not done so correctly these can underestimate the moisture content if the initial stage is too rapid because of case hardening of the shells. Another common method is based on the use of a humidity meter. A sample of capsules is placed in a small stainless steel box, which has the meter sensor set in its base. The RH of the air surrounding the capsules is measured and calibration curves are used to determine the moisture content.

Sorting for defects

After the capsules have been inspected either electronically or manually, they are sampled by quality control inspectors and the results compared with an inspection plan (see Chapter 13 for more information). Capsules that fail to meet this plan can either be re-sorted or rejected depending on the type or frequency of the faults found.

Printing inspection

After the capsules have been printed, they are sampled by quality inspectors. The samples are inspected for the quality of print. As with capsule faults, the capsule manufacturers, in consultation with the users, have agreed-upon quality levels for the nature and frequency of defects in printing. The results of the examination are compared with the inspection plan. Capsules that fail this can either be re-sorted or rejected. It is difficult to remove capsules with printing faults by sorting, as the faults are not easy to see except if image analysing equipment is used.

Final inspection

When the capsules have been placed in the final container, a further sample is taken for comparison with the inspection plan for dimensions, physical defects and colour. Capsules can be rejected even at this late stage in the process. During the whole of the manufacturing process, capsules are being continuously monitored for microbiological quality and are only dispatched if this is satisfactory. Samples of each approved lot are stored following standard GMP rules. The final step is to check all the paperwork that is associated with a batch and when this is found to be complete the batch is approved for dispatch to the customer. At this step many manufacturers prepare a certificate of analysis or conformity to send to the customer. It is also common practice for the manufacturers to take a random sample of empty capsules to be used by the incoming quality department of their customers, after they have been validated to do it. The reason is that it avoids the need for the sealed bags of capsules to be opened for sampling for approval. This maintains the integrity of the original manufacturers packaging and ensures that all of the capsules in a lot are stored with the same degree of protection, thus removing another potential cause of variation within a lot that could interfere with the filling process.

Types of hard capsules

The basic shape of the hard two-piece capsule is the same as that proposed by Lehuby in his patent in 1847. Since that time, however, specialist capsules have been made to meet the demands of certain applications; self-locking capsules, capsules for liquid filling, capsules for administration to animals and capsules used for certain clinical trials. For current applications there are certain design features that all hard capsules must possess; a feature to hold the empty capsule shells together, a self-locking feature, an air venting system and a feature to allow accurate rejoining after filling. These features will be described with reference to capsules of the two companies that are used as examples in earlier chapters. They illustrate the basic concepts. However, there are other producers around the world that have solved the problems in different ways, either because of the need to avoid patent infringement or by finding a new solution. The mould pins on which the capsules are formed have these indentations made in their surfaces using presses.

Capsules are made to be self-locking by moulding into the inner surface of the capsule cap and the outer surface of the capsule body certain features that when positioned correctly after the filled capsule has been closed, prevent it from coming apart. The original designs for these features were described in the first edition of this book (Cole, 1987). These features are all distinctive and they enable the manufacturer of the empty capsule to be identified. The trademarks for standard self-locking capsules are Posilok for Shionogi Qualicaps and Coni-snap for Capsugel.

When self-locking capsules were first made on a commercial scale in the late 1960s and early 1970s they came apart more readily than regular capsules in shipment and in handling prior to filling. This was because on the manufacturing machines the caps and bodies could not be closed together as much as previously, otherwise the locking features would have become engaged and the filling machines would not have been able to separate them. To overcome this problem a set of indents were moulded onto the cylindrical part of the cap, commonly called a prelock feature, whose function is to hold the body within the caps and keep the parts together until they were separated by the filling machine. The prelock feature on Capsugel capsules consists of a series of circular depressions and on the Shionogi Qualicaps capsules there are four rectangular-shaped indents aligned around the long axis.

After filling, capsules are conveyed through a transport system to high-speed automatic packaging machines and during this process they are subjected to many forces, principally vibratory and mechanical impact. The self-locking features on capsules are thus designed to prevent the capsules coming apart owing to these stresses. They cannot prevent the capsule being opened manually and capsules in the USA that are required to have a 'tamper-evident' feature have to be sealed in some way after filling (see Chapter 9). The self-locking mechanisms have features on both the cap and body of the capsules. The Shionogi Qualicaps Posilok system has a circular groove near the open end of the body and a corresponding grove in the cap close to the dome that is divided into three parts by unindented areas called lands. The purpose of these lands is to act as air vents. These are required particularly on a high-speed filling machine. When the capsule parts are being rejoined, the filled body is lifted at speed up into the cap, which is restrained from above and because of this movement air is forced into the cap. As the body enters the cap the gap between them is rapidly being closed so that without a special feature to allow air to escape it would be trapped inside the capsule. This would result in the air pressure inside the capsule being greater than that outside and this could result in the caps being easily dislodged during subsequent handling. The Capsugel Coni-snap system has a circular groove on the open end of the body and another on the inside of the open end of the cap. It has two flat areas on either side of the body to act as an air vent. When both types of capsule are closed to their correct closed joined lengths (see Tables 4.1 and 4.2), the features on the caps and bodies are in line and an interference fit is produced that is strong enough to hold the parts together during mechanical handling.

Capsule shells are made to close dimensional tolerances and the gap between the open end of the body and the inside of the open end of the cap is about 0.1 mm. The rejoining of capsules on filling machines is done at high speeds. The walls of empty capsules are flexible and they can flex and distort slightly during mechanical handling. The bushes that carry the capsule parts must also have a clearance allowance to enable the shells to move easily in and out of them. Therefore problems can occur at this point unless everything is perfectly aligned. If the body wall comes into contact with the cap wall the latter will split and a sliver of body will be left outside of the cap instead of inside it. This capsule closing fault, a damaged poorly joined capsule, is commonly known in the industry as a 'telescope'. The reason why such an odd name was chosen that in no way describes the fault has now been lost in time. To avoid this problem the bushings and alignment parts of the filling machines must be well maintained. The empty capsules have a feature designed to minimise this problem. The Capsugel Coni-snap has a tapered rim at the open end of the body. This helps to guide the body into the cap even if it is slightly out of line as it emerges from the body bushing. Shionogi Qualicaps Posilok has tackled the problem in another way; it has a circular groove indented in the cap close to its open end whose function is to maintain the circularity of the opening. This ensures the optimum condition for the entry of the body.

The liquid filling of hard capsules with either liquids mobile at ambient temperatures or with liquefied semi-solid matrix formulations is becoming more popular. The main challenge in this usage is to prevent the leakage of low viscosity mixtures from the capsules before they can be sealed (see Chapter 9). The way to reduce the potential for this kind of leakage is to ensure that the open end of the body comes into contact with the dome of the capsule. Capsugel have made a special grade of capsule for this use, the Licaps (see Table 4.4). The differences from their standard Coni-snap capsules are that the body lengths are longer for sizes 1 and 2, the closed joined lengths for these sizes are shorter and the larger sizes have a more hemispherical-shaped end. The standard Posilok capsule from Shionogi Qualicaps functions satisfactorily when filled with liquids because their bodies make contact with the base of the cap dome.

There is a demand for capsules for veterinary use and a wide range of special sizes have been produced for this purpose ranging from the very small for use with rodents up to the very large for administration to cattle. Capsules for large animals have been available since the late 19th century. Currently they come in a range of sizes with capacities from 3 mL up to 28 mL, diameters from 15 to 24 mm and lengths from 3 to 9 cm. Capsules for administration to rats and other small rodents were developed comparatively recently by some German workers in collaboration with Elanco Qualicaps (Shionogi) (Stanislaus et al., 1979; Lax et al., 1983). Lax et al. (1983) devised a method using a stainless steel tube to administer filled size 9 capsules directly into the stomachs of rats who were conscious. This method was thought to be an improvement on those used previously because it causes less shock and tissue damage to the animals. These capsules are mainly used for toxicity studies. For these 'size 9' capsules, the Shionogi Qualicaps version have

Table 4.4   Dimensions of Licaps capsules made by Pfizer Inc. Capsugel Division, for liquid and semi-solid filling, measured at a moisture content between 13 and 16% w/w (Capsugel 1999, 2000)

Capsule size	Cap length (mm)	Body length (mm)	Cap diameter (mm)	Body diameter (mm)	Closed joined length (mm)
00el	12.95	22.20	8.53	8.18	25.3
00	11.73	20.22	8.53	8.18	23.8
0	10.72	18.44	7.64	7.73	21.7
1	9.78	17.01	6.91	6.63	19.7
2	8.94	15.57	6.35	6.07	17.9

a capacity of 0.025 mL and are 8.4 mm long with a diameter of 2.65 mm, whereas the Capsugel PC caps have a capacity of 0.02 mL and a length of 7.18 mm.

Hard capsules are the ideal dosage form to use in preparing for double blind trials because it is a comparatively easy task to fill tablets into opaque capsules to disguise them. The standard sizes of capsules are capable of containing a good proportion of the tablets in use, however, problems are caused by tablets that have diameters larger than 7 mm, which is the largest diameter tablet that can easily fit inside a size 0 capsule. Capsugel have produced some special capsule sizes for this usage that are called DB, which stands for double blind. The sizes are labelled, AA, A and B, their capacities are 0.94, 0.68 and 0.50 mL and internal diameters are 9.08, 7.81 and 7.85 mm, respectively.

A gelatin/PEG (polyethylene glycol) capsule was one of the outcomes of Shionogi Qualicaps development work that tried to make a hard capsule whose physical properties were not so dependent on its moisture content. The objective for this was to try and make the shell more flexible at lower than normal moisture contents. The result was a gelatin capsule that contains 4% PEG 4000 and it has been shown to exhibit less brittleness than gelatin capsules at the same moisture content (Ogura *et al.*, 1998). However, they are also very brittle at moisture contents less than 10%.

Properties of hard capsules

Moisture content

Gelatin capsules

The capsule when supplied to the user should have a moisture content in the range 13.0–16.0% w/w. This moisture content can differ depending upon the atmospheric conditions to which the capsules have been exposed. Empty capsules should be supplied in moisture-proof containers and it is when the seals of these are broken that unwanted changes in moisture content may occur. The way in which water is held in the

gelatin wall and the quantity that may be transferred has been the subject of study.

There is a pronounced hysteresis in the sorption–desorption isotherms for water in gelatin, which is particularly pronounced at the desorption stage (Strickland and Moss, 1962; York, 1981). To measure the lag time of water vapour diffusion through the capsule wall, Strickland and Moss devised an apparatus in which sealed capsules were suspended on quartz springs in a controlled atmosphere. They found a four-hour lag time before water vapour penetrated through to the contents, which were a hygroscopic powder. York showed that the separation of the absorption and desorption isotherms was due to the amount of water located internally in the material, which cannot easily be replaced.

The moisture transfer between the capsule shell and its contents has been studied by examining the behaviour of several insoluble macromolecular substances, such as starch and microcrystalline cellulose, filled into capsules (Ito *et al.*, 1969). Sorption–desorption isotherms were obtained both for these materials and for empty capsules and the final equilibrium moisture contents of the components of the filled capsules could be calculated from these values. From this work it would appear that, to prevent moisture from passing either to or from the capsule shell or contents on storage, each component should be used at its equilibrium moisture content for the storage conditions. This has been found to be valid for most substances and, in particular, it is true for the preparation of a stable formulation for a capsule of cefalexin (Bond *et al.*, 1970). However, some anomalous behaviour has also been reported (Bell *et al.*, 1973; York, 1981). Bell *et al.* in a study of capsules of sodium cromoglicate found that when filled capsules were recycled through hygrostats at varying humidities they did not return to the initial equilibrium moisture contents. The sodium cromoglicate in the capsules was acting as a moisture sink and was continuously withdrawing water from the capsule shell. York found a similar effect with maize starch and sodium barbitone, which he thought was due to a higher amount of water held internally in these powders. He speculated that increase in moisture content of the encapsulated materials was the cause of reported ageing effects.

The water present in the gelatin film is chemically bound to varying extents. Under normal ambient conditions, capsules cannot be dried to less than 4% w/w of water. This residual water is strongly bound to the gelatin molecule and if it is removed down to a level of 0.3% w/w, the gelatin will not reconstitute to the same physicochemical state (Yannas and Tobolsky, 1967). Above 4% w/w the water molecules are less strongly bound and can easily be added to or removed. The moisture content of the capsule wall strongly affects its physical properties, the optimum being in the 13–16% w/w range. When it falls below 10%, the capsules become brittle and will easily fracture on handling and when it rises above 18% the capsules soften and distort (Bond *et al.*, 1970; Kontny and Mulski, 1989; Liebowitz *et al.*, 1990). The change in moisture content of the capsule is reflected by changes in its physical dimensions. There is a change of about 0.5% in dimensions for each 1% change in moisture content over the range 13–16% w/w (Cole, 1972).

The heat absorption, tensile strength and moisture content of capsules exposed to different humidities have been measured simultaneously by using a specially adapted balance and thermistors (List and Schenk, 1974, 1975). The results were affected by the type of gelatin, the overall composition of the capsules and even the colourants used. A non-destructive method for determining the physical properties of liquid-filled hard gelatin capsules has been made using a texture analyser (Kuentz and Röthlisberger (2002). The objective of this study was to measure the changes in mechanical properties on storage. The authors proposed that the formulations should contain a balanced amount of water to make them compatible with the capsule shell. They showed that the stiffness of the shells changed with time, water content and the excipients used. They thought this could be a useful technique for the rapid evaluation of the effect of formulations on capsule shell properties and hence physical stability.

The equilibrium moisture content (EMC) of hard gelatin capsules has been determined by several groups of workers. The values will vary depending upon the different types of gelatin and colourants used. An empty capsule will have its optimum performance on high-speed filling machines if it is handled in an atmosphere with a RH between 35 and 55%. If the conditions are outside this range the capsules could be adversely affected, but provided only the minimum number are exposed for fairly short times no practical difficulties will ensue, because the lag time to reach the EMC is several hours. Bond *et al.* (1970) used the EMC values of hard gelatin capsules and the cefalexin that they were filling to predict both the best starting moisture contents for the raw materials and the probable stability of the finished product. Zografi *et al.* (1988) used a special apparatus to measure the sorption–desorption transfer between an empty capsule and microcrystalline cellulose, maize starch and silica gel. They used a mathematical model to predict the moisture redistribution between the components and were able to simulate the effect of headspace in a container and temperature. Bremecker (1987) devised an apparatus that was able to measure very small changes in the diameter of capsule bodies when they were exposed to fill materials with different amounts of free water. He proposed this as a rapid method to predict product stability when capsules were filled with a wide variety of different excipients, solid, semisolid and liquid (Bremecker, 1988).

HPMC capsules

HPMC capsules have a lower moisture content specification compared to gelatin capsules. Shionogi Qualicaps Quali-V capsules contain 4–6% and Capsugel Vcaps contain 5–7%. HPMC films are less permeable to water vapour and moisture plays a different role to that in gelatin films. It does not act as a plasticiser, which means that if the capsules lose their moisture for whatever reason, e.g. exposure to low humidities or are filled with hygroscopic formulations, they do not become brittle. Ogura *et al.* (1998) dried gelatin and HPMC capsules down to below their standard moisture content and subjected them to a brittleness test that involved dropping a 50 g weight onto them from a height of 10 cm. The results showed that gelatin capsules below about 11% moisture content become very brittle and that HPMC capsules showed no signs of brittleness even when dried down to below 1%. Podczeck (2002) compared gelatin, gelatin/PEG and

HPMC capsules by measuring indentation forces after storage for up to a week at a range of humidities. HPMC capsules were found to be less resistant to indentation but maintained their mechanical properties at both low humidities and elevated temperatures whereas gelatin and gelatin/PEG capsules did not.

Solubility and disintegration

There is a difference in the solubility in aqueous media between gelatin and HPMC capsules and this affects the way in which the capsules disintegrate and dissolve. This is due to the difference in their permeability to water vapour, gelatin being more permeable (Nagata, 2002). This influences the way in which they dissolve because the polymers first need to absorb water and hydrate before they can dissolve, which is the beginning of capsule break up and the start of release of their contents.

Gelatin is readily soluble in biological fluids at body temperature, which is one of the main reasons that it has been used for so many years to make capsules. However, if the temperature falls much below 37°C then the rate of solution declines. The change in the rate of gelatin dissolution in the range 35–39°C, which is the range used in many pharmacopoeias for disintegration and dissolution testing, has been found to be about 30% (Jones and Cole, 1971). Thus temperature of the test solution becomes significant for the *in vitro* testing of gelatin capsules.

The way in which gelatin capsules disintegrate and dissolve has been studied (Jones and Cole, 1971; Ludwig *et al.*, 1979; Ludwig and Ooteghem, 1980a, 1980b, 1980c, 1981). One of the problems in measuring the dissolution of empty capsules is that they float and if filled with material to make them sink this will influence the gelatin solubility (see section on disintegration in Chapter 13). Jones and Cole (1971) overcame this by putting a metal ball bearing inside a capsule and suspending the body in an aqueous medium. The time for the capsule shell to rupture could then be determined by observing when the ball bearing fell from the capsule. They showed that at 30°C gelatin capsules were insoluble. In their extensive study Ludwig and her co-workers used various

visualisation techniques to observe the break-up of gelatin capsules (Ludwig *et al.*, 1979; Ludwig and Ooteghem, 1980a, 1980b, 1980c, 1981). The walls of gelatin capsules are not completely uniform in thickness because they have been formed by a dipping process. The thickness and structure of gelatin was studied by using isoelectric focusing and microscopy. It was found that the structure of the walls was uniform but that the thickness varied and the thinnest areas were on the shoulders of the cap and body domes (Ludwig *et al.*, 1979). Optical microscopy was used to track the stages in the mechanism of disintegration of capsules in 0.1N HCl with 0.001% Tween 80 (Ludwig and Ooteghem, 1980a). After 40 s of immersion the shell walls begin to wrinkle and the capsules first split at their shoulders. They used a scanning electron microscope to track the movement of the test solution into the shell and found that after only 30 s 75% of the shell had been penetrated (Ludwig and Ooteghem, 1980b). They next studied the rupture time for capsules by using various dissolution apparatuses. They found that the best results were obtained by using the rotating paddle apparatus with the capsule held in a sinker, which consisted of a wire spiral with legs to hold the capsule off the bottom of the vessel (Ludwig and Ooteghem, 1980c). They filled capsules with several model crystalline compounds and measured the time for the first crystal to escape. The results were related to the hydrophilicity and hydrophobicity of the fill materials. In a final paper, the influence of the composition of the test fluids on capsule disintegration was measured (Ludwig and Ooteghem, 1981). The rupture time of the capsules was decreased in acid media and by the addition of pepsin, sodium chloride and Tween 80.

A comparative study of shell dissolution of gelatin, gelatin/PEG and HPMC capsules has been made (Chiwele *et al.*, 2000). The ball-bearing method was used and the test carried out at a range of temperatures from 10 to 55°C, which were chosen because they represent the normal range of temperatures for cold and hot drinks. The dissolution of HPMC capsules was not affected by changes in temperature but that of the gelatin-based capsules was. The release time of the ball bearing in the range 35 to 39°C was less than 100 s for gelatin-based capsules and greater than 300 s for HPMC capsules. Gelatin capsules were found to be insoluble at temperatures below 30°C. Various test solutions were used and it was found that in phosphate buffers the release times for all the types of capsule were prolonged and more variable. The authors suggested that because of their findings it would be better to recommend that gelatin capsules were taken with a warm drink. HPMC capsules could be taken with cold or hot drinks and that carbonated cola-type drinks should be avoided for all types of capsule.

The difference in the time to the first rupture of the capsule shell and hence the start of release of the contents could have an influence on the dissolution and availability of products from capsules. Work published to date does not support this conclusion when comparing gelatin and HPMC capsules. The *in vitro* studies that have been performed have shown that there is a longer lag time before the HPMC capsules start to open but after that the dissolution of the active ingredient is not affected (Honkanen *et al.* 2001; Podczeck and Jones, 2002). Honkanen *et al.* used ibuprofen as a model active compound and compared its release from HPMC and gelatin capsules in two phosphate buffer systems. In trisodium phosphate buffer after 15 min the amounts released were identical, whereas in a potassium phosphate buffer it took longer to reach the point when the results were equal and there was more variability in their results. Potassium ions are known to increase the strength of carrageenan gels. The bioavailability of these two capsules was compared by administering them orally and rectally. They found that by the oral route there was no significant difference between the capsules in any of the standard pharmacokinetic parameters measured, whereas by the rectal route the T_{max}, C_{max} and AUC (area under the curve) values were significantly higher at the 5% level for HPMC. An early *in vivo* trial by Shionogi Qualicaps using capsules filled with cefalexin showed that there was no significant difference between gelatin and HPMC (Ogura *et al.*, 1998). Podczeck and Jones (2002) carried out a study using theophylline as a model compound filled into gelatin, gelatin/PEG and HPMC capsules. Different formulations were prepared using lactose, as an example of a soluble diluent, and microfine cellulose, as an example

of an insoluble diluent, and 0.5% magnesium stearate was added as a lubricant. Capsules were filled with the active alone and the two mixtures by making powder plugs at known compression forces using an Höfliger & Karg powder plug rig. An analysis of variance of the results showed that the formulation and the capsule shell materials were the most influencing factors. It was pointed out that the pharmacopoeias for standard release preparations only specify a percentage released at a specific time and that is important to understand the rate at which dissolution occurs. This was done by calculating the mean dissolution time (Podczeck, 1993), which showed that the rate from the HPMC capsules was faster. It was thought that this was due to the way in which the HPMC capsule dissolves/disintegrates. The time to the first rupture is longer but then the shell disperses evenly over all of its surfaces thereby exposing all the surface of the plug to the dissolution medium, unlike a gelatin capsule where the central tube left by the ends splitting open remains longer before it disperses and unmasks the plug (Podczeck and Jones, 2002).

The assessment of the solubility behaviour of HPMC capsules must be considered in a different way from gelatin capsules. The reason for this is that the formulation of all gelatin capsules is identical, to all intents and purposes, and as a result capsules from any source will have very similar properties in terms of solubility, disintegration and release. The same is not true for HPMC capsules because each manufacturer has his own system, mostly patented, for making a gelling system. Their formulations are not identical and they have different properties. A study of dissolution profiles of acetaminophen, filled into three different makes of HPMC capsules that were available on the market in Japan, was made in two test media of the Japanese Pharmacopoeia (JP) and water (Sakaeda *et al.*, 2002). This showed that only one of the makes of capsule gives satisfactory release in all media, the other two makes had poor release in JP test fluid no. 1, which is 0.1 N HCl with 0.03 M NaCl.

Some workers have shown that gelatin capsules have a tendency to adhere to the oesophagus. Work has been published that compares the performance of gelatin and HPMC capsules (Ponchel and Degobert, 1999; Honkanen *et al.*,

2002). A comparative *in vitro* study was carried out using pieces of porcine oesophageal mucosa fixed to the top of a cylinder (Ponchel and Degobert, 1999). The capsule was fixed on the end of a cylinder that could be raised and lowered, which was attached to a force-measuring device. The assembly was lowered until the capsule contacted the mucosa and the force to remove it was measured after 15 seconds, equivalent to the time of normal oesophageal transit. The adhesivity to the oesophageal mucosa was the same for both types of capsule. A similar study was carried out by Honkanen *et al.* (2002). They used a piece of pig oesophagus kept in a special organ bath. Capsules that had been filled with lactose were placed on the surface of the tissue, left for 1.5 min and the force needed to remove them was measured. They found that the force to detach gelatin capsules was nearly 2.5 times greater than that for HPMC capsules. They concluded that this fact was a reason to use HPMC capsules and commented that the difference in their results from earlier work may have been caused by using a different source of HPMC capsule.

Gas permeability

The filled hard capsule is not gas tight unless it has been banded and gases can readily reach the contents by passage through the space between the cap and body in the region of overlap. Czetsch-Lindenwald (1967) was the first to publish work on the rates of diffusion of both oxygen and carbon dioxide into the contents of filled capsules. He filled capsules with materials that readily absorb the relevant gas, e.g. the catalyst bithional sulfoxide for oxygen or calcium oxide for carbon dioxide. To determine the quantity of gas that passes through the wall relative to that passing through the gap between the cap and the body, he used normally filled capsules and compared them to filled capsule bodies the open ends of which had been sealed. The rate of uptake will be dependent upon the concentration of gas external to the capsule. Hence, to relate the measurements to pharmaceutical practice, he placed capsules in closed containers, e.g. standard glass bottles with a metal screw cap. Packing in a sealed container significantly reduces

the problem. Shah and Augsburger (1989) used an oxygen-sensitive probe, which they placed inside a capsule, to measure the diffusion into standard and banded capsules. They demonstrated that there was an approximate 60-fold decrease in oxygen diffusion into banded capsules. Thus most of the gases enter the capsules through the gap in the cap body overlap region. The quantity of gas travelling by this route will only present a problem if the contents are very susceptible to oxidation. If the contents of banded capsules are still susceptible to oxygen or carbon dioxide they can be protected by formulation procedures, such as suspending the actives in an oily medium.

The diffusion of gases is greater into HPMC capsules than gelatin capsules and this is thought to be due to the structure of the shell wall, which in scanning electron micrographs appears to have a more open structure (Nagata, 2002). If this presents a problem to the formulator then antioxidants can be added to the formulation or aluminium foil blister packaging could be used that would limit the exposure of the capsule.

Stability and storage

Gelatin capsules are relatively inert packages and have been successfully used over more than 100 years to make stable products. However, they do present certain problems in use. They undergo chemical reactions with a few compounds that are well documented (Price, 2000). When capsules are subjected to ICH-accelerated storage conditions, of 40°C and 75% RH for 6 months, the gelatin undergoes a cross-linking reaction that significantly changes their solubility and the dissolution rate of products filled into them. The latter dissolution phenomenon has been shown to be an artefact of the conditions.

An extensive study was carried out in the 1990s by the Capsule Group set up by the American FDA and representatives of academia and industry. It involved both hard and soft gelatin capsules. Hard gelatin capsules were cross-linked by filling them with lactose to which formaldehyde solution has been added (Cadé *et al.*, 1994). Two levels of treatment were used to give capsules with different levels of cross-linking, which were called moderate and severe (Malinowski, 1997) The capsules were filled with acetaminophen and their dissolution rate was tested in water and US Pharmacopoeia (USP) simulated gastric fluid (SGF) with and without enzymes (Gray, 1997). The moderately cross-linked capsules only passed the dissolution test in the SGF with enzymes whereas the severely stressed capsules failed in all three fluids. The capsules were used in a bioavailability study and the usual pharmacokinetic parameters measured (Mhatre *et al.*, 1997). The AUCs from all three capsules were the same and the C_{max}, T_{max} and lag time could be ranked in the order, untreated > moderate stress > severe stress. There was no significant difference between the untreated capsules and moderately stressed capsules, but the severely stressed capsules were outside the 80–125% confidence limits. This result was confirmed in an *in vivo* scintigraphic investigation (Brown *et al.*, 1998), which showed that there was no significant difference in the disintegration time in the stomach between the unstressed and moderately stressed capsules.

It was suggested by Digenis *et al.* (1994) that the best way in which to get an estimate of the probable *in vivo* dissolution rate of gelatin capsules would be to carry out the test using a medium containing enzymes. An initial test should be performed using water and if the capsule product did not comply then it should be repeated using enzymes. Either USP SGF or intestinal fluid could be used with the addition of either pepsin or pancreatin, respectively. The USP has now adopted these test conditions that are used for hard and soft gelatin capsules and gelatin-coated tablets (see Chapter 13). HPMC does not undergo cross-linking reaction, therefore products filled into HPMC capsules when stored under ICH conditions or even more stressful conditions are unaffected (Nagata *et al.*, 2001).

Substances that contain aldehyde groups such as ascorbic acid will discolour on storage at high temperatures and humidities. This is thought to be caused by a reaction between ascorbic acid and an α-amino group in the gelatin molecule and the reaction is catalysed by water (Ogura *et al.*, 1998). The problem can be reduced by controlling the moisture content of gelatin capsules and by using moisture-proof packaging or alternatively it can be avoided by filling the ascorbic acid into HPMC

capsules. Shionogi & Company Ltd. has such a product, Cinal-A, on the market in Japan.

References

Bell, J. H., Stevenson, N. A. and Taylor, J. E. (1973). A moisture transfer effect in hard gelatin capsules of sodium cromoglycate. *J. Pharm. Pharmacol.*, 25, Suppl., 96P–103P.

Bond, C. M., Lees, K. A. and Packington, J. L. (1970). Cephalexin: A new oral broad spectrum antibiotic. *Pharm. J.*, 20S, 210–214.

Bremeker, K.-D. (1987). Modell zur Prüfung der Formstabilität und Versprödung von Hartegelatinekapseln. *Pharm. Ztg*, 132, 1076–1080.

Bremeker, K.-D. (1988). Stabilitätsprüfungen im Zeitraffer: Füllgutformulierungen für Hartgelatinekapseln. *Pharm. Ind.*, 50, 487–490.

Brown, J., Madit, N., Cole, E. T., Wilding, I. R. and Cadé, D. (1998). The effect of cross-linking on the in vivo disintegration of hard gelatin capsules. *Pharm. Res.*, 15, 1026–1030.

Cadé, D., Madit, N. and Cole, E. T. (1994). Development of a test procedure to consistently cross-link hard gelatin capsules with formaldehyde. *Pharm. Res.*, 11 (10) Suppl., p. S-147, PT 6067.

Capsugel (1999). Multistate file, 1st edition, Capsugel Library, pp 78.

Capsugel (2000). General worldwide specifications for Capsugel hard gelatin capsules, revised August 2000, Capsugel Library, pp 20.

Chiwele, I., Jones, B. E. and Podczeck, F. (2000). The shell dissolution of various empty hard capsules. *Chem. Pharm. Bull. Tokyo*, 48, 951–956.

Cole, G. C. (1987). Capsule types, filling tests and formulation. In: Ridgway, K. (Ed.), *Hard Capsules, Development And Technology*, London: Pharmaceutical Press, 165–175.

Cole, W. V. J. (1972). Personal communication.

Czetsch-Lindenwald, H. v. (1967). Sinn und Grenzen der 'in vitro'- Versuche (1. Mitteilung). *Pharm. Ind.*, 29, 145–149.

Digenis, G. A., Gold, T. B. and Shah, V. P. (1994). Cross-linking of gelatin capsules and its relevance to their *in vitro–in vivo* performance. *J. Pharm. Sci.*, 83, 915–921.

Eli Lilly & Co. (1966). *Capsule with an integral locking band*. British Patent No. 1 108 629.

EMEA (1997). ICH Topic Q 3 C, Impurities: residual solvents. *Note For Guidance On Impurities: Residual Solvents* (CPMP/ICH/283/95). London: European Agency for the Evaluation of Medicinal Products, pp 18.

Gray, V. A. (1997). Two-tier dissolution testing. In: *Session Hard and Soft Capsules: Issues, Research and Outcome*. Boston: AAPS meeting.

Hannon, J. T. and Markowski, M. J. (1996). Using techniques for reducing system validation time and cost. *Pharm. Tech.*, 20(11), 40, 42, 44, 46, 48, 50, 52, 54.

Honkanen, O., Eerikäinen, S., Tuominen, R. and Marvola, M. (2001). Bioavailability of ibuprofen from orally administered hydroxypropyl methylcellulose capsules compared to corresponding gelatine capsules. *S. T. P. Pharma Sci.*, 11, 181–185.

Honkanen, O., Laaksonen, P., Marvola, J., Eerikäinen, S., Tuominen, R. and Marvola, M. (2002). Bioavailability and *in vitro* oesophageal sticking tendency of hydroxypropyl methylcellulose capsule formulations and corresponding gelatine capsule formulations. *Eur. J. Pharm. Sci.*, 15, 479–488.

Hostetler, V. (1986). Hard gelatin capsules. In: L. Lachman, H. A. Lieberman and I. L. Kanig (Eds.), *The Theory and Practice of Industrial Pharmacy*, 3rd edition. Philadelphia: Lea & Febiger, 374–398.

Ito, K., Kaga, S.-I. and Takeya, Y. (1969). Studies on hard capsules I. Water vapor transfer between capsules and powders. *Chem. Pharm. Bull., Tokyo*, 17, 1134–1137.

Jones, B. E. (1974). The manufacture and printing of hard gelatin capsules. *Ann. méd. Nancy*, 13, 191–200.

Jones, B. E. (1982). The manufacture of hard gelatin capsules. *Chem. Eng, London*, No. 380, 174–177.

Jones, B. E. and Cole, W. V. J. (1971). The influence of test conditions on the disintegration time of gelatin capsules. *J. Pharm. Pharmacol.*, 23, 438–443.

Kontny, M. J. and Mulski, C. A. (1989). Gelatin capsule brittleness as a function of relative humidity at room temperature. *Int. J. Pharm.*, 54, 79–85.

Kuentz, M. and Röthlisberger, D. (2002). Determination of the optimal amount of water in liquid-fill masses for hard gelatin capsules by means of texture analysis and experimental design. *Int. J. Pharm.*, 236, 145–152.

Lax, E. R., Militzer, K. and Trauschel, A. (1983). A simple method for oral administration of drugs in solid form to fully conscious rats. *Lab. Animals*, 17, 50–54.

Liebowitz, S. M., Vadino, W. A. and Ambrosio, T. J. (1990). Determination of hard gelatin capsule brittleness using a motorised compression test stand. *Drug Dev. Ind. Pharm.*, 16, 995–1010.

List, P. H. and Schenk, G. D. (1974). Untersuchungen an Hartgelatinekapseln. Teil 1: Messungen des hygroskopischen Verhaltens mit Thermistoren. *Arch. der Pharm.*, 307, 719–726.

List, P. H. and Schenk, G. D. (1975). Untersuchungen an Hartgelatinekapseln. Teil 2: Messungen des hygroskopischen Verhaltens bei kontinuierlicher Steigerung der Feuchtigkeit. *Pharm. Ind.*, 37, 91–96.

Ludwig, A. and van Ooteghem, M. (1980a). Disintegration of hard gelatin capsules, Part 2: Disintegration mechanism of hard capsules investigated with a stereoscopic microscope. *Pharm. Ind.*, 42, 405–406.

Ludwig, A. and van Ooteghem, M. (1980b). Disintegration of hard gelatin capsules, Part 3: Penetration and diffusion of liquid through the capsule wall investigated by scanning electron microscopy. *Pharm. Ind.*, 42, 1040–1043.

Ludwig, A. and van Ooteghem, M. (1980c). Disintegration of hard gelatin capsules, Part 4: Method for measuring the time of rupture of the capsule. *Pharm. Ind.*, 42, 1140–1141.

Ludwig, A. and van Ooteghem, M. (1981). Disintegration of hard gelatin capsules, Part 5: The influence of the composition of the test solution on the disintegration time of hard gelatin capsules. *Pharm. Ind.*, 43, 188–190.

Ludwig, A., van Ooteghem, M. and Delva, A. (1979). Disintegration of hard gelatin capsules, Part 1: composition and structure of shell wall. *Pharm. Ind.*, 41, 796–798.

Malinowski, H. (1997). Overview of the process: Issues, research and outcome. In: *Session Hard and Soft Capsules: Issues, Research and Outcome*. Boston: AAPS meeting.

Martyn, G. W. (1974). The people computer interface in a capsule molding operation. *Drug Dev. Comm.*, 1, 39–49.

Mhatre, R. M., Malinowski, H., Nguyen, H., Meyer, M. C., Straughn, A. B., Lesko, L. and Williams, R. L. (1997). The effects of crosslinking in gelatin capsules on the bioequivalence of acetaminophen. *Pharm. Res.*, 14(11), S–251.

Nagata, S. (2002). Advantages to HPMC capsules: A new generation's hard capsule. *Drug. Delivery Tech.*, 2 (2) 34–39.

Nagata, S., Tochio, S., Sakuma, S. and Suzuki, Y. (2001). Dissolution profiles of drugs filled into HPMC capsule and gelatin capsule. *AAPS Pharm. Sci.*, 3 (No. 3).

Norris, W. G. (1959). Hard gelatin capsules – how Eli Lilly make 500 million a year. *Manuf. Chem.*, 30, 233–236.

Norris, W. G. (1961). P.D.'s new capsule plant. *Manuf. Chem.*, 32, 249–252, 258.

Ogura, T., Furuya, Y. and Matsuura, S. (1998). HPMC capsules – An alternative to gelatin. *Pharm. Technol. Eur.*, 10(11), 32, 34, 36, 40, 42.

Parke Davis & Co. (1968). *Hard shell capsules*. British Patent No. 1 108 629.

Podczeck, F. (1993). Comparison of *in vitro* dissolution by calculating mean dissolution time (MDT) or mean residence time (MRT). *Int. J. Pharm.*, 97, 93–100.

Podczeck, F. (2002). The strength and brittleness of hard shell capsules made from different materials. In: *Business Briefings, PharmaTech 2002*, World Market Research Centre Ltd, London, April 2002, 128, 130, 132, 134, 135.

Podczeck, F. and Jones, B. E. (2002). The *in vitro* dissolution of theophylline from different types of hard shell capsules. *Drug Dev. Ind. Pharm.*, 28, 1163–1169.

Ponchel, G. and Degobert, G. (1999). Capsules dures d'hydroxypropyl méthylcellulose: une alternative, à la gélatine, étude comparative de leur adhésivité. *S. T. P. Pharma Pratiques*, 9, 12–17.

Price, J. C. (2000). Gelatin, 11, Incompatibilities. In: Kibbe, A. H., (Ed.), *Handbook of Pharmaceutical Excipients*, 3rd edition. Washington, D.C.: American Pharmaceutical Association and London: Pharmaceutical Press, 216.

Sakaeda, T., Nakamura, T., Okumura, K., Tochio, T. and Nagata, N. (2002). Dissolution properties of hydroxypropylmethylcellulose (HPMC) capsules (in Japanese). *Jpn J. Pharm. Health Care Sci.*, 28, 594–598.

Shah, R. and Augsburger, L. L. (1989). Measurement of oxygen permeation through band-sealed and unsealed hard gelatin capsules. Poster PT642, AAPS meeting.

Shionogi Qualicaps (2002). *The Two-Piece Gelatin Capsule Handbook*. Shionogi Qualicaps, S. A., Alcobendas, Spain, SQ 014/ 0 – 2002 E, pp 81.

Stanislaus, F., Schneider, G. F. and Hofrichter, G. (1979). Zur Technik der Verabreichung von Arzneistoffen in fester Form an der Ratte. *Drug Res.*, 29, 186–187.

Strickland, W. A. and Moss, M. (1962). Water vapor sorption and diffusion through hard gelatin capsules. *J. Pharm. Sci.*, 51, 1002–1005.

USDHH (US Department of Health and Human Services), Food and Drug Administration, Center for Drug Evaluation and Research (CDER) Center for Biologics Evaluation and Research (CBER) (1997). *Guidance for Industry*, Q3C impurities: residual solvents, December 1997, pp 13.

Yannas, I. V. and Tobolsky, A. V. (1967). Cross-linking of gelatine by dehydration. *Nature*, 215, 509–510.

York, P. (1981). Analysis of moisture sorption hysteresis in hard gelatin capsules, maize starch and maize starch: drug powder mixtures. *J. Pharm. Pharmacol.*, 33, 269–273.

Zografi, G., Grandolfi, G. P., Kontny M. J. and Mendenhall, D. W. (1988). Prediction of moisture transfer in mixtures of solids: transfer via the vapor phase. *Int. J. Pharm.*, 42, 77–88.

5

Powder, granule and pellet properties for filling of two-piece hard capsules

Fridrun Podczeck

Introduction

The manufacture of two-piece hard capsules on modern fully automated machinery under industrial conditions requires powders and granules to flow, yet also to pack and to arch to form a sufficiently strong plug. Filling the pellets requires them to flow, to pack and to be mechanically strong enough not to be chipped or broken by the filling tools. Factors that require consideration during the development of two-piece hard capsules are the capsule size in relationship to the required fill weight and dose of drug, the filling method, the powder, granule or pellet properties, the bioavailability of the active ingredient and the stability of the formulation. The required powder and granule properties will vary with the filling method employed. They have to be monitored carefully to ensure that the final dosage form meets all criteria and specifications listed in regulatory documents and the pharmacopoeias.

Powder properties

Particle size and size distribution

For the development of powder formulations for two-piece hard capsules a knowledge of the mean or median particle size as well as the width and shape of the particle size distributions of the active ingredient(s) is important. Any excipient added in large quantities (e.g. fillers, bulking agents) should match the particle size distribution of the drug(s) closely to avoid particle segregation and to promote the homogeneity of the powder mixture. Usually, the filling properties of powders possessing a narrow monomodal particle size distribution are fairly easy to understand. However, the filling behaviour of powders comprising a wide distribution of particle sizes and/or more than one modal fraction is more difficult to predict and such powders often give problems during filling. Typically, such powders tend to segregate and hence there will be a continuous increase in fill weight with time. Whilst for pharmaceutical powders the problem can usually be solved by granulation, the filling of herbal powders often suffers from these problems.

Various methods of particle size analysis can be used, although to describe these is beyond the aims of this chapter. However, the following information is based on weight distributions and thus in order to benefit from these principal findings, the method chosen should also provide a weight distribution or a conversion into a weight distribution should be undertaken.

In general, a larger number of fines, i.e. more than 20% of particles with a diameter of less than 50 µm, leads to poor flow properties of the powder and an increased variability in fill-weight results. In tamp-filling machines the powder bed inside the powder bowl is often angled and varies greatly owing to the discontinuous powder feeding mechanism. At the other extreme, coarse powders with the majority of particles being larger than 150 µm usually show excellent flow properties. However, this will also result in an increased variability in fill weight. In tamp-filling machines the so-called 'water effect' can be seen. When the tamping pins come into contact with the powder bed, particles are displaced sideways and are not pushed into the dosing bores. This phenomenon is similar to that observed when a stone hits the surface of a water reservoir, i.e. the water splashes away from the surface. This water effect can be reduced by increasing the powder bed height so that the tamping pins are immersed in the powder even at their highest position when retracted from the dosing bores. Excellent powder flow is accompanied by a smooth, but not angled powder bed and hence the powder bowl can be filled to a greater extent. When using a dosator nozzle machine, the problems will occur during the plug transfer, because excellent powder flow is usually accompanied by the inability of the powder to form an arch. As a consequence, the powder plug will not be fully retained in the nozzle, powder will be spilled and the fill weight will vary grossly. Similarly, the poor plug-forming abilities of free-flowing powders can influence the fill-weight variability negatively when using tamp-filling machines to try to overfill the capsule body. Overfilling of the capsule body is common practice, where large doses have to be accommodated, for example, when filling antibiotics or non-steroidal anti-inflammatory drugs. Between plug transfer into the capsule body, which is immersed in the bushings, and joining the cap to the body, the part of the plug protruding from the capsule body will collapse, if sufficient plug strength was not achieved, and hence the final fill weight will be reduced in an irregular fashion.

If the median particle size of a powder mixture is larger than 150 µm, it will usually not be possible to fill such powder on a dosator nozzle machine. (Note that this does not apply to granules, if these contain sufficient binding capacity; see under section 'Filling of granules and pellets'). Filling on tamp-filling machines can be achieved by increasing the bed height, but overfilling of the capsule body must be avoided for the reasons discussed above. The plug strength for powders with a median particle size between 100 and 150 µm will still be variable, as the plug produced will often not be strong enough to withstand the transfer. However, in many cases this problem can be solved by adding a filler of similar particle size, which possesses adequate dry binding capacity, for example, 10–20% of microcrystalline

or microfine cellulose. The ideal median particle size range for capsule filling lies between 50 and 100 µm. In this case, the plug formation can be controlled by the tamping pressure or the compression setting when filled on either a tamp-filling or a dosator nozzle machine. Minor flow problems might occur, which can be removed by the addition of an optimal amount of a lubricant. Powders with a median particle size between 20 and 50 µm show an increased tendency to adhere to the metal parts of the filling machinery and also exhibit extremely reduced flowability. The latter is reflected in an increase in fill-weight variability, while the former in the extreme case can lead to blockage of all moving parts and the machine ceases to function. A typical remedy for these problems adopted in Europe is to granulate the powder. However, the addition of excipients such as pregelatinised starch with a small particle size, yet self-lubricating properties and thus better flow properties should also be considered. The addition of optimised quantities of lubricants and glidants might also be sufficient to solve the problem. Typically, magnesium stearate and silicon dioxide in a ratio of 4:1 or magnesium stearate and talcum powder in a ratio of 1:5 are recommended. In any case a low tamping force or compression setting should be used to avoid an increase in wall friction and to prevent an increase in plug disintegration times. Finally, powders with a median particle size below 20 µm can usually not be filled successfully owing to excessive adhesion and friction and extremely poor powder flow. In this case, granulation appears to be the only suitable option.

Particle shape

The shape of powder particles influences all stages of powder handling, in particular mixing, packing and flow. For capsule filling, one critical particle shape is the needle- or rod-shape and this includes all fibrous powders and those with a clearly elongated shape. For crystalline materials with a median particle size above 20 µm, particle size reduction could be considered, but problems resulting from the reduced particle size might be counterproductive for success. Granulation is hence often the only successful method to overcome problems in capsule filling caused by unfavourable particle shape. An extremely irregular particle shape is also unfavourable. However, here the addition of excipients with roundish particle shape of similar particle size, for example, some grades of microcrystalline or microfine cellulose and dried maize starch, is sometimes helpful.

Stickiness

Stickiness of powders to all metal parts of capsule-filling machinery is a serious problem, which is often difficult to solve. It causes a large variability in fill weight and in the extreme case the machine will cease to function after a prolonged running time. Reasons for sticky formulations are mainly large adhesion forces caused by a fine particle size (see above), a low melting point and the tendency of powders to adsorb moisture. The critical melting point is about 100°C. The movement of the machine parts and metal–metal friction can lead to the softening and spot-melting of such powders. It has been shown that for lower dose formulations, i.e. when at least 50% of excipient is added, a mixture of pregelatinised starch and microfine cellulose can be used to prevent stickiness owing to the active substance having a melting point as low as 70°C (Gohil, 2002). Powders, which tend to adsorb moisture, should not be mixed with excipients that contain larger amounts of water such as celluloses or starches. Moisture-regulating excipients such as mannitol or anhydrous lactose can be included. These excipients prevent moisture adsorption and hence reduce the chance of capillary forces, which cause excessive powder stickiness, to develop. However, hygroscopic powders should not be filled into gelatin capsules, as these materials pose severe stability problems.

When using dosator nozzle filling machinery, stickiness caused by adhesion can only be overcome by modification of the formulation. In tamp-filling machines, the early signs of stickiness are a slightly reduced velocity of the machine and an increase in noise, which can be heard by experienced users. Here, stickiness is much more detrimental as not only will the product properties such as fill-weight variability be poor, but also

serious damage to machine parts such as the dosing disk, tamping ring, motor and gears will eventually occur. Increasing the gap size can prevent powder stickiness in the gap between dosing disk and tamping ring in some cases. However, this is usually accompanied by a larger loss of powder owing to the suction employed to keep this gap free of powder. Some machine manufacturers offer to widen the dosing bores in a conical fashion to reduce friction between the tamping pin and the wall of the dosing bores. Although this is sometimes successful in reducing the variability in fill weight and the chance of the machine ceasing to function, this measure does not remove the problem of the stickiness of the powder.

Stickiness is difficult to quantify prior to test runs on capsule-filling machinery. The most sophisticated test is the use of a modified shear cell to measure powder wall friction (Tan and Newton, 1990a; Podczeck and Newton, 2000). Here, the shear plane is between the powder and a polished stainless steel plate. The stainless steel plate can be replaced by other materials relevant to the capsule-filling machinery, for example, brass or steel plates with different surface roughness. Values for the angle of wall friction below 10° are found for non-sticky powder. Other large-scale tests are rather empirical but can also help to evaluate stickiness. Two round plates of about 5 cm in diameter are made from the material(s) to be investigated, for example, brass or brushed stainless steel to simulate the gap between dosing disk and tamping ring. A small amount of powder is distributed between these plates, which are then pressed together with a defined force and twisted against each other. After separating the plates all loosely adhering powder should be blown off using a mild air stream. The stickiness of the powder is then evaluated by examining the remaining material using the following scale:

0 no visible amount of powder remained adhered
1 very little powder remained adhered
2 powder adhered on some places in the form of little spots
3 powder adhered on many places in the form of little spots
4 powder adhered in the form of larger plaques
5 powder adhered completely, cake-like.

If the powder is rated on the above scale with a value of 0 or 1, no major problems caused by stickiness should be experienced. A rating of 2 or 3 indicates stickiness, whereby the optimisation of lubricant and glidant/anti-adhesive concentrations should be sufficient to prevent major filling problems. However, experience has shown that powder rated as 4 or 5 cannot be filled on automated machinery. A less detailed small-scale test involves the use of a metal funnel with a large opening to enable powder flow. The funnel should be filled with 100 g or more of powder and the funnel wall should be inspected after the powder has been emptied by gravitational force. A fine film of dust covering the funnel wall is often observed. If this film can be easily removed, for example, by a gentle air stream, major problems caused by stickiness are less likely to occur, unless stickiness is caused by a low melting point material. The latter cannot be assessed using such a crude method. On the small scale, stickiness can be measured directly as an adhesion force, employing methods such as the centrifuge technique (Podczeck, 1999a), or other methods (Podczeck, 1998).

Powder flow and packing properties

Any powder can expand or contract over a limited range by changing its porosity. It can remain in this expanded or contracted state if not disturbed. Shear forces can deform a powder so that it can flow like a viscous liquid, but the shear force required to produce deformation is a function of the load applied in perpendicular direction to the shear plane. Powder behaviour such as flow or packing is therefore difficult to quantify and predict (Ridgway and Shotton, 1972).

The flowability of powder has a major influence on capsule filling and insufficient flow is usually accompanied by a large variability in fill weight. Sufficient powder flow will ensure that the powder bed height in the powder bowl remains constant, that the powder bed structure is even and that grooves and holes in the powder bed, which are the result of dosing, are refilled immediately. The powder packing properties are similarly important for a constant bed height and even powder bed structure.

The transfer of the powder from the hopper into the powder bowl is often controlled by a force-feeding mechanism and hence is less dependent on powder flow. However, the powder packing properties are again important as the powder might be densified inside the hopper over time owing to machine vibration and the weight of the powder itself. Consequently, a more densified powder might be fed into the powder bowl changing the powder bed structure and resulting in an increase in capsule fill weight. At the end of a filling process, the opposite is often observed, because as the hopper starts to be less full, less densified powder enters the powder bowl. It is therefore important not only to determine the flow properties, but also the dynamic packing properties of a powder.

In tamp-filling machines poor powder flow will also be reflected in the shape of the powder bed as is shown in Figure 5.1a–c. The more angled the powder bed is, the more difficult it is to adjust and control the capsule fill weight, because the first and/or second tamping station will play less or no part in the plug formation. As the increase in the average powder bed height is limited by the size of the powder bowl and spillage of powder from the powder bowl must be avoided in line with the current GMP (good manufacturing practice) standards, a powder bed shape as shown in Figure 5.1c is unsuitable for capsule filling. In such a case the amount of powder at each tamping station will vary continuously and thus the amount of powder tamped at each event will also vary considerably. A large variability in fill weight will be the result.

Different methods are available to measure powder flow and packing. Many of these are empirical and there is also only limited correlation between such flow properties and powder filling owing to simultaneous influence of the packing and plug-forming properties. Experimental values are extremely dependent on the moisture content of the powder and the environmental properties such as relative humidity of the air and, to some degree, temperature. Hence, it is very important to control precisely the experimental conditions for any tests carried out.

(a)
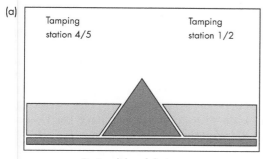
Dosing disk and dosing cone

(b)
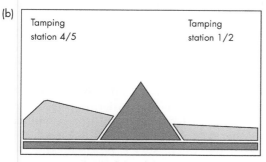
Dosing disk and dosing cone

(c)
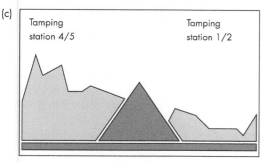
Dosing disk and dosing cone

Figure 5.1 Powder bed behaviour in the powder bowl of a tamp-filling machine. (a) Flat, smooth powder bed, characteristic of excellent flow properties; (b) angled, smooth powder bed, characteristic of good or moderate powder flow; (c) angled, erratic powder bed, characteristic of poor flow properties.

Angle of repose

The angle of repose can be determined by various methods (Train, 1958). All these methods suffer from the fact that they can be applied only to free-flowing powders. The reproducibility of results is extremely poor, depending upon the operator and the method used. The methods can be divided into static and dynamic ones, whereby the angle of repose obtained from a static method tends to be slightly smaller than that from a dynamic one. This might be the result of the collapse of the heap (static methods) or because some degree of inertia needs to be overcome (dynamic methods).

The commonly used static methods are the 'fixed funnel and free standing cone' and the 'fixed base cone.' In both cases the powder needs to flow out of a funnel to form a powder heap, from which the angle of repose α can be estimated as follows:

$$\tan \alpha = \frac{2h}{d} \tag{5.1}$$

$$\tan \alpha = \frac{2h}{d - D/2} \tag{5.2}$$

with h is the height of the cone, d is the diameter of the cone and D is the diameter of the funnel outlet. In the case of the fixed base cone method, d equals the base diameter onto which the powder flows and which restricts the heap by its size. The height of the heap must be determined for both methods. In addition, the cone diameter must be measured when using the fixed funnel and free standing cone method. It is obvious that the fixed base cone method is the more accurate. To avoid heap collapse, the funnel should be positioned so that at the end the cone that has formed has its tip just inside the funnel opening. Sometimes, researchers vary the funnel outlet diameter in order to accommodate less well flowing powders. However, the outlet diameter has an influence on the angle of repose owing to its influence on the arch strength of the powder (see critical orifice diameter). Therefore, Equation 5.2 should preferably be used to calculate the angle of repose.

The dynamic methods embrace the 'tilting box', the 'rotating cylinders' and those methods that give a 'drained angle of repose'. The tilting box is a rectangular box lined with sandpaper to enhance the adhesion and friction forces between the powder and the box walls to a maximum. The box is filled with powder and then tilted until the top powder layer starts to move downwards.

Particles can start to roll or slide down, depending on their properties. The rotating cylinder method has been further developed (see powder avalanching), but is still based on the same principle, i.e. a powder bed can be tilted until the particles overcome friction forces and start to roll or slide. The drained angle of repose is obtained by allowing the powder to pass through the central hole or a side hole of a large cylindrical container. Depending on the method, the angle of the powder heap then needs to be measured directly, or both Equations 5.1 or 5.2 can be employed. In Table 5.1 limited guidelines for the use of the angle of repose as an indicator for capsule-filling properties on tamp-filling and dosator nozzle machines are provided.

Electronic powder flow meters

Electrical or electronic powder flow meters were first developed by Gold et al. (1966). They are used to observe the flow velocity of powders continuously. Figure 5.2a and b shows the typical set-up for such a system and the mass–time profiles obtained from these measurements. An electronic recording balance registers the mass flowing out of the hopper as a function of time. By differentiation of the mass–time curve, the mean flow velocity (in $g\ s^{-1}$) and its standard deviation can be obtained. The ideal powder is characterised by a small mean flow velocity and a standard deviation close to zero. Larger flow velocities and/or larger standard deviations are indicative of flow problems. To overcome arching and blocking of the hopper orifice, some flow meters are equipped with a vibration machine to encourage flow. This appears a sensible approach as on most capsule-filling machinery there is some vibration. So far, no relationships between flow velocity and capsule-filling performance have been reported in the literature.

Powder avalanching

The initial tool used to study the avalanching behaviour of powders was based on feeding powder continuously onto a ramp with a defined tilting angle. It was found that the powder did not immediately and continuously slide down the ramp, but that a powder heap was formed, which gained weight until the accumulated powder bulk suddenly moved down avalanche-like (Kaye, 1993). Kaye et al. (1995) later developed an advanced system, which used a rotating drum and is available commercially. Here, a fixed amount of powder is kept in motion. Ideally, the powder should remain stationary and hence be transported upwards in the direction of the drum rotation, until an angle of repose similar to that obtained in the tilting box method is reached. At that moment the powder heap would collapse avalanche-like and move downwards in the opposite direction to the drum rotation. However, powder movement in a rotating cylinder depends

Table 5.1   Rough guidelines for the use of the angle of repose (fixed base cone method) as an indicator of capsule-filling properties on tamp-filling and dosator nozzle machines

Approximate range (°)	Flow type	Tamp-filling machines	Dosator nozzle filling machines
<25	Very good	Limited filling; water effect, no overfilling possible; use higher powder bed	No filling owing to insufficient plug strength (no arching)
25–30	Good	General range for capsule filling	General range for capsule filling
30–45	Limited	Problems on high-speed machines; fill-weight variability larger at low compression settings, but acceptable at higher compression settings	Good plug strength and arching, but larger fill-weight variability
>45	Poor	No filling within acceptable fill-weight variability	No filling within acceptable fill-weight variability

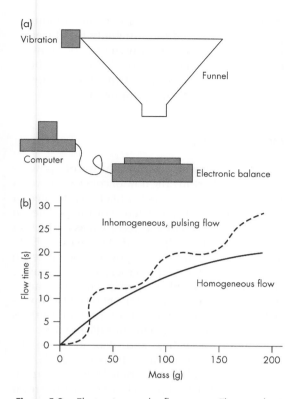

(a) Vibration

Funnel

Computer

Electronic balance

(b)

Figure 5.2 plot: Flow time (s) vs Mass (g)

Inhomogeneous, pulsing flow

Homogeneous flow

Figure 5.2 Electronic powder flow meter. The powder is filled into a funnel, which can be vibrated if necessary. The powder flows out of the funnel onto a recording balance, which is connected to a computer. (a) Set-up of the equipment; (b) flow time as a function of the powder mass flowing out of the funnel onto the recording balance.

on interparticulate forces and the powder–wall friction. Henein *et al.* (1983) consequently identified five different types of movement that can be observed (see Figure 5.3). Boateng and Barr (1996) reported that the rolling bed motion (Figure. 5.3c) would be a desirable powder property for powder mixing. Rolling is a sign of very good powder flow. Cascading (Figure. 5.3d) and cataracting (Figure. 5.3e), however, are signs of poor powder flow owing to large interparticulate forces. Slipping (Figure. 5.3a) and slumping (Figure. 5.3b) indicate that the interparticulate forces are considerably larger than forces between powder and drum wall and hence the powder cannot be transported far enough upwards during drum rotation to reach the critical level of the angle of repose. Occurrence of slipping and

slumping are hence not useful in the determination of powder flow. The drum surface can, however, be modified to increase the wall friction, for example, by fitting a sieve-like metal collar or by coating the drum wall with a thin polymer film. This might result in a change in the type of movement to one more characteristic of the powder flow behaviour.

In the APT AeroFlow® system the rotating drum is placed into a light box, which is equipped with electronic photocells that can record the position of the powder bulk in relation to the drum movement. A real-time two-dimensional picture of the powder bulk is recorded and events such as those shown in Figure 5.3 can be detected and processed. The time between individual events is used to characterise the powder flow properties. A so-called 'strange-attractor plot' is produced by joining the points for the time of two consecutive events (e.g. t_1 and t_2) and the next two events (t_3 and t_4) (see Figure 5.4).

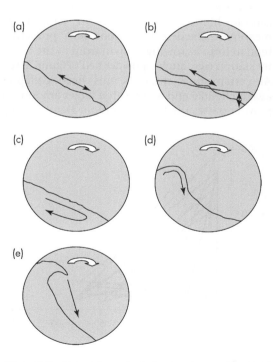

Figure 5.3 Types of powder behaviour in a rotating drum. (a) Powder slipping; (b) slumping; (c) rolling; (d) cascading ('avalanching'); (e) cataracting. Modified from Boateng and Barr (1996).

The centroid of this plot is equivalent to the mean time to avalanche (MTA), i.e. the average time occurring between two events as defined in Figure 5.3. The expansion of the strange-attractor plot in the two dimensions of space illustrates the scatter of times between each individual event and is a reflection of the regularity in flow behaviour. A powder with good flow properties is characterised by a low value of MTA and low scatter. However, these values alone can be very deceptive as, for example, powders that slip provide values of MTA and scatter patterns as small as those observed for rolling motion. However, slipping powders do not necessarily flow well (Trowbridge *et al.*, 1997). Lee *et al.* (2000) hence recommended a dual approach, which is based on a combined evaluation of type of motion and numerical data. In this way, powders with good flow properties could be clearly distinguished from powders with poor flow properties and powders could be ranked in a sensible order with respect to practical experience.

Avalanching behaviour can be observed during capsule filling, for example, in tamp-filling machines. Powders with poor flow properties tend to form a growing heap in front of the transfer station (see 'angled powder bed', Figure. 5.1c). At certain time intervals this heap collapses avalanche-like and as a result a larger amount of powder passes the transfer station. Hence, filling

stations 1 and 2 vary grossly in the amount of powder present and the variability in fill weight is increased. The measurement of powder avalanching behaviour could therefore be a useful tool to predict problems during capsule filling when using a tamp-filling machine.

Powder rheology

One researcher who described the use of a powder 'rheometer' to study the flow of cohesive powders was Cole (1987). A propeller blade was rotated in the powder and the resistance of the powder to the blade movement was recorded as a function of time or blade tip speed. Podczeck (1999b, 1999c) used an automated powder rheometer to study the resistance of microcrystalline cellulose powders against blade movement. The automated equipment enabled the choice of a combination of up- and downward rotor path settings and various blade tip velocities (Figure 5.5). The rotor path was helical and, depending on the direction of movement of the rotor blade, the position of the blade face could be either perpendicular to the helical path (1), or so that the rotor blade angle and the helix angle are similar (2). Depending on whether the torque exerted by the powder onto the blade was measured during the downward (3) or upward (4) movement of the blade, four modes of operation could be distinguished. In the slicing mode (combination of 2 and 3) the impeller blade cuts knife-like through the powder bed, whereas in the compression mode (combination of 1 and 3) the impeller blade pushes the powder behind its face. Similarly, combinations of 2 and 4 and 1 and 4 were defined as 'aeration' and 'moderate compaction and shearing of the powder', respectively. The aeration mode was used to condition the powder back into its original packing state only. Podczeck (1999c) concluded that the most discriminatory operation mode was 'compaction' (1 + 3), whereas 'slicing' (2 + 3) appeared to be the least sensitive procedure, particularly at higher helix angles. Preconditioning of the powder in the measuring vessel was also found to be important, as the packing state of the powder will change from loose packing towards more dense packing during the first few up and down cycles. The 'zero

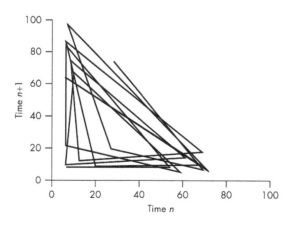

Figure 5.4 Strange-attractor plot: the time points of two consecutive 'avalanche' events and the next two events are joined to give a two-dimensional illustration of the scatter of times between individual events.

torque limit' (ZTL), i.e. the intercept of the torque–displacement curves was used to characterise the flow properties of the powder.

Freeman (2000) used the same equipment to define a series of flow parameters. He based his investigations on the area under the torque–displacement curves, which he related to the energy consumed by the powder during the test. He followed Podczeck's findings in using only the compaction mode for data collection. The compaction flowability index (CFI) was defined as the ratio of the energy consumed during the test by a sample densified to its maximum bulk density

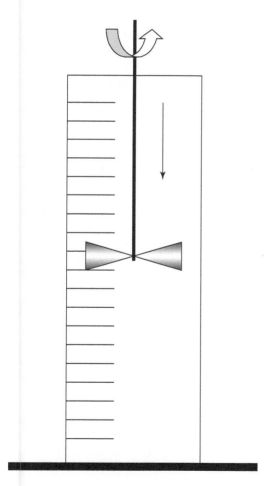

Figure 5.5 Principle of the powder rheometer. Powder is filled into a graduated vessel and a stirrer rotates in the powder bed, moving upward or downward at the same time.

and the energy consumed by a sample that was conditioned by a preceding set of up and down conditioning cycles only. The flow rate flowability index (FRFI) was defined as the energy required for a test traverse at a blade tip speed of 10 mm s^{-1} and the energy obtained using a blade tip speed of 100 mm s^{-1}. Other indices defined concerned powder attrition and segregation. In each case, the ratio between the energy required for a test traverse after a multitude of conditioning steps and the energy consumed initially was sought.

From the above it becomes clear that the powder rheometer allows a large number of different tests to be applied to powder samples. While this is its strength in terms of the adaptability to various needs of the samples, it is also its weakness as comparability between samples and laboratories is only obtained by using exactly matching test conditions. Also, the fill weight of the powder, its apparent particle density and the degree of packing achieved during the preconditioning and test cycles influence the energy values obtained. It is hence difficult to compare the flowability of powders that differ in their general particle properties.

Podczeck and Newton (2000) used the powder rheometer to predict the filling properties of lubricated granulated cellulose powder in a tamp-filling machine. A relationship between the ZTL values and the change in the mean capsule fill weight as a function of the magnesium stearate concentration was found, when filling under higher compression settings.

Determination of powder packing properties

Powders are a mixture of solid particles in air and, depending on the number of contact points between individual particles and the extent of air entrapped in the interparticulate voids, powders can take up a variety of packing states. These are limited by their two extremes, namely 'loose' and 'tight' packing. Loosely packed powders contain a maximum of air space between the particles and a minimum of contact points, whereas tightly packed powders contain a minimum of air entrapped and demonstrate a maximum of interparticulate contact. The two extremes are illustrated for spherical and cubical particles in Figure 5.6. The powder porosity values presented in the figure quantify the amount of air entrapped.

If a defined amount of powder (in grams) is investigated, the powder porosity P can be determined from the volume the powder bulk occupies (V_b):

$$P = 1 - \frac{V_p}{V_b} \qquad (5.3)$$

where V_p is the volume the powder would occupy if no air was entrapped between the particles. The latter can be obtained from the apparent particle density ($V_p = 1/\rho_p$). The apparent particle density is theoretically a material constant and can be measured using an air or a helium pycnometer. However, the major sources of error for this quantity are closed pores inside the particles. It is hence advisable to measure the apparent particle density on micronised samples. The value of V_b can be determined simply using a measuring cylinder.

To obtain a value for the upper extreme powder volume, i.e. the poured or loose bulk volume (V_a) the powder has to be poured into the measuring cylinder without any disturbance, which requires considerable skill. The reciprocal value of V_a is the so-called poured, minimum or loose bulk density (ρ_a).

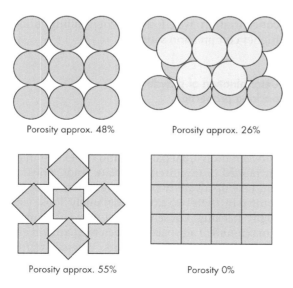

Porosity approx. 48%   Porosity approx. 26%

Porosity approx. 55%   Porosity 0%

Figure 5.6 Loose and tight packing of spherical and cubic particles.

To obtain an estimate of the tightest packing arrangement, the powder bulk is usually subjected to 'tapping', i.e. a measuring cylinder containing the powder is dropped from a certain height onto a surface. The drop height must be sufficient for the particles to have no contact during the enforced jump, so that no adhesion or friction forces are present and the particles can rearrange their relative position to each other. Tapping has to be repeated until the volume occupied by the powder bulk does not reduce any further. The reciprocal of the tapped bulk volume V_t is called the tapped or maximum bulk density (ρ_t).

Tapping can be undertaken employing automated 'tap volumeters'. Hausner (1967) employed the ASTM Standards B-243-61 and B-329-6, which define the tap volumeter as an equipment with a free fall height of 1 inch (about 25 mm) and a tapping frequency of 30 taps per minute. Modern pharmacopoeias (e.g. the European, British and United States Pharmacopoeia), however, describe tap volumeters with a free fall height between 1.5 and 3 mm only and the tapping frequencies are 300 or 150 taps per minute. Podczeck and Sharma (1996) pointed out that with these modern tap volumeters no constant tap density could be reached in a reasonable time span and that the values were significantly different from those that can be predicted from densification models such as that reported by Furnas (1931) or Staple (1975). Abdullah and Geldart (1999) compared a tapped density volumeter meeting Hausner's specification (Hosakawa Powder Tester) with that described in the European Pharmacopoeia (Copley/Erweka Tap Density Volumeter). They concluded that the latter did not provide correct and reproducible values. Thus it appears necessary to change back to the original equipment specifications, i.e. a free fall height of 25 mm and a tapping frequency of no more than 30 taps per minute. It should be noted that all relationships with respect to capsule filling discussed in the following text are based on findings using a tap volumeter meeting the Hausner specification.

Carr's compressibility index and Hausner's ratio

Powder packing and powder flow are closely related bulk properties. Similar to packing, during powder flow, particles will rearrange their relative position to each other in the powder bed by slipping over each other. Also during powder flow, such particle rearrangement will be inhibited by interparticulate friction. As a consequence of the forces acting between particles, there will be a larger difference between the poured and the tapped powder bulk density and the number of taps required to transform the powder bulk from its loosest to its tightest packing state will also increase with an increase in interparticulate friction. Similarly, powder flow will become more and more disturbed. Carr (1965) and Hausner (1967) used the relationship between powder packing and powder flow to define a simple measurement technique to estimate powder flow from powder packing.

Carr's compressibility index CI (Carr, 1965) is defined as follows:

$$CI = \frac{\rho_t - \rho_a}{\rho_t} = \frac{V_a - V_t}{V_t} \qquad (5.4)$$

with ρ_t and ρ_a being the tapped and poured bulk density and V_t and V_a being the tapped and poured bulk volume, respectively. Usually, the values are multiplied by 100 to obtain a percentage value. Hausner's ratio H (Hausner, 1967) is defined as:

$$H = \frac{\rho_t}{\rho_a} = \frac{V_a}{V_t} \qquad (5.5)$$

Carr's compressibility index and Hausner's ratio are equivalent measures and hence only one, usually Carr's compressibility index, is determined. The relationship between the values of CI and the filling of powders and granules into hard shell capsules is summarised in Table 5.2.

Dynamic packing behaviour

The dynamics of the packing process can be determined if the change in packing density during tapping experiments is drawn as a function of the number of taps employed. This is illustrated for three common excipients in Figure 5.7. The process of capsule filling involves the powder bulk being held for a longer time both in the powder storage and dosing hopper. In particular in the storage hopper the powder is subjected to mechanical vibration and hence, the powder will change its packing state. Powders, which reach

Table 5.2 Guidelines for the use of Carr's compressibility index (CI) as an indicator of capsule-filling properties on tamp-filling and dosator nozzle machines. (The findings are based on tap volumeter data obtained with equipment following the original Hausner specification, i.e. 25 mm free fall height and 30 taps min^{-1})

CI (%)	Flow type	Tamp-filling machine	Dosator nozzle machine
<15	Free flowing	Water effect, larger variability in fill weight, filling with largely increased powder bed height required	No plug formation owing to lack of ability to arch; flowability must be reduced (grinding) or addition of binder
15–25	Good	Ideal filling range; powder bed angled but smooth	Ideal filling range; problems with plug strength can be overcome by adding small amounts of dry binder
25–35	Acceptable/poor	Powder bed very angled, but still relatively smooth; addition/ optimisation of lubricant and/or glidant; alteration of machine settings; mechanical flow aids	Good plug formation but variability of fill weight owing to poor flow; adjust machine settings, add/optimise lubricant and/or glidant concentration
>35	Extremely poor/no flow	Powder bed very angled, erratic; no successful filling; reformulation required (granulation or exchange of direct filling excipients)	Good plug formation, but no reformation of smooth powder bed; no successful filling; reformulation (granulation or exchange of excipients)

their maximum bulk density only slowly, will change their packing state over a long period and hence a larger variability of fill weight might result. The dynamic packing profiles such as those shown in Figure 5.7 are helpful in identifying powders that could pose such problems. As can be seen, microcrystalline cellulose (median particle size 50 µm) changes its packing state much more slowly than lactose monohydrate of similar particle size.

Dynamic packing profiles can also be used to define another set of flow parameters that have proved useful in the manufacture of powder- or granule-filled hard capsules. The principles of the first model were derived by Kawakita (1956) and were used for dynamic packing studies by Lüdde and Kawakita (1966):

$$c = \frac{V_a - V_N}{V_a} = ab\frac{N}{1 + bN} \qquad (5.6)$$

where N is the number of taps and a and b are constants. V_a is the poured bulk volume and V_N is the bulk volume after N taps. More commonly, the packing volume is normalised and then drawn as a function of the number of taps employed:

$$\frac{N}{c} = \frac{1}{ab} + \frac{1}{a}N \qquad (5.7)$$

The constant b has no real physical meaning in

tapping experiments, yet seems to reflect to some degree the influence of particle shape and surface texture on powder flow. The reciprocal value of b was found to increase with increasing adhesion forces between the particles in the powder bulk (Yamashiro et al., 1983). It has been shown in theory (Lüdde and Kawakita, 1966) and practice (Podczeck and Lee-Amies, 1996) that the Kawakita constant a and Carr's compressibility index are identical in magnitude and meaning. However, as the determination of the value of V_a is often erroneous, the Kawakita constant a is often a more accurate estimate than Carr's compressibility index. Owing to repeated tapping the error involved in the calculation cancels out, because the slope of the complete function is used to determine its value. Hence, if dynamic packing profiles have been obtained, the Kawakita constant a should be used instead of Carr's compressibility index. The interpretation follows that provided in Table 5.2.

The next model based on dynamic packing experiments was first described by Varthalis and Pilpel (1976). The model was initially the result of empirical data treatment. However, the authors reported a close relationship of their model constant θ (see below) with the angle of internal friction obtained from shear cell experiments. The model describes the change in powder porosity P as a function of the number of taps N:

$$K = \frac{NP^2}{1 - P} = f(N) \qquad (5.8)$$

The slope (i.e. θ) of this linear function is termed 'angle of internal flow' and is regarded as a measure of the cohesive properties of the powder bulk. Increased adhesion, autoadhesion and/or friction forces between the particles in a powder bulk are reflected in an increase in the value of θ. Newton and Bader (1987) reported a linear relationship between the weight of capsules filled on a hand-filling machine simulator and the angle of internal flow, whereby the fill weight was found to decrease with an increase in the value of θ. Tan and Newton (1990b), however, using an instrumented continuous dosator nozzle filling simulator, did not find a relationship between the angle of internal flow and the coefficient of fill-weight variability. Podczeck et al. (1999a)

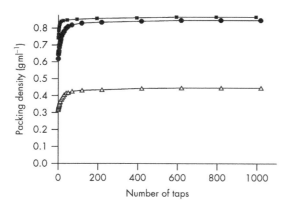

Figure 5.7 Dynamic packing profiles of lactose monohydrate (■), pregelatinised starch (●) and microcrystalline cellulose with a median particle size of 50 µm (△).

found, when filling granules of different size distribution, a direct relationship between the co-efficient of fill-weight variability and the value of θ both on a tamp-filling and on an intermittent dosator nozzle machine. Such a relationship was also reported for filling powders (Podczeck and Newton, 1999) and powdered herbs (Podczeck et al., 1999b) into hard capsules on a tamp-filling machine. The use of this model to characterise powders for capsule filling is hence recommended. However, at this stage it does not appear possible to give a threshold value for θ that should not be exceeded in order to produce capsules with a low variability in fill weight.

Mohammadi and Harnby (1997) proposed the following equation to model dynamic packing profiles:

$$\rho_N = \rho_t - (\rho_t - \rho_a)\exp^{-N/T} \qquad (5.9)$$

where ρ_N is the packing density reached after N taps and ρ_t and ρ_a are the theoretical tapped and poured bulk density, respectively. It should be noted that these two values are fitted by non-linear regression to remove errors that have occurred during filling of the powder into the measuring cylinder and during tapping. The con-stant T or 'compaction constant' is supposed to describe the ease of densification. The value of T appears also to be related to powder flow and an optimum value of 20–25 with respect to capsule filling was reported (Podczeck and Newton, 1999). Values of 15 or less were found to be typical for the flooding tendency ('water effect'), whereas values above 35 indicated that the powder could not be filled on a tamp-filling machine.

Hauer et al. (1993) described the dynamic packing properties as a bi-exponential function, whereby the first exponential term covered the initial rapid densification process and the second term was used to model the slow densification close to reaching the tapped density. The authors found a direct correlation of the first exponent to powder flow and an indirect proportionality to capsule filling on a dosator nozzle machine. The better the flow properties were, the larger was the coefficient of fill-weight variability when filling lactose monohydrate powders of different particle size distribution. Podczeck and Newton (1999) reported an optimum range of 3.5–3.7 for the first

exponent when filling powders on a tamp-filling machine. The second exponential term appeared in both publications as being unimportant.

Critical orifice diameter

The measurement of the critical orifice diameter is not a measure of powder flowability, but of the ability of a powder to form an arch. The values are hence related to the attractive forces acting between the particles in the powder bed.

The critical orifice measurement equipment consists of a stainless steel container of cylindri-cal shape with a capacity of about 200 mL. The container has a set of variable bottom plates with a hole in the centre. The hole size can be varied between 1 and 30 mm. The critical orifice is the hole diameter through which the powder does not flow, i.e. the largest diameter of the hole, where an arch is formed.

The determination of the critical orifice is a helpful tool when working with dosator nozzle machines. Here, the powder must flow well enough to allow reshaping of the powder bed immediately after an individual dose has been removed from the bed by the nozzle. On the other hand, the forces acting between the particles and the nozzle wall must be large enough for the powder to form an arch at the open end of the nozzle. The powder arch must be able to support the powder inside the nozzle and hence prevent the powder from running out during plug transfer to the capsule body. Thus, the critical orifice diameter should be at least of the size of the diameter of the dosator nozzles being used, or even larger. In any case, such measurement should always be accompanied by a measurement of powder flow in order to have the correct balance between flow and the ability to form an arch.

Shear cell measurements

Shear cell measurements aim to assess the strength of a powder bed when it is subjected to a shearing force. For a detailed description of the physical background of such measurements

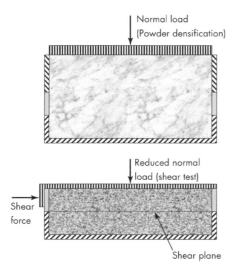

Figure 5.8 Schematic diagram of a Jenike shear cell. Top: the base ring, shear ring and loading ring are assembled and filled with powder. A normal load is applied, which densifies the powder bed. Bottom: the loading ring has been removed and replaced by the shear lid. The normal load acts now on top of the shear lid, but has been reduced by a certain amount. A shear force applied will cause the powder layer contained in the shear ring to slide over the powder bed contained in the base ring.

special literature should be consulted (Pilpel, 1969, 1971; Podczeck, 1998). Shear testers differ in their design. The Jenike shear cell (Jenike, 1961) represents the oldest design (Figure 5.8). It is a linear shear cell and difficult to handle. For the determination of each point of a yield locus (see below), fresh powder has to be used. The more easily operated shear cell designs are the rotational shear testers such as the annular shear cell (Carr and Walker, 1967–8) and the split-level shear tester (Peschl, 1984). In the rotational shear testers, one complete yield locus can be obtained on one powder sample. Measurements are less operator dependent and the reproducibility of the results is very good. The variability of the results here is an expression of the inhomogeneity of powder samples rather than a reflection of the experimental errors.

To carry out a shear cell measurement in a rotational shear tester, the powder is packed into the two halves of the shear cell and consolidated with a defined normal stress. A shear stress is then applied across the two halves of the shear cell to cause failure of the powder bed at the interface of the two halves. The powder is reconsolidated using the initially applied normal stress. The stress is then reduced by a defined amount to give a reduced normal load and shearing is repeated. The complete procedure of reconsolidation and shearing is repeated several times, always using a further reduced normal load. The shear stress measured as a function of the (reduced) normal load applied is called the 'yield locus'. The slope of the yield locus, tan δ, is proportional to the dynamic coefficient of friction and the value of δ is called the internal angle of friction. The larger the value of δ, the more and larger frictional forces are acting between the particles during the shear test. The intercept of the yield locus is called 'cohesion coefficient' or stickiness. It describes the shear strength of a powder at failure for a hypothetical zero normal stress. This value is related to adhesion between the particles in the powder bed.

As both friction and adhesion forces are relevant for powder flow it appears useful to obtain Jenike's flow factor (Jenike, 1961) from the values of the angle of internal friction and the cohesion coefficient. For this reason, several yield loci are obtained employing different normal loads. The unconfined yield strength f_c and the major principal stress σ_m are determined for all yield loci either on the basis of the Mohr semi-circles (Pilpel, 1971), or via numerical evaluation (Jolliffe, 1980; Tan, 1987; Podczeck, 1998). The unconfined yield strength can be regarded as the strength required in a powder bed for arching to occur, whereas the major principal stress is equal to the stress along the major principal plane at zero shear stress. This stress consolidates the powder and is present at any point in the powder bed during gravitational flow (Richards, 1966). The Jenike flow factor is the reciprocal of the slope of the values of f_c as a function of σ_m. For free-flowing powder, the flow factor exceeds a value of 10; easy flowing powders are characterised by a flow factor between 4 and 10; poor powder flow results in values between 1.6 and 4 and values below 4 are typical for non-flowing powders.

Shear cell measurements have been employed to study capsule filling, both on dosator nozzle

(Patel and Podczeck, 1996) and on tamp-filling machines (Gohil, 2002). The best range for filling appears to be a flow factor between 4 and 10.

Jolliffe *et al.* (1980) used Jenike's flow factor to calculate the minimum stress that must be applied to a powder bed in order for it to form a stable arch. In this way the authors could calculate the minimum stress required for the powder to be retained in a dosator nozzle. Using this approach the maximum nozzle diameter for capsule filling can also be obtained (Tan and Newton, 1990c).

A wall yield locus can be measured if one half of the shear cell is replaced by a solid plate that bounds the powder bed. The shear plane is hence not between the two powder layers, but between powder and plate surface. The plate material could be stainless steel, aluminium, plastic or the like. An angle of wall friction is measured instead of the internal angle of friction and the intercept of the yield locus now quantifies adhesion forces between powder and wall. Jolliffe *et al.* (1980), Jolliffe and Newton (1982) and Tan and Newton (1990a) employed this technique to study the influence of wall materials on powder arching in dosator nozzles. Tan and Newton (1990d) also studied the influence of dosator nozzle wall texture on capsule fill-weight variability. They found that the angle of wall friction and the value for powder wall adhesion are functions of the powder properties such as particle size and size distribution. Powder retention problems were found to be associated with low values of the angle of wall friction. When using lactose mono-hydrate powders of different particle size, Jolliffe and Newton (1983) again could clearly identify the influence of wall texture on capsule filling with dosator nozzle machines. However, when using self-lubricating powders such as pregela-tinised starch and microcrystalline cellulose, no effect was observed (Tan and Newton, 1990d). Podczeck and Newton (2000) found a decrease in the angle of wall friction with an increase in the concentration of magnesium stearate, which was proportional to a reduction in the tamping force measured.

It can be concluded that shear cell measurements are a very sophisticated technique to study powder flow. The complexity of the theoretical background should not hinder this technique being employed more extensively in the determination of powder flow properties and their relationship to capsule filling.

Filling of granules and pellets

In principle, the filling of granules into hard shell capsules should obey the same rules as outlined for powders except that for wet granulation products the concentration of binder ensures good plug formation even for larger particle sizes. Granulation usually results in better flow properties, but the granule surfaces are rougher than those of crystalline powders. Hence, the values for the angle of internal flow are much larger than those observed for powders. Typically, for powders these values do not exceed 40° (Podczeck and Newton, 1999), whereas for granules values of 50° or more were observed (Podczeck *et al.*, 1999a; Podczeck and Newton, 2000), yet the granules were all free flowing. Similarly, values for the compaction constant T for granules were found to be between 75 and 140, thus not being in the optimum range proposed for powders (see above). As for powders there appears to be a direct relationship between the angle of internal flow and the coefficient of fill-weight variation (Podczeck *et al.*, 1999a). An acceptable filling performance can be achieved on all types of capsule-filling machines. However, the tamp-filling principle appears to be the better choice for coarser granules, as this method does not rely so much on the formation of a firm plug. Podczeck and Newton (1999a) suggested that in situations where low plug density is required to ensure good drug dissolution and bioavailability, a tamp-filling machine should be employed. On the other hand, if a greater extent of compression was required in order to fill larger doses of drugs, the dosator nozzle principle appeared to be more suitable. However, granulation is not always able to reduce the total powder bulk volume and the choice of the granulation method has also to be considered in this respect (Podczeck and Lee-Amies, 1996).

To fill pellets into hard shell capsules, their size and size distribution, their shape and, if applicable, the film coating properties are important factors to consider. Especially critical is the pellet size distribution (Marquardt and Clement, 1970). As the individual pellets are usually larger than 200 µm, pellets are more prone to rapid segregation. Hence, it is advisable to restrict the range of pellet sizes by removing smaller and larger ones by sieving. The critical difference between smallest and largest pellet size is, unfortunately, dependent on the median pellet size. While for pellets of a median size of about 1000 µm an interquartile range of 100–200 µm is still acceptable, for pellets of a medium diameter of 500 µm only, the interquartile range should be well below 100 µm. A deviation of the pellet shape from being spherical will be detrimental for the filling of pellets if their aspect ratio, i.e. the ratio between length to breadth, exceeds a value of 1.2 (Chopra *et al.*, 2002).

If pellets are film coated, there is a high probability of these pellets acquiring electrostatic charges during the filling process. As a result, the filling mechanism will be blocked and either no filling or filling with large variability in fill weight will occur. Chopra *et al.* (2002) showed that this problem could be solved by blending the pellets with 1% talcum powder prior to filling. The film coating should be elastic and mechanically resistant to avoid damage by the sliders or the dosator nozzle edges. Dynamic mechanical analysis (DMA) offers the possibility to measure the mechanical properties of films *in situ* on the pellets (Podczeck and Almeida, 2002).

What cannot be filled into hard shell capsules?

Powders with increased content of moisture tend to be very sticky and poorly flowing owing to the existence of capillary forces. However, if these materials are not thermolabile, they can be dried prior to filling and filling can be carried out under dry conditions. A classical example is maize starch. Care has to be taken when packing such products, as these have to be protected from moisture. The capsule shell material chosen should not become brittle because of moisture uptake by the powder during storage. Thus hydroxypropyl methylcellulose capsule shells should be used. Powders with a low melting point

also tend to stick to the machinery, because the movement of the powders, the resulting friction and the compression involved in the filling process lead to the generation of heat.

Herbal powders containing stone cells (typically from fruit, seeds or hard bark) destroy the tooling because of the particles scratching the metal surfaces. The released metal dust can contaminate the powder and should hence be regarded as a potential impurity, for which the final product should be checked. Hardened tooling and frequent tool replacement are advisable to overcome this problem.

Large fibrous powders such as herbal powders from roots (e.g. liquorice root) block the filling mechanism and produce capsules with a large variability in fill weight. The large fibrous residues should be removed by sieving the powder and, if necessary, the grinding procedure should be changed.

Any powder that is incompatible with gelatin should be filled into hydroxypropyl methylcellulose capsules, for example, strong acids or bases, proteolytic ingredients, aldehydes, anionic and cationic polymers and astringents such as tannins.

Hygroscopic powders should not be filled into capsules at all. Such substances try to reach their equilibrium with the environment in terms of humidity by water uptake from a chemical reaction, for example, the formation of hydrates. There is no true moisture equilibrium. Such substances remove the water from gelatin capsule shells, which makes them brittle. As they take up moisture from the environment, they become more and more sticky during the capsule-filling process. Hence, these products vary greatly in their fill weight and their stability is impaired.

References

Abdullah, E. C. and Geldart, D. (1999). The use of bulk density measurements as flowability indicators. *Powder Technol.*, 102, 151–165.

Boateng, A. A. and Barr, P. V. (1996). Modelling of particle mixing and segregation in the transverse plane of a rotary kiln. *Chem. Eng. Sci.*, 51, 4167–4181.

Carr, J. F. and Walker, D. M. (1967–8). An annular

shear cell for granular materials. *Powder Technol.*, 1, 369–373.

Carr, R. L. (1965). Evaluating flow properties of solids. *Chem. Engrg.*, 72, 163–168.

Chopra, R., Podczeck, F., Newton, J. M. and Alderborn, G. (2002). The influence of pellet shape and film coating on the filling of pellets into hard shell capsules. *Eur. J. Pharm. Biopharm.*, 53, 327–333.

Cole, G. C. (1987). Powder characteristics for capsule filling. In: Ridgway, K. (Ed.), *Hard Capsules Development and Technology*. London: The Pharmaceutical Press, 80–86.

Freeman, R. E. (2000). The flowability of powders – an empirical approach. *I. Mech. E.*, 566, 545–556.

Furnas, C. C. (1931). Mathematical relations for beds of broken solids of maximum density. *Ind. Eng. Chem.*, 23, 1052–1058.

Gohil, U. C. (2002). *Investigations into the Filling Properties of Powder Mixtures into Hard Shell Capsules.* PhD Thesis, University of London (UK), 184, 202–212.

Gold, G., Duvall, R. N. and Palermo, B. T. (1966). Powder flow studies I: Instrumentation and applications. *J. Pharm. Sci.*, 55, 1133–1136.

Hauer, B., Remmele, T., Züger, O. and Sucker, H. (1993). Gezieltes Entwickeln und Optimieren von Kapselformulierungen mit einer instrumentierten Dosierröhrchen–Kapselabfüllmaschine I. *Pharm. Ind.*, 55, 509–515.

Hausner, H. H. (1967). Friction conditions in a mass of metal powder. *Int. J. Metall.*, 3, 7–13.

Henein, H., Brimacombe, J. K. and Watkinson, A. P. (1983). An experimental study of segregation in rotary kilns. *Metall. Trans. B*, 16, 763–775.

Jenike, A. W. (1961). Gravity flow of bulk solids. *Utah Eng. Exp. Stn. Bull.*, 108, 1–294.

Jolliffe, I. G. (1980). *The Filling of Hard Gelatine Capsules.* PhD Thesis, University of Nottingham (UK), 91.

Jolliffe, I. G. and Newton, J. M. (1982). Practical implications of theoretical consideration of capsule filling by the dosator nozzle system. *J. Pharm. Pharmacol.*, 34, 293–298.

Jolliffe, I. G. and Newton, J. M. (1983). The effect of dosator nozzle wall texture on capsule filling with the mG2 simulator. *J. Pharm. Pharmacol.*, 35, 7–11.

Jolliffe, I. G., Newton, J. M. and Walters, J. K. (1980).

Theoretical considerations of the filling of pharmaceutical hard gelatine capsules. *Powder Technol.*, 27, 189–195.

Kawakita, K. (1956). (in Japanese). *Science*, 26, 149.

Kaye, B. H. (1993). Fractal dimensions in data space; new descriptors for fine particle systems. *Particle & Particle Systems Characterization*, 10, 191–200.

Kaye, B. H., Liimatainen, J. and Faddis, N. (1995). The effect of flow agents on the rheology of a plastic powder. *Particle & Particle Systems Characterization*, 14, 53–66.

Lee, Y. S. L., Poynter, R., Podczeck, F. and Newton, J. M. (2000). Development of a dual approach to assess powder flow from avalanching behaviour. *AAPS Pharm. Sci. Tech.*, 1 (3), article 21, www.pharmscitech.com (accessed 22 September 2003).

Lüdde, K.-H. and Kawakita, K. (1966). Die Pulverkompression. *Pharmazie*, 21, 393–403.

Marquardt, H. G. and Clement, H. (1970). On the dosage accuracy of pellets using a high speed hard gelatin capsule filling and sealing machine. *Drugs Made in Germany*, 13, 21–33.

Mohammadi, M. S. and Harnby, N. (1997). Bulk density modelling as a means of typifying the microstructure and flow characteristics of cohesive powders. *Powder Technol.*, 92, 1–8.

Newton, J. M. and Bader, F. (1987). The angle of internal flow as an indicator of filling and drug release properties of capsule formulations. *J. Pharm. Pharmacol.*, 39, 164–168.

Patel, R. and Podczeck, F. (1996). Investigation of the effect of type and source of microcrystalline cellulose on capsule filling. *Int. J. Pharm.*, 128, 123–127.

Peschl, I. A. S. Z. (1984). New developments in the field of shear test equipment and their application in industry. In: Iinoya, K., Beddow, J. K. and Jimbo, G. (Eds.), *Powder Technology*. Washington D.C.: Hemisphere Publishing, 150–164.

Pilpel, N. (1969). La cohésivité des poudres. *Endeavour*, 28, 73–76.

Pilpel, N. (1971). Cohesive pharmaceutical powders. In: Bean, H. S., Beckett, A. H. and Carless, J. E. (Eds.), *Advances in Pharmaceutical Sciences*, Vol. 3. London: Academic Press, 173–219.

Podczeck, F. (1998). *Particle–Particle Adhesion in Pharmaceutical Powder Handling*. London: Imperial College Press, 99–104, 141–159.

Podczeck, F. (1999a). Investigations into the reduction of powder adhesion to stainless steel surfaces by surface modification to aid capsule filling. *Int. J. Pharm.*, 178, 93–100.

Podczeck, F. (1999b). Rheological studies of the physical properties of powders used in capsule filling I. *Pharm. Technol. Eur.*, 11(9), 16–24.

Podczeck, F. (1999c). Rheological studies of the physical properties of powders used in capsule filling II. *Pharm. Technol. Eur.*, 11(10), 34–42.

Podczeck, F. and Almeida, S. M. (2002). Determination of the mechanical properties of pellets and film coated pellets using Dynamic Mechanical Analysis (DMA). *Eur. J. Pharm. Sci.*, 16, 209–214.

Podczeck, F. and Lee-Amies, G. (1996). The bulk volume changes of powder by granulation and compression with respect to capsule filling. *Int. J. Pharm.*, 142, 97–102.

Podczeck, F. and Newton, J. M. (1999). Powder filling into hard gelatine capsules on a tamp-filling machine. *Int. J. Pharm.*, 185, 237–254.

Podczeck, F. and Newton, J. M. (2000). Powder and capsule filling properties of lubricated granulated cellulose powder. *Eur. J. Pharm. Biopharm.*, 50, 373–377.

Podczeck, F. and Sharma, M. (1996). The influence of particle size and shape of components of binary powder mixtures on the maximum volume reduction due to packing. *Int. J. Pharm.*, 137, 41–47.

Podczeck, F., Blackwell, S., Gold, M. and Newton, J. M. (1999a). The filling of granules into hard gelatine capsules. *Int. J. Pharm.*, 188, 59–69.

Podczeck, F., Claes, L. and Newton, J. M. (1999b). Filling powdered herbs into gelatine capsules. *Manuf. Chem.*, 70(2), 29–33.

Richards, J. C. (1966). Bunker design. In: Richards, J. C. (Ed.), *The Storage and Recovery of Particulate Solids*. London: Institution of Chemical Engineers, 95–137.

Ridgway, K. and Shotton, J. B. (1972). Aspects of pharmaceutical engineering. *Pharm. J.*, 208, 574–576.

Staple, W. J. (1975). The influence of size distribution on the bulk density of uniformly packed glass particles. *Soil Sci. Soc. Am. Proc.*, 39, 404–408.

Tan, S. B. (1987). *Filling of Hard Gelatine Capsules with Powders of Different Flow Properties*. PhD Thesis, University of London (UK), pp. 112.

Tan, S. B. and Newton, J. M. (1990a). Influence of capsule dosator nozzle wall texture and powder properties on the angle of wall friction and powder–wall adhesion. *Int. J. Pharm.*, 64, 227–234.

Tan, S. B. and Newton, J. M. (1990b). Powder flowability as an indication of capsule filling performance. *Int. J. Pharm.*, 61, 145–155.

Tan, S. B. and Newton, J. M. (1990c). Minimum compression stress requirements for arching and powder retention within a dosator nozzle during capsule filling. *Int. J. Pharm.*, 63, 275–280.

Tan, S. B. and Newton, J. M. (1990d). Capsule filling performance of powders with dosator nozzles of different wall texture. *Int. J. Pharm.*, 66, 207–211.

Train, D. (1958). Some aspects of the property of angle of repose of powders. *J. Pharm. Pharmacol.*, 10, 127T–135T.

Trowbridge, L., Williams, A. C., York, P., Worthington, V. L. and Dennis, A. B. (1997). A comparison of methods for determining powder flow: correlation with tabletting performance. *Pharm. Res.*, 14, S–415.

Varthalis, S. and Pilpel, N. (1976). Anomalies in some properties of powder mixtures. *J. Pharm. Pharmacol.*, 28, 415–419.

Yamashiro, M., Yuasa, Y. and Kawakita, K. (1983). An experimental study on the relationship between compressibility, fluidity and cohesion of powder solids at small tapping numbers. *Powder Technol.*, 34, 225–231.

6

Dry filling of hard capsules

Fridrun Podczeck

Introduction

In 1970, the production of hard capsules was estimated to be between 25 and 30 billions per year (Clement and Marquardt, 1970). Meanwhile this figure has certainly doubled even if only pharmaceutical manufacture is considered. However, hard capsules are also an integrated part of the health supplement industry, where in particular powdered herbs are filled into hard capsules (Podczeck *et al.*, 1999).

In the pharmaceutical sector, the majority of formulations filled into hard capsules are dry powder mixtures or granules. Pellets and tablets are also filled into hard capsules (see Figure 6.1). A combination of all three types of formulation can be achieved, but more commonly, combinations of two formulation types are required. Liquids and semi-solids can also be filled alone, or in combination with dry formulations. Such hard capsule products may require special sealing technology to prevent leakage. The topic of filling of liquids/ semi-solids into hard capsules is discussed in Chapter 9.

The advantages of filling powder or granules into hard capsules include the often improved bioavailability owing to a higher porosity of the filled mass, the less stringent requirements on powder flow (see Chapter 5) when compared to tabletting and the ability to fill formulations that are not compressible to the extent required for tabletting. Pellets coated with delicate films, which would rupture in any tabletting process, can be filled into capsules. Controlled release can be achieved by mixing pellets with different release characteristics prior to filling, or filling two or three types of pellets in a consecutive fashion into the capsules. This would also apply to filling of mini-tablets or two to three small tablets into one capsule. Patient compliance would be improved, as they would have to take only one capsule, rather than two or three different tablets at a time.

The use of hard capsules is also a common feature in clinical trials, as the filling of tablets or even capsules themselves (see Figure 6.1) will blind the dosage forms studied. However, this is not easily achieved, as the weight and sensory characteristics have to be matched. For this reason, additional excipients are needed to fill the empty spaces remaining in the capsule shell, which requires careful development and adequate machinery.

In the following text, the general concepts of filling powders/granules, pellets and tablets are explained only from the viewpoint of modern

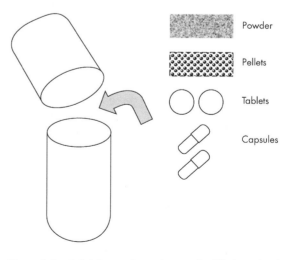

Figure 6.1 Solid dosage forms that can be filled into hard capsules. Up to three combinations of fillings are possible.

filling machinery and bench-top devices. All machines and devices whether automatic or manually operated have to follow the sequence of events illustrated in Figure 6.2.

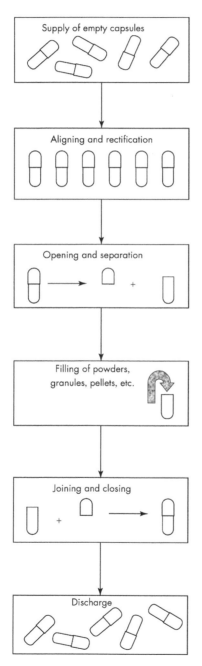

Figure 6.2   General steps in the filling of hard shell capsules.

Manually operated bench-top equipment

Manually operated bench-top equipment is used in hospitals and in the early development phase of capsule formulations, when only 50–100 (max. 300) capsules are produced as a batch. The classical apparatus consists of a series of stacked plates with sets of holes matching either the capsule body or cap in diameter. These plates can be exchanged in order to handle different capsule sizes. Empty capsules are poured on top of the upper plate. A special mechanism helps in rectification, although some manual help to position all shells correctly into the holes provided and to remove the surplus of capsule shells might be needed. Afterwards, capsules are pushed down into the filling plate ('dosing disk') and the cap and body of each capsule shell are separated. The powder must now be evenly filled into the capsule bodies. This can be achieved by pouring a predetermined weight of the powder bulk on top of the plate containing the open capsule bodies and using a stroking device to spread the powder evenly over the plate. The powder will thus fill into the capsule bodies on the basis of powder flow only. If the flow properties are not good enough, air pockets inside the capsule bodies could hinder an even distribution of the formulation and cause a deviation in fill weight between individual capsules. Hence, often such apparatus will be provided with a 'tamping device' (see Figure 6.3). The tamping device consists of a plate set with pins matching the inner diameter of the capsule bodies. When the pins are lowered inside the capsule bodies, they compress the powder by removing the entrapped air. If required, more powder can then be distributed on the plate. A further advantage of the tamping device is the possible increase in fill weight owing to densification of the powder beyond its poured bulk density. Finally, capsule caps and bodies are rejoined and the capsules are closed. The capsules can then be ejected by pushing the capsules out of the dosing plate. All plates must be cleaned carefully before filling a further batch of capsules. Manufacturers and suppliers of such equipment include Anchor Mark (India), Feton (Belgium), Grovers International (India), Labocaps (Denmark), Torpac (USA), T.U.B. Enterprises

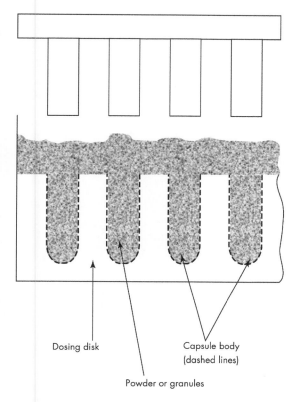

Dosing disk

Capsule body
(dashed lines)

Powder or granules

Figure 6.3 Schematic of manual capsule filling with tamping device.

(Canada) and Westeam Enterprises (Philippines). Torpac provides an extensive handling description plus video on its website.

Capsule rectification, opening and closing on fully automated capsule-filling machinery

The majority of manufacturers of fully automated capsule-filling machines employ a rectification mechanism as illustrated in Figure 6.4. The capsules are supplied from magazines, whereby in the sorting tubes the capsule can face either cap up- or downward. The rectification pin will always touch the capsule cap, independent of orientation. When this pin is pushed forward, the capsule has to give way, but because the pin is in contact with the cap it will align horizontally, with the cap either in the upper or lower position of the rectification mechanism (see Figure 6.4). The capsule is then pushed forward until its body has left the segment. The sorting pin pushes downwards, thereby aligning the capsule vertically, always with the cap facing upward.

A different principle has been patented by Shionogi Qualicaps (Japan) and is shown in Figure 6.5a. Here capsule shells are moving in a drum reservoir, which releases the capsules into a second, counter-rotating drum. Capsules are oriented so that if the cap faces upwards they drop into a cavity deep enough for the capsule to be immersed to its full length. The cap cannot enter the slightly smaller lower part of the cavity (Figure 6.5b). If, however, the cap faces downwards, the capsule body will protrude into the outside (Figure 6.5c). As the drum rotates further,

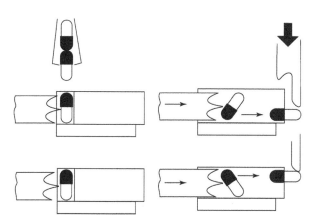

Figure 6.4 Standard capsule rectification mechanism in automated capsule-filling machines. (Reproduced from Bosch Training Manual GKF 400S, with permission of Robert Bosch GmbH, Germany.)

(a)

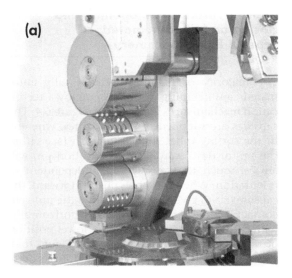

Figure 6.5 Capsule rectification system of LIQFIL^{Super} machines. For a full explanation refer to the text. (Figure 6.5a reproduced from Customer Leaflet, with permission of Shionogi Qualicaps, Japan.)

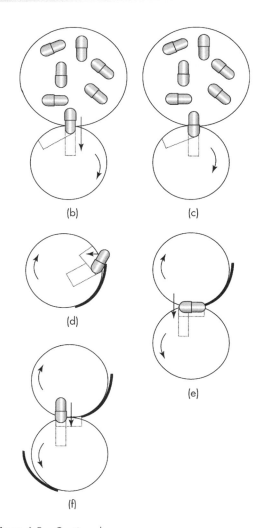

Figure 6.5   Continued.

a finger moves the protruding capsules, so that they lie flat in the preformed segments (Figure 6.5d). The capsules are released from the scraper plate when in contact with the next, counter-rotating drum (Figure 6.5e). Capsules previously fully immersed vertically in the drum now drop into the next drum with their bodies protruding, whereas the capsules positioned vertically in the previous segment will simply drop into the next segment (Figure 6.5f). Another finger will realign the protruding capsules, so that now all capsules are held vertically in the segments with the cap facing in the direction of rotation. The capsules are then released into the segment plate below the drums, caps facing upwards.

Capsule opening is usually vacuum assisted. The general principle is outlined in Figure 6.6. After rectification the capsules are positioned in the segments or 'bushings' that are in pairs. The top segment contains the capsule cap, to which initially the capsule body is still attached, resting on a guide pin inside the lower segment. Suction is applied to separate the cap and body. In order to avoid damage to the capsule body when hitting the base of the lower segment, the guide pin is slowly lowered during the application of the suction, always keeping contact with the

shoulder of the capsule body. After the separation process is completed, the upper and lower segments are separated to allow filling.

The segments are repositioned on top of each other after filling has been completed. They are mechanically joined by pushing the lower closing pins upwards. This is illustrated schematically in Figure 6.7. Before closing can occur, faulty capsules, i.e. unopened capsules or capsule bodies with two caps, are removed usually by a suction mechanism applied to the upper segment, in which these defective capsules would have remained during the filling process.

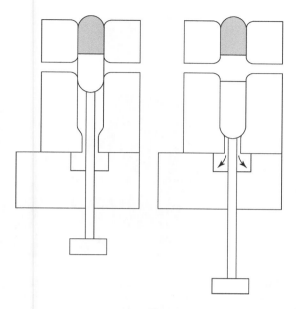

Figure 6.6 Vacuum-assisted capsule opening in automated capsule-filling machines. (Reproduced from Bosch Training Manual GKF 400S, with permission of Robert Bosch GmbH, Germany.)

Direct filling methods for powders and granules

Auger-filling principle

Capsule filling using the auger principle is the oldest semi-automatic filling technique, which was developed by Arthur Colton in the first decade of the 20th century. Modern machines can, however, also be fully automatic. The general filling principle is illustrated in Figure 6.8. The powder or granules are contained in a hopper similar in design to mass-flow hoppers. Inside these is a rotating auger to which a stirrer is attached. Powder is fed continuously out of the hopper outlet owing to the rotation of the auger. The capsule bodies are placed into the bushings of a rotating turntable ('pick-up ring') and will pass the hopper outlet during the rotation process. The amount of powder fed directly into the capsule body depends on the time the capsule body spends underneath the hopper outlet, on the auger speed and on the twist angle of the auger. Slower rotation of the turntable increases the fill weight, because the capsule bodies remain underneath the hopper outlet for a longer time

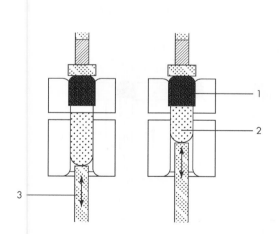

Figure 6.7 Capsule closing mechanism in automated capsule-filling machines. 1 = capsule cap, positioned in the upper segment; 2 = capsule body, positioned in the lower segment; 3 = closing pin. (Reproduced from Bosch Training Manual GKF 400S, with permission of Robert Bosch GmbH, Germany.)

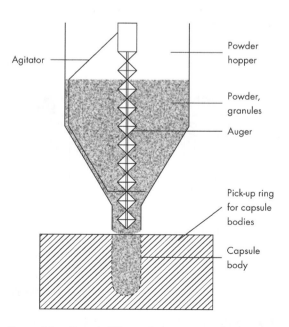

Figure 6.8 Capsule filling with the auger method.

span. An increase in the auger speed and a steeper twist angle will also increase powder fill weight.

A fully automated auger capsule-filling machine is the Liqfil^{super} JCF 40/80, manufactured by Shionogi Qualicaps (Japan). This machine uses a screw-type single or double auger filling system (Figure 6.9). Here, the individual segments of the pick-up ring are cleaned between individual filling cycles by compressed air and vacuum. The machine can produce 40 000 or 80 000 capsules per hour and can fill capsule sizes between 00 and 5, except for non-standard capsules (Supro). It has to be considered that powder in a hopper of a machine is changing its bulk density with time, whereby it gradually reaches its tapped density. The change in bulk density can lead to a proportional increase in capsule fill weight, unless auger speed and turntable rotation are adjusted accordingly. Such potential problems can be identified employing dynamic packing studies (see Chapter 5). The rotation of the stirrer can occasionally lead to powder compaction onto the hopper walls rather than powder transport out of the hopper.

Vibration-assisted filling

Vibration-assisted filling is a method introduced in 1960 by the Osaka Company (Japan). Figure

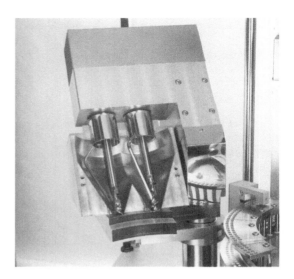

Figure 6.9 Double screw auger used in LIQFIL^{Super} machines. (Reproduced from Customer Leaflet, with permission of Shionogi Qualicaps, Japan.)

6.10 illustrates the individual steps of the filling procedure. Also here, the powder is directly filled into the capsule bodies. The capsule bodies are again positioned into a rotating turntable and pass underneath the powder bowl. The powder bowl has a mesh floor connected to a mechanical vibrator. Powder or granules pass the mesh floor,

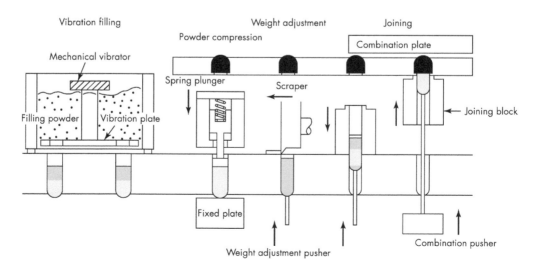

Figure 6.10 Vibration-assisted filling of hard capsules. For a full explanation refer to the text. (Reproduced from Customer Leaflet, with permission of Shionogi Qualicaps, Japan.)

whereby the vibration assists with desagglomeration and flow. The capsule bodies are filled to the maximum extent plus more powder on top, filling the cavities of the turntable completely. Further rotation of the turntable positions the capsule bodies so that a spring-loaded plunger can compress the powder inside the dosing bores to a more or less firm plug. A fixed counter plate underneath the turntable assists with this. The capsule body now contains a powder plug usually considerably longer than the size of the body itself. The complete capsule body is now pushed upwards so that the powder plug is partially outside the turntable bores. A scraper removes the powder outside the turntable bores and the capsule bodies are pushed further upwards to be joined to the corresponding cap and closed. The final fill weight is hence adjustable via the following mechanisms:

- rotation speed of the turntable and extent of vibration
- compression setting of the plunger
- final plug length defined by the amount of lift and the action of the scraper.

It has been reported that the above described system works well for dense, free-flowing powders (Jones, 2002). The system is commercially available as Liqfil^{super} 40 (Shionogi Qualicaps, Japan), producing up to 40 000 capsules per hour. All sizes between 00 and 5 can be filled except for non-standard capsules. One obvious problem is the recycling of the parts of the plug scraped off. This powder is more densified and hence an increase in powder fill weight might be expected with time.

A recently developed bench-top equipment (Xcelodose, Meridica, Melbourn, Herts., UK) uses gravimetric filling of powders or granules rather than volumetric filling. A microbalance with an accuracy of ± 1 ng is used here and allows the filling of individual doses between 0.1 mg and 100 mg directly into the capsule body. The powder is filled into the dispensing head equipped with a sieve of defined mesh size. Currently, 32 different sieve mesh sizes are available and the mesh size has to be chosen so that the powder forms an arch above the meshes. When the dispensing head is tapped mechanically, some of the powder bridges will collapse and a small quantity of powder (about 2% of the

overall weight) falls through the sieve into the capsule body. The system weighs the capsule body after each dispensing step. Initially, the system dispenses the powder using a series of rapid mechanical taps with a tapping frequency of about 25 Hz. The system then slows down to achieve maximum fill weight accuracy. The precision of fill weight that can be achieved for powders with good flow properties appears to be below 2%, even for single doses as low as 500 µg. Currently, two models are available, which can fill either 120 or 600 capsules per hour. The machines can accommodate capsule sizes 00 to 3.

Indirect filling methods for powders and granules

Tamp-filling machines

In tamp-filling machines the powder is contained in a powder bowl ('dosing hopper') that rotates by indexing and six rotational events are required for the bowl to perform a full 360° cycle. The bottom of the powder bowl is formed by a removable dosing disk, which contains six sets of dosing bores, corresponding to the indexing of the rotation. The powder bowl is fed with powder from a powder feed hopper fitted with an auger mechanism. The amount of powder released from the powder feed hopper into the powder bowl is controlled by a conductive sensor positioned in the powder bowl at a defined height above the dosing disk. Powder flowing from the powder feed hopper is further distributed throughout the powder bowl with the help of a dosing cone, i.e. an angled central insert that aids powder flow away from the centre towards the walls of the powder bowl. The exact positions of the hopper feed outlet, the dosing cone and the bed height sensor are illustrated in Figure 6.11. The powder is tamped into the dosing bores by compressing pins ('tamping fingers'). The tamping is repeated five times to form an individual powder plug and at the sixth tamping event, the final plug is pushed into the capsule body (Figure 6.12). To avoid the powder being pushed through the dosing bores, the dosing disk rotates over a brass support ring ('tamping ring'). The gap between the tamping ring and the dosing

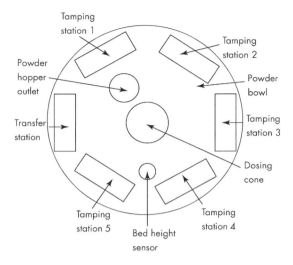

Figure 6.11 Top view and principal position of the filling elements used in tamp-filling machines.

disk must be chosen very carefully, so that there is no increased friction between the two metal parts, but also no larger gap for powder to enter. The gap width is kept usually between 20 and 50 μm. Larger gap sizes might be required for sticky powder. The gap is permanently cleaned by suction. This means that powder entering the gap is removed to a filter bag and the larger the gap size the more powder is lost.

As can be seen from Figure 6.12, the final plug consists ideally of five individual segments of similar size and density. However, in practice the height of the individual segments depends very much on the powder flow properties. The segments formed in the dosing bores during the first tamping event are often large compared to those formed in later tamping events. However, if the flow properties are poor, the first one or two tamping stations are starved of powder and segments are either not formed at all, or are only small.

In order to avoid excessive pressure on the powder during tamping, the tamping pins are spring loaded. The springs deflect when the tamping force exceeds a given value. In modern tamp-filling machines the exchangeable springs give maximum forces between 20 and 200 N only, in contrast to older machines. In this way, the plug density is kept low, which aids

powder plug disintegration, drug dissolution and bioavailability. Older machines had rigid springs allowing loads to form highly densified plugs at >500 N, which defeated the common reasons for producing capsules. Powder plugs produced by such older machines could even be subjected to bending strength tests (Shah *et al.*, 1986). A plug produced by a modern tamp-filling machine is fragile and easily destroyed between two fingers. If not, the tamping forces allowed by the springs must have been grossly exceeded, damaging the machine in the long term.

Greater changes in fill weight can be achieved by varying the powder bed height. However, this may be restricted to a certain level by the powder flow properties. An increase in powder bed height can lead to spillage of the powder between tamping station 5 and the transfer station, if the powder flow is poor. Smaller adjustments in fill weight can be achieved by modification of the settings of the penetration depth of the tamping pins. In contrast to common belief, the tamping pins do not push all powder situated below the pin surfaces inside the dosing disk bores. The majority of powder forming the plugs enters the dosing disk bores during rotation of the dosing disk owing to flow (Podczeck, 2000). The thickness of the dosing disk also influences the capsule fill weight. The ideal thickness for a given powder can be found experimentally by means of the 'Höfliger & Karg plug simulator', which is nowadays produced by Robert Bosch GmbH (Germany). A weighed amount of powder, i.e. the required fill weight, is tamped manually in a single event with the chosen tamping force into a die and the plug length is measured. The equipment and its use are described in more detail by Jones (1988, 1998). Alternatively, the dosing disk can be selected via a numerical approach from the tapped density of the powder (Davar *et al.*, 1997).

Tamp-filling machines are produced by a variety of manufacturers. One of the larger machine manufacturers is Robert Bosch GmbH (Germany). Their current range of machines includes GKF 400S, GKF 700S, GKF 2000ABS and GKF 2500S. The S-models are 'standard' machines in terms of filling control and they are cheaper than the ABS-versions, which offer 100% weight check and feedback control. The GKF 400S is a small machine for laboratory use and production

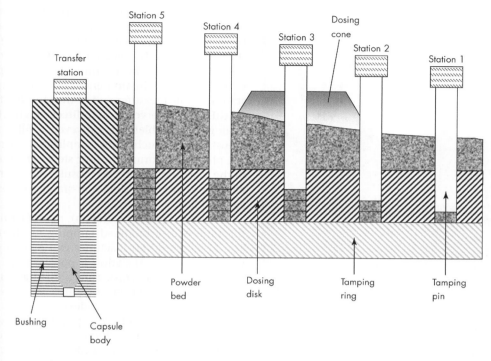

Figure 6.12 *Schematic drawing of the tamp-filling mechanism. The five tamping stations plus transfer station are positioned on a circular path (see Figure 6.11). For a full explanation refer to the text.*

of clinical trial samples. Each tamping station has three tamping pins. The machine produces 24 000 capsules per hour and standard and so called 'tamper-proof' (Coni-Snap Supro) capsules between sizes 00 and 5 can be filled. The GKF 700S fills 48 000 capsules per hour, i.e. is a medium-scale production machine with five pins per tamping station. The GKF 2000ABS is able to fill capsule sizes 00–5 plus tamper-proof capsules with an output of 150 000 capsules per hour. Each tamping station has 18 pins positioned in two rows. The ABS-system performs a fully automatic 100% online weighing, ejects over- and underfilled capsules and initiates automatic readjustment of the fill weights. The latter is achieved with pneumatic tamping heads on stations 3 and 4. Here, the springs are replaced by a cushion of compressed air. If the air pressure is increased, the fill weight will also increase as the tamping pins will compress the powder to a larger extent and hence create more space inside the dosing bores. The weight checker is connected to the tamping heads and persistent increase or

drop in weight will trigger a feedback pressure valve to increase or decrease the air pressure inside the tamping head accordingly. The GKF 2500S (Figure 6.13) has a revised powder station. A blocking slide is fitted between dosing disk and tamping ring. This rotates with the dosing disk and hence avoids shearing the plug at the gap between the rotating dosing disk and the stationary tamping ring. At the transfer station, the blocking slide is opened to allow for plug transfer. This design has improved the weight uniformity of powder-filled capsules on this type of machine.

Bosch GKF 2000ABS and GKF 2500S machines can be equipped with a 'height-adjustable dosing disk' (Figure 6.14). The dosing disk consists of a base dosing disk and an upper dosing disk. The bores of the two disks are connected by integrated tubes, which define the space between the bottom of the base and the top of the upper dosing disk and hence the length of the powder plug formed. The tubes can be expanded or collapsed like a telescope, which allows in-place changes of the disk thickness.

Figure 6.13 View into the GKF 2500S capsule-filling machine. Capsules are fed from a hopper into the rectification system (left) and opened. Powder plugs are formed in the tamping unit (right) and transferred into the lower segments, which follow a circular path for opening, filling and closing (front). (Reproduced from Bosch Customer Leaflet VT/VFW 04/02-1E, with permission of Robert Bosch GmbH, Germany.)

Index, a sub-group of Romaco (Italy) produces machines of the K-series, which can produce between 40 000 and 150 000 capsules per hour. These machines can fill all capsule sizes between 000 and 5. The index 'i' indicates 'intelligent choice', i.e. these machines are equipped with automatic 100% weight control. Also Harro Höfliger (Germany) provides the tamping mechanism as one possible filling process on their KFM-III series. These machines can produce between 9000 and 24 000 capsules per hour, depending on the powder properties. All capsule sizes between 00 and 5 can be filled.

IMA (Italy) also produces tamp-filling machines with its IMPRESSA series. On these machines, the four tamping stations are not mounted to form a full circle. Instead, a semi-circle with a large radius is produced. Opening the empty capsules, filling and closing form one large processing ring. As a consequence, the

powder bowl is also large and the powder remains inside the powder bowl for a longer time. These machines have no tamping ring and the capsule body segments serve as lower tamping support. The machines can be equipped with strain gauge instrumented pins on tamping station 4. The measurement of the tamping force on this station serves as an indirect weight check analogous to tabletting. However, a fill-weight adjustment is much more complicated owing to the plug being formed in four steps. The output of the machines is 130 000 capsules per hour and all capsule sizes including non-standard capsules can be filled.

Shionogi Qualicaps (Japan) manufactures a machine with a modified tamping mechanism. The Liqfil^{super} 40/80 machines can be equipped with this system. At station 1, powder is fed into the bores of a dosing disk using a stirring agitator. The powder is then tamped only twice, i.e. on stations 2 and 3. A further indexing of the dosing disk positions the so-formed powder plug above the 'moulding slag', which is a lower punch underneath the dosing disk. This punch is pushed upwards, which will cause the collapse of any powder bridge or hole inside the plug. Excess powder is then scraped off on station 5. As in their vibration machines, the excess powder is recycled. Finally, the plugs are transferred into the capsule bodies. As mentioned before, the recycling of compressed powder might result in a

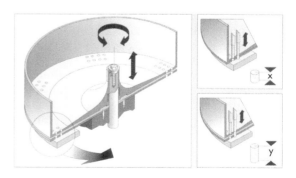

Figure 6.14 'Height'-adjustable dosing disk* for GKF 2000ABS and GKF 2500S machines. (Reproduced from Bosch Customer Leaflet VT/VFW 04/02-1E, with permission of Robert Bosch GmbH, Germany.) *Name as given in leaflet; it means thickness-adjustable.

gradual increase in fill weight. Certainly, to prevent the scraping resulting in uneven fracture planes (i.e. uneven plug lengths), the formation of strong plugs must be avoided.

Dott. Bonapace & C.S.R.L. (Italy) produce bench-top tamp-filling machines (In-cap). These machines use very strong springs similar to old-fashioned capsule-filling machines and the powder movement is not circular, owing to space restrictions. Filling experiences reported are sparse, but indicate some difficulties, in particular with powder filling in contrast to filling of granules. The maximum output is 3000 capsules per hour. The machine can fill powders, pellets and tablets into all capsule sizes including non-standard capsules, but special change parts are required for the latter.

Dosator nozzle machines

Dosator nozzle machines are available in both intermittent motion and continuous filling operation versions. The working principle of a typical intermittent machine is illustrated in Figure 6.15. The powder is fed via a powder hopper into a dosing hopper which rotates and a metal plate positioned behind the hopper outlet inside the dosing hopper controls the powder bed height. The dosators are linked in pairs for their operation. A dosator nozzle consists of a hollow dosing tube, inside which is a piston. The initial position of the piston inside the tube defines the volume and consequently the weight of the powder plug to be formed. The volume of this chamber is controlled by the piston position. The dosator nozzle is lowered into the powder bed and powder enters the dosing tube and is compressed. The dosator nozzle is then lifted up from the powder bowl. At the same time, the second nozzle is positioned over the capsule body and the powder plug is pushed out of the dosing tube by the piston being moved down to the end of the tube. After the nozzles have been lifted, they swap position so that the newly formed plug can be released into a new capsule body and at the same time the second nozzle is lowered into the powder bed. In contrast to tamp-filling machines, the powder plug is hence produced in one compression step. It will thus be more homogeneous

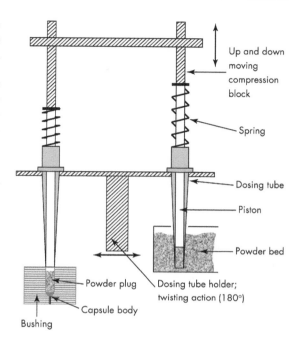

Figure 6.15 Intermittent dosator nozzle filling principle. For a full explanation refer to text.

in density. However, for the powder plug to be retained in the nozzle during the transfer, i.e. the switch of the nozzle positions, the powder must be able to form an arch, as otherwise it would drop out of the dosing tubes during its transfer. The powder properties required for arching to occur are discussed in Chapter 5.

In dosator nozzle machines, the fill weight of the capsules can be adjusted first by varying the powder bed height in the dosing hopper. Assuming a constant dosator volume, an increase in the bed height will increase the fill weight as more powder is compressed. As mentioned above, the fill weight can also be adjusted via the volume inside the dosator, i.e. by adjusting the initial piston position.

Modern dosator nozzle designs are streamlined, easy to clean and simple in design. Figures 6.16 and 6.17 show dosator nozzles used by Harro Höfliger (dissembled) and MG2 (assembled), respectively.

Intermittent dosator nozzle machines are manufactured mainly by IMA (Italy). These are divided into the Zanasi and the Zanasi-Plus series.

Figure 6.16 Dissembled dosator nozzles used in KFM-III machines. (Reproduced from Harro Höfliger Customer Leaflet HH 1.0/02 WA-D, with permission of Harro Höfliger, Germany.)

All Zanasi machines can be obtained in F or E version. The E-version ('electronic') provides an automatic sampling unit with check weigher in addition to the standard fittings ('F'). Samples are drawn on a statistical basis and the information obtained from the weights is used to adjust the piston settings automatically. If several capsule weights are out of range, the system will stop the machine. The different machine types, i.e. Zanasi 6, 12, 25 and 40 can fill between 6000 and 40 000 capsules per hour. Zanasi plus machines come similarly as F or E versions and can fill between 8000 and 85 000 capsules per hour.

Continuous dosator nozzle machines are more comparable with rotary tablet machines and hence have larger output rates. Here, the powder bowl is a circular trough fed with powder from a supply hopper. The dosator nozzles are fixed in a turret, which has a smaller diameter than the powder bowl and is slightly off-centre (Figure 6.18). Both powder trough and turret rotate in the same direction. In position 1 the dosator nozzle dips into the powder bed and compresses a plug. In positions 2–4 the nozzle is withdrawn from the powder bed and because of the smaller diameter of the turret it is shifted to a position outside of the powder trough. In positions 5 and 6 the plug is released into the capsule body. The nozzle is then retracted and moves again over the powder bed as the turret rotates further. Figure

Figure 6.17 Assembled dosator nozzle used in MG2 machines. (Reproduced from Customer Leaflet, with permission of MG2, Italy.)

6.19 shows the turret and general machine of the G140 produced by MG2 (Italy).

Continuous dosator nozzle machines are produced by MG2 (Italy). The SUPREMA models can fill 24 000 or 48 000 capsules per hour into capsule sizes 00–5. The FUTURA range can be equipped with different numbers of dosator nozzles to give 6000, 12 000, 24 000, 48 000 or 96 000 capsules per hour. As the turret and powder trough dimensions remain the same, problems with scale-up of the formulations are kept to a minimum. All capsule sizes and types between 000 and 5 can be filled, which also applies to all machines of the G-series. The G38/N and G60 fill up to 60 000 capsules per hour, G37/N and G100 fill 100 000 capsules per hour and G120/N and G140 fill up to 120 000 capsules per hour. G100 and G140 include an automatic capsule weight control system.

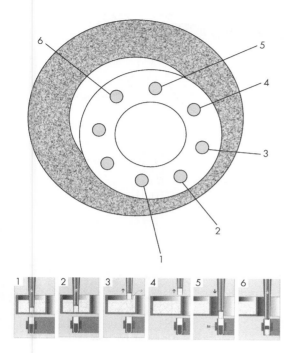

Figure 6.18 Continuous capsule filling with dosator nozzles. Position and relative size of the ring-shaped powder bowl and turret (top). (Schematic drawings to illustrate steps 1 to 6 reproduced from MG2 Customer Leaflet, with permission of MG2, Italy).

IMA (Italy) also produces continuous dosator nozzle machines. The MATIC series machines can be installed with rotating powder trough or with a suction tank. The suction tank aims to precompress the powder to higher bulk density values. This is particularly advantageous for light or aerated powder and increases the dosing precision. Machines can produce 60 000, 90 000 or 120 000 capsules per hour. A statistical weight control system tests all capsules from each dosator in turn and is able to self-adjust all working parameters of the machine automatically. Machines of the IMATIC series produce 100 000, 150 000 or 200 000 capsules per hour. They are equipped with an automatic 'clean-in-place' system, which performs prewashing, washing with detergent, rinsing with demineralised water and drying with hot air. The cleaning parameters can be adjusted onto the product properties. In-place cleaning is enabled by a complete sealing of the powder dosing area from the remainder of the machine.

Harro Höfliger (Germany) also offers a continuous dosator nozzle system as an option for the KFM-III series. Their machine can fill up to 24 000 capsules per hour into capsule sizes 00–5.

Vacuum-filling method

The vacuum method is based on a continuous dosator nozzle principle. The moving piston has been replaced by a piston fitted with a porous plate at the lower end. Its position remains fixed during the complete filling process. The upper end of the dosing tube is connected to an airflow system (Figure 6.20). When the nozzle is dipped into the powder bed, powder is sucked into the cavity formed by dosing tube and piston plate. When the nozzle is positioned over the capsule body, the suction is reversed into a mild air blow, which transfers the powder into the capsule

Figure 6.19 View of the rotary dosing unit of a G140 continuous dosator nozzle capsule-filling machine. (Reproduced from Customer Leaflet, with permission of MG2, Italy.)

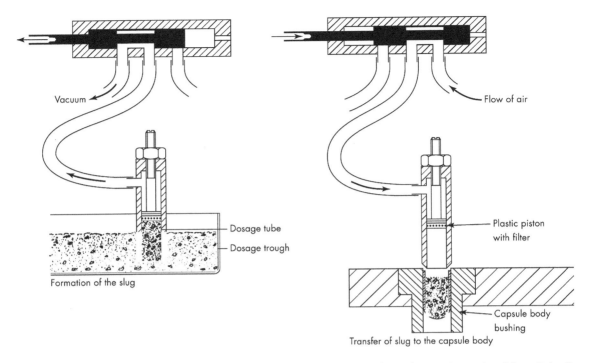

Figure 6.20   Vacuum-assisted filling of hard capsules. For full explanation refer to the text. (Reproduced from Cole, G. C. (1987). The mechanical operations of filling hard capsules. In: Ridgway, K. (Ed.), *Hard Capsules Development and Technology*. London: The Pharmaceutical Press, 92–103, with permission of publisher and author.)

body. Powder filled with this mechanism is much less densified, very small doses can be filled and owing to the fact that there are no moving parts inside the dosing tube, the need for a lubricant in the formulation is limited and often no lubricant is required. This makes such a system ideal for filling low doses and inhalation products.

Romaco (Italy) produces the 'Macofar' series of filling machines, which are equipped with vacuum dosators. The range comprises CD-5, CD-20, CD-40 and CD-60, which fill between 5000 and 60 000 capsules per hour, respectively. Capsule sizes 00 to 5 can be filled. The KFM-III produced by Harro Höfliger (Germany) also offers vacuum-assisted dosator nozzles as one option. Their nozzles were developed particularly for filling of low doses between 3 and 20 mg. The vacuum filters inside the nozzles require the particles to be larger than 15 µm in order not to pass the filter mesh (Information leaflet HH 1.0/02 WA-D). Therefore, these machines cannot be used for filling dry powder inhalations, as in

dry powder inhalations the particle size of the drug is mainly below 5 µm. Their claim that the filling system is suited for micronised powder (Information leaflet HH 1.0/02 WA-D) is thus not justified. Dependent on the powder properties, between 9000 and 24 000 capsules can be filled per hour.

Filling of pellets

Filling of pellets into hard capsules requires separate techniques from those used for powders and granules, because pellets are usually larger in size (800–1400 µm), they cannot be compressed and they are often coated with a film to control drug release. Uncoated pellets appear to pose few problems during filling as long as they are reasonably round, i.e. their aspect ratio should not exceed a value of 1.2, their surface is fairly smooth (Chopra *et al.*, 2002) and they should be non-friable (Pfeifer and Marquardt, 1986). Film-coated pellets, however, pose many problems for filling technology. The filling mechanism should not damage the coating, which is often fragile and delicate. The development of electrostatic charging must be prevented, which can be achieved by blending of the pellets with 1% talcum powder (Chopra *et al.*, 2002).

Capsule machines manufactured by Robert Bosch GmbH (Germany) provide two filling mechanisms for pellets. The currently most used is the double slide method, which is illustrated in Figure 6.21. In the first step, pellets flow from the pellet magazine into the dosing chamber. The size of the dosing chamber can be varied by moving the left chamber slide closer or further apart from the right wall of the dosing chamber. The dosing slide is then moved to the left, whereby it closes the dosing chamber from the pellet magazine. At the same time, the outlet slide moves to the right and opens the dosing chamber at the bottom end, so that the pellets can flow through the conical outlet into the capsule body. Reversing the movement of the dosing and outlet slide starts a new filling cycle. For particularly delicate film coatings the company suggests the use of a dosing disk method, which is illustrated in Figure 6.22. The method is comparable to that used for powder filling (Figure 6.12). The tamping pins are removed and a blocking slide is installed between

transfer station and the capsule segments. Pellets fill the dosing bores of the dosing disk during the indexing rotation of the dosing disk controlled by flow only. The blocking slide opens between two indexes to release the pellets from the dosing disk bore into the capsule body and the movement of the transfer pins aids complete emptying. Problems with this method have been reported for smaller pellet sizes, where the blocking slide was fracturing pellets remaining in the dosing bore during closing.

IMA (Italy) and Harro Höfliger (Germany) offer a method based on the use of vacuum-assisted dosator nozzles (Figure 6.23). Initially the nozzles are dipped into the pellet bed and owing to the suction applied, pellets are retained in the dosing tubes when these are lifted out of the pellet bowl. Pellets outside the dosing tube are removed by a scraper (metal bar or soft brush) or, for extremely delicate film coatings, by an air-blow mechanism. When the dosing tubes are positioned over the capsule body, the pellets are released by reversing the direction of the airflow inside the nozzle. Although the scraper technique has been improved by using a brush or the air-blow system, dipping the nozzle into the pellet bed can still damage the coating.

The filling system available in MG2 machines (Italy) is illustrated in Figure 6.24. Here, the dosing chamber is formed by a piston moving in

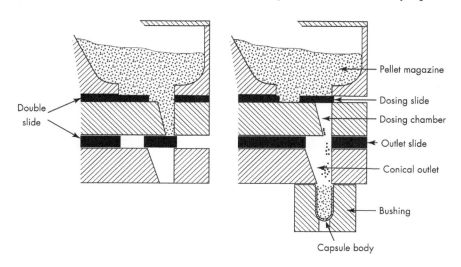

Double slide

Pellet magazine

Dosing slide

Dosing chamber

Outlet slide

Conical outlet

Bushing

Capsule body

Figure 6.21 Filling of pellets with the double slide mechanism. (Reproduced from Bosch Training Manual GKF 400S, with permission of Robert Bosch GmbH, Germany.)

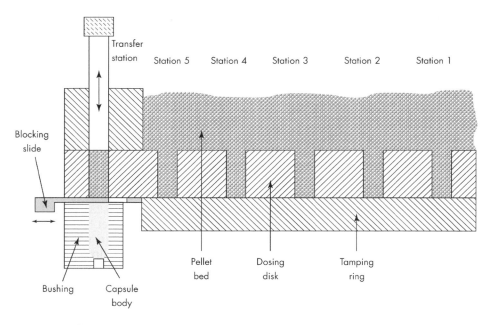

Figure 6.22 Filling of pellets with a modified dosing disk system. For full explanation refer to text.

a dosing tube and the open end of the dosing tube is positioned underneath the pellet hopper. Pellets enter the dosing chamber formed by the piston in the upper end of the dosing tube by flow. The dosing tube rotates with the machine turret and hence the dosing chamber passes under the pellet hopper. On further rotation the piston moves down and opens the outlet tube. Pellets will flow through the outlet tube into the capsule body, whereby a mild air-blow assists complete emptying. Finally, the piston moves fully upward to remove all remaining pellets from the tube and is then retracted to its original position.

The pellet-filling system provided by Shionogi Qualicaps machines from type Liqfil^super 40 has some similarities to the double slide method described above. Here the pellets are in a hopper, where they are agitated to prevent agglomeration. They are separated from a set of measuring chambers by an upper shutter. When the upper shutter is opened, the pellets can flow into the measuring chambers, which are sealed off from the capsule body containing segments by a lower shutter. The upper shutter is then closed and the lower shutter opened, so that the pellets leave the measuring chambers and enter the capsule bodies by flow.

Filling of tablets

Tablet-filling mechanisms provided by Robert Bosch GmbH (Germany), MG2 (Italy) and Harro Höfliger (Germany) are very similar and the principle is outlined in Figure 6.25. A dosing slide, which can accommodate exactly one tablet standing on its edge, moves underneath the tablet feeder. The slider including the tablet moves over the capsule body, where the tablet simply drops into it. At the same time, a pin goes down into the capsule body. If a tablet has been filled correctly, the travel path of the pin is limited and the horizontal bar connected to the pin just touches the sensor. If, however, no tablet has been filled, the pin travels down to the bottom of the capsule body. In this case, the horizontal bar will switch the sensor, indicating an incorrect filling. Empty capsules can be detected and eliminated from the product.

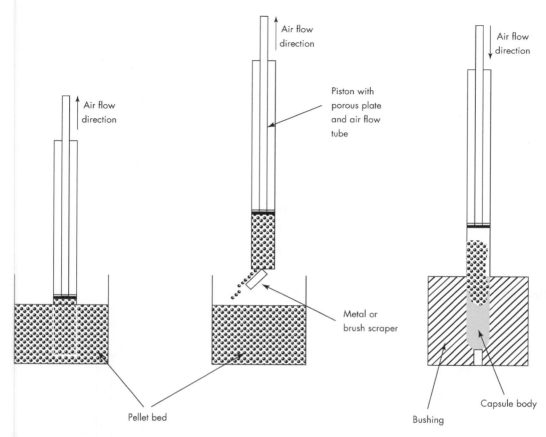

Figure 6.23 Filling of pellets with a vacuum-assisted dosator nozzle system. For full explanation refer to text.

IMA (Italy) offers a slightly different filling mechanism, which is illustrated in Figure 6.26. Here, the tablets are resting on their faces rather than on the outer diameter. Also, in this system a slider transports one tablet over the capsule body. The tablet pushes the sensor slightly up and thus provides an electrical signal indicating successful filling.

Filling of combinations of powders, pellets and tablets

Detailed information about filling combinations of two or three dosage forms into a single capsule is not available from all machine manufacturers.

MG2 machines of the FUTURA series can combine up to three different dosage forms into one capsule, whereby the first one to be filled can be a liquid or semi-solid. The G-series machines can also fill up to three dosage forms in combination, but combinations with liquid or semi-solid fillings are not possible.

Zanasi 6 and 12 (IMA, Italy) can fill combinations of up to three solid dosage forms, whereas Zanasi 25 and 40 machines can fill combinations of two solid dosage forms only. The Zanasi-Plus 8, 16, 32 and 48 machines can fill combinations of up to three dosage forms, including liquids or semi-solids. Here, at the first filling station, powders, pellets or liquids/semi-solids are filled. At the second filling station, tablets or pellets are added and on the third filling station again

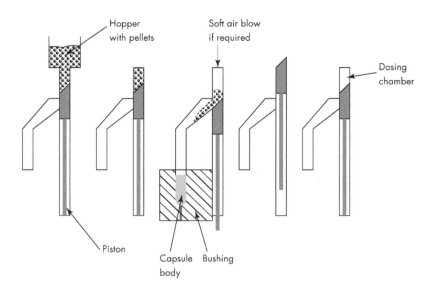

Figure 6.24 Filling of pellets on MG2 capsule-filling machines.

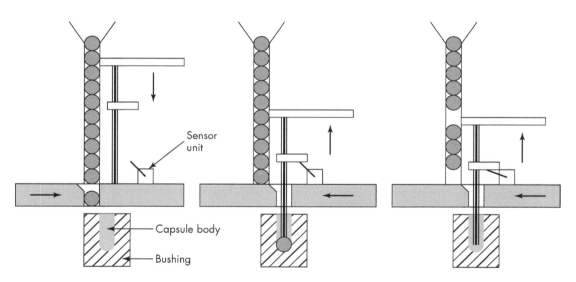

Figure 6.25 Schematic drawing of the most often used filling mechanism for tablets into hard capsules.

powders, pellets or liquids/semi-solids are filled. Zanasi-Plus 70 and 85 use two filling stations fed by one hopper, hence filling one and the same product. This doubles the number of capsules filled per hour, but restricts the machines in terms of possible combinations, i.e. no combined filling is possible.

All machines produced by Robert Bosch GmbH (Germany) can fill any combination of two dry dosage forms. The GKF 400S and 700S can also fill one powder plus two different batches of pellets or the triplet of powder, pellets and tablet. The GKF 2000ABS and 2500S can fill three different batches of pellets and two pellet batches plus

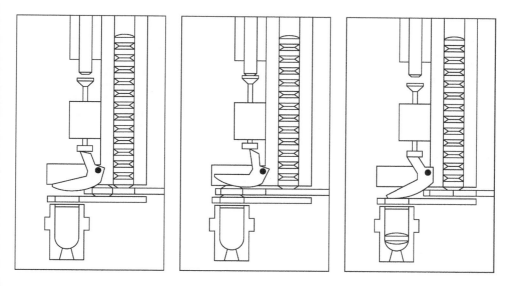

Figure 6.26 Filling of tablets into hard capsules on IMA machines. (Reproduced from Customer Information material, with permission of IMA, Italy.)

tablet. GKF 400S and 700S can also fill one powder plus two tablets. The older GKF 800 series could fill three solid dosage forms in all possible combinations plus one liquid.

Summary

A series of different filling mechanisms is available for filling solid dosage forms into hard capsules. The mechanism offered is machine manufacturer dependent. Only one manufacturer offers machines that can be equipped with almost all known techniques used for the filling of powders and granules. The machine manufacturer dependence of filling mechanisms often limits the possibilities for companies during the development of new formulations, as the filling principle employed should ideally match that used in production to avoid scale-up problems. When making changes in filling principle for the manufacture of licensed products, SUPAC regulations (see, for example, Hileman, 1997) have to be considered.

References

Chopra, R., Podczeck, F., Newton, J. M. and Alderborn, G. (2002). The influence of pellet shape and film coating on the filling of pellets into hard shell capsules. *Eur. J. Pharm. Biopharm.*, 53, 327–333.

Clement, H. and Marquardt, H. G. (1970). Erfahrungen über die maschinelle Verarbeitung von Hartgelatinekapseln. *Pharm. Ind.*, 32, 169–176.

Davar, N., Shah, R., Pope, D. G. and Augsburger, L. L. (1997). The selection of a dosing disk on a Höfliger–Karg capsule filling machine. *Pharm. Technol.*, 21(2), 32–48.

Hileman, G. A. (1997). Regulatory issues in granulation processes. In: Parikh, D. M. (ed.), *Handbook of Pharmaceutical Granulation Technology*. New York: Marcel Dekker, 483–502.

Jones, B. E. (1988). Powder formulations for capsule filling. *Manuf. Chem.*, 59(6), 28–33.

Jones, B. E. (1998). New thoughts on capsule filling. *S.T.P. Pharma Sci.*, 8, 277–283.

Jones, B. E. (2002). Capsules, hard. In: Swarbrick, J. and Boylan, J. C. (eds.), *Encyclopedia of Pharmaceutical Technology*. 2nd edition. New York: Marcel Dekker, 302–316.

Pfeifer, W. and Marquardt, H. G. (1986). Investigations of the frequency and causes of dosage errors during the filling of hard gelatin capsules. 2nd Communication: Dosage errors during the filling of pellets into hard gelatin capsules. *Drugs Made in Germany*, 29, 217–220.

Podczeck, F. (2000). The development of an instrumented tamp-filling capsule machine. I: Instrumentation of a Bosch GKF 400S machine and feasibility study. *Eur. J. Pharm. Sci.*, 10, 267–274.

Podczeck, F., Claes, L. and Newton, J. M. (1999). Filling of powdered herbs into gelatine capsules. *Manuf. Chem.*, 70(2), 29–33.

Shah, K. B., Augsburger, L. L. and Marshall, K. (1986). An investigation of some factors influencing plug formation and fill weight in a dosing disk-type automatic capsule-filling machine. *J. Pharm. Sci.*, 75, 291–296.

7

Instrumented capsule-filling machines and simulators

Norman Anthony Armstrong

Introduction

Instrumentation in the context of this chapter means the fitting of measuring devices, also known as transducers, to a piece of capsule-filling equipment, measuring the output of the transducers and relating this to some parameter connected with the performance or operation of the equipment.

Tablet presses fitted with suitable transducers have been available for several decades. For reasons which will be discussed later, instrumentation of capsule-filling equipment has lagged considerably behind that of tablet presses. Nevertheless it is instructive to consider briefly tablet press instrumentation, since in many cases the parameters measured in both types of equipment are qualitatively identical though quantitatively very different. Furthermore, approaches which have been successful in tablet press instrumentation have often proved equally successful in the instrumentation of capsule-filling machines.

The term 'instrumented tablet machine' was introduced by Higuchi *et al*. (1954). Building on the pioneering work of Brake (1951), Higuchi's group described an excentric tablet press equipped with transducers for the measurement of force and distance, both with respect to time. This work was made possible by the development and ready availability of strain gauges for measuring force, displacement transducers for measuring distance and associated electronic devices for data recording and storage. Thus it was now possible to measure accurately both the force applied by the punches to particulate material contained in a die and also the punch movement

which gave rise to these forces. Hence techniques became available to measure the fundamental parameters of the tabletting process, namely force and punch movement and to relate these to each other and also to time.

The importance of this development in the study of powder compression and tablet formulation cannot be overestimated. Virtually all properties of compressed tablets – porosity, physical strength, disintegration, dissolution, friability – are dependent on the compressive force to which the constituent powder particles have been subjected. It follows that unless the magnitude of the force is known, meaningful studies on tablet properties are difficult, if not impossible. Instrumentation was soon extended to rotary tablet presses, though the rotation of the turret, punches and dies of such presses posed challenges, both in the fitting of and collecting the output from the transducers.

Instrumented methods have been increasingly applied to all aspects of tabletting technology, aided by the availability of powerful yet relatively cheap computing power. A wide range of parameters associated with the compression of powders can be measured and considerable insight has been gained into the fundamentals of the tabletting process. Tablet presses are now manufactured with integral instrumentation packages, usually for the automatic control of tablet weight. Methods used for the instrumentation of tablet presses and the information that can be obtained from them have been comprehensively reviewed by Watt (1988) and Celik (1992, 1996).

Whether attached to tablet presses, capsule-filling equipment or indeed any other type of machinery, certain principles of instrumentation apply. A transducer of appropriate sensitivity must be fitted to a suitable location on the equipment and a calibration method must be available, as must some method of recording transducer output. It is also vitally important to be confident that the collected data are a measure of the intended parameter and do not constitute an artefact introduced by the measuring device or its attachment, an error in data collection or manipulation or some uncontrolled feature of the overall system. The prospective user of instrumentation will receive scant guidance from the published literature in this respect, since, almost exclusively, only details of successful experiments are published.

Though the remainder of this chapter will be devoted to the instrumentation of capsule-filling equipment, there will be several instances where reference is made to instrumented tablet presses. There are similarities, as well as important differences, between fitting of transducers to tablet presses and capsule-filling equipment and challenges faced in fitting transducers to tablet presses are often of direct relevance to capsule filling.

Instrumentation of capsule-filling equipment

Even though the capsule-filling process involves the same variables of force, distance and time as tabletting, capsule-filling equipment has received much less attention than tablet presses from the instrumentation point of view. There are two possible reasons for this. First, the force applied to the particulate mass of a capsule fill is much lower than that applied to a tablet in a press – at most a few hundred newtons compared to tens of kilonewtons. Hence a more sensitive measuring system is needed. Second, the fitting of transducers, and especially displacement transducers, demands that they be fixed to a secure and massive fixing point. Such points are readily available in tablet presses, but are not so abundant on capsule-filling equipment. How these two problems have been overcome will be described later in this chapter.

Although there are two types of tablet press – rotary and excentric – they have basically an identical mechanism of tablet formation. A particulate solid is compressed in a die between two punches. In contrast there are several types of capsule-filling equipment, each with its own mode of operation. The fill material is treated in different ways in each case and hence the challenges of fitting instrumentation are also different.

Capsule-filling equipment is described in detail by Jones (2001) and elsewhere in this volume, but it is necessary to describe briefly the operating principles of the two most popular types here. These are machines that make use of the dosator tube principle and those based on a tamping mechanism into a dosing disk.

In the first machine, the dosator consists of a tube open at one end, inside of which is a moveable piston (see Chapter 6). The dosator is plunged into a bed of particulate solid contained in a hopper. Powder enters the open end of the dosator and is then consolidated by downward movement of the piston to form a plug. The dosator is then withdrawn from the hopper, taking the plug with it, and is positioned over the body of the capsule shell. The piston moves down and ejects the plug into the capsule body and the upper part of the shell is then fitted. A very free-flowing powder is not required, as a cohesive plug would not then form inside the dosator tube. However, the powder cannot be too cohesive since the powder bed must be maintained at a relatively uniform depth. A lubricant such as magnesium stearate may be required. Examples of machines using this technique are the mG2, Zanasi and Macofar (see Chapter 6).

The second type of filling equipment – the dosing disk – in some ways resembles a tablet press (see Chapter 6). The dosing disk has a number of holes bored through it, all except one being closed off by a stop plate. Powder flows into the first hole and is compressed by a tamping finger. This hole is then moved to position two, further powder flows in and is again tamped. This is repeated until the last hole is reached when, after excess powder is scraped off, the dosing disk positions the plug over a capsule body. The plug is then ejected by a piston. The overall arrangement is analogous to the die cavity of a tablet press in which the contents are compressed several times. This type of machine is exemplified by those made by Robert Bosch GmbH and Harro Höfliger (see Chapter 6).

As stated earlier, fixing measuring devices to tablet presses constituted a major advance in pharmaceutical technology and the ability to measure compression forces, perhaps related to punch position, brought about major advances in our understanding of tablet formation. It is therefore not surprising that similar efforts have been made with capsule-filling equipment. Since the contents of a hard shell capsule also consist of an aggregate of particles that have been subjected to a compressive force, then a quantitative knowledge of that force is obviously desirable.

Instrumentation of dosator tube capsule-filling equipment

The first published reports of instrumented capsule-filling equipment of this type were made by Cole and May (1972, 1975), using a Zanasi LZ64 machine. Foil strain gauges (120 Ω) were mounted on flat surfaces ground on to opposing sides of the dosator piston shank. The wiring from the strain gauges was led out through a hole drilled along the length of the piston to a recording oscillograph. Compression forces were thus measured along the axis of the dosator piston. Since the dosator on this type of machine constantly rotates during operation, Cole and May were faced with a problem similar to that encountered by users of instrumented rotary tablet presses, namely how to prevent twisting and rupture of the electrical cables leading to and from the transducers. They overcame this by fitting a planetary gear to the dosator head, which caused the dosator to make a complete clockwise rotation for each anticlockwise revolution of the dosator support arm.

Using this device, Cole and May were able to record for the first time the compression and ejection forces generated during plug formation and transfer for lactose, microcrystalline cellulose and pregelatinised starch. These powders were used either unlubricated or after the addition of 0.5% magnesium stearate. Cole and May noted that the small values of the forces (typically 20–30 N)

made measurement difficult, since a high degree of signal amplification was needed, with the attendant problems of signal-to-noise ratio.

They reported that up to four regions could be distinguished on the oscillograph:

- A force, tens of newtons in magnitude, which represented the compression force, is generated as the dosator was pushed into the powder bed. It is worth noting that the compressive force in a tablet press would be of the order of tens of kilonewtons.
- A force of a few newtons, termed the retention force, which was detected while the dosator was being raised from the powder bed and positioned over the empty capsule shell. The retention force showed that the plug remained in contact with the face of the piston during transfer.
- An ejection force as the plug was pushed out of the dosator into the capsule shell. This force was dependent on lubrication. For example, with unlubricated lactose, it could progressively rise to several hundred newtons, but addition of 0.5% magnesium stearate virtually abolished it.
- A 'drag force' which resisted the full retraction of the dosator piston after ejection of the plug was complete, indicating that the dosator rod was in tension. The drag force, the magnitude of which was also dependent on the degree of lubrication, was attributed to particles lodging between the sides of the dosator rod and the inner surface of the nozzle. It was most marked with pregelatinised starch, which had the smallest particle size of the powders studied.

Shortly after the full publication of the pioneering work of Cole and May, work describing the instrumentation of a Zanasi LZ64 machine was published by Small and Augsburger (1977). This was the first part of a major body of work in the field of instrumented capsule-filling machines to come from a group led by Professor Augsburger at the University of Maryland. Over a period in excess of 20 years, their instrumentation programme has involved the progressive development of instrumentation for both dosator tube and dosing disk machines.

One of the aims of Small and Augsburger was to modify the original equipment as little as possible. Four strain gauges were bonded to flattened areas of the piston shank to give a complete Wheatstone bridge. They of course faced the same problem as Cole and May in getting the electrical supply to the transducer and the signals out, owing to rotation of the dosator head. They solved this by using a mercury contact swivel between the instrumentation and the amplifier. Signals were then fed into an oscilloscope or a recorder. This particular piece of equipment was used to generate a considerable body of useful information on plug formulation.

Small and Augsburger detected the compression, retention, ejection and drag forces reported by Cole and May. However, they showed that the compression event could be divided into two stages. The first stage occurred as the dosator plunged into the powder bed. No force was detected until the dosator had penetrated to a depth equal to the height of the piston in the dosator. Then as the dosator continued downwards, a force built up, maximum force coinciding with maximum penetration. This they termed precompression. Then the main compression event took place, which was caused by downward movement of the piston inside the dosator, providing further consolidation. The adjustable movement of the piston is a feature of the Zanasi design but this was not used by Cole

and May in their work. They had not used the platen because they had removed it from the machine in order to accommodate their mechanism for turning the dosator.

Retention forces were not observed with lubricated powders and the authors surmised that this was because the lubricant had allowed the plug to slip inside the dosator, contact between the plug and the piston tip thereby being lost. The negative force after ejection, which had been reported by Cole and May was also noted by Small and Augsburger and attributed to the same cause.

Thus from the work of Cole and May (1975) and Small and Augsburger (1977), the sequence of changes in force which occur when a plug is formed in a dosator tube machine can be envisaged (Figure 7.1). The instrumentation consists of force transducers mounted on the piston shank. It must be borne in mind that Figure 7.1 shows all the events that can occur. The magnitude of these events will depend on the formulation, especially the degree of lubrication and some events might not be detected at all.

- Point A. The dosator tube descends into the powder bed. Powder enters the tube and comes into contact with the piston tip, where a force is detected.
- Point B. The dosator tube continues to descend

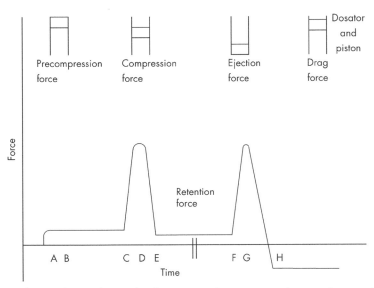

Figure 7.1 Changes in force with time during plug formation and ejection in a dosator tube capsule-filling machine.

and hence an increasing force is detected at the piston tip. There is no *relative* movement between the piston tip and the end of the dosator tube. At B, the tube descends no further and a constant force – the precompression force – is detected.

- Point C. The piston now moves down the dosator tube, compressing the powder in it. Force now increases to a maximum at D, after which it decreases rapidly as the dosator is drawn out of the powder bed. However force does not drop to zero as the plug remains in contact with the piston tip (point E).

- Between points E and F, the dosator assembly is rotated so that the tube containing the plug is positioned over the base of an empty capsule shell.

- Point F. The piston moves downwards, thereby ejecting the plug from the dosator. The magnitude of the ejection force (point G) is dependent on lubrication.

- Point H. After the plug has been ejected, there is now no contact between it and the piston tip. Hence force should fall to zero, but in fact it may fall to below zero. This means that an extensive stress rather than a compressive stress is detected by the transducers. This has been termed the 'drag force' and is attributed to powder on the inside of the dosator tube preventing full retraction of the piston.

In a subsequent study, Small and Augsburger (1978) examined the effects of powder bed height, piston height, lubricant type, lubricant concentration and compression force on the force needed to eject the plug from the dosator, using three fillers (microcrystalline cellulose, pregelatinised starch and anhydrous lactose). As might be expected, ejection force increased with compression force. However ejection force was also directly proportional to powder bed height and piston height. As these two factors are increased, the plug length is also increased and hence there is a greater area of contact between the plug and the inside of the dosator tube. It follows from this work that for reproducible plug properties, consistency in powder bed height and piston settings are necessary.

Ejection force minima were achieved with 1% magnesium stearate for anhydrous lactose, 0.5%

for microcrystalline cellulose and 0.1% for pregelatinised starch. It is interesting to note that despite the much lower forces used for compression, the ejection forces are comparable to those encountered in tablet presses and the levels of lubricant are very similar to those used in tablet formulations containing these three diluents.

When forces in the capsule-filling process can be measured, it follows that they can also be controlled. Thus those properties of the powder plug such as physical strength and release characteristics which depend on the compression force can be investigated meaningfully. Mehta and Augsburger (1981) were able to study the effect of magnesium stearate on plug strength and drug dissolution at a constant compressing force of 15 kg (about 150 N). They measured the physical strength of plugs by a three-point bending test and found that this could be correlated with dissolution rate. With microcrystalline cellulose, the hardness of the fills decreased markedly as the magnesium stearate concentration was increased. A similar reduction has been noted with microcrystalline cellulose tablets lubricated with magnesium stearate (Bolhuis and Hölzer, 1996).

The effect on release characteristics was more complex. At low lubricant concentrations, the reduction in physical strength of the plugs permitted easier water penetration, but as the level of lubricant was raised, increased hydrophobicity inhibited drug release. With lactose, hardness was not significantly reduced and only the retarding effect on dissolution was noted. This too has parallels in tablet formulation.

The instrumented Zanasi LZ64 filling apparatus was modified by Botzolakis *et al.* (1982) who replaced the mercury swivel contact with slip rings, an approach previously used with rotary tablet presses (Watt and Rue, 1979). They studied the effect of disintegrating agents on capsule fills, pointing out that in many previous studies on drug release from hard shell capsules, the capsules had been hand filled, with a resulting high porosity. It was not surprising that disintegrating agents had been found to have little effect, since there was no structure for them to press against. Hence wettability and water penetration would be the more important factors. Botzolakis *et al.* were able to keep piston height, powder bed height and compression force constant and examined the effects of a range of disintegrating agents on capsule fills made from dicalcium phosphate dihydrate and microcrystalline cellulose, both lubricated with magnesium stearate. All disintegrating agents improved drug release with sodium starch glycolate and croscarmellose sodium being the most effective.

In a more elaborate study, Botzolakis and Augsburger (1984) used the same two fillers in a three-factor, three-level factorial design, the three factors being disintegrant concentration, lubricant concentration and compression force. The responses were fill weight variation, plug strength, disintegration and dissolution. They found that disintegration times did not always have the same rank order as dissolution, but that all three factors and their interactions had significant effects, the magnitude of which differed according to the solubility of the filler.

The fitting of displacement transducers (LVDTs – linear variable differential transformers) to tablet presses and their use to quantify the compression process was pioneered by de Blaey and Polderman at the University of Leiden (see for example, de Blaey and Polderman (1970)). By plotting force as a function of punch position, the work expended in the compression event could be calculated, since work is dimensionally equal to the area enclosed by the force–displacement curve. Fitting LVDTs to capsule-filling equipment was first reported by Mehta and Augsburger (1980a). They monitored movement of the instrumented piston described by Small and Augsburger (1977), a task complicated by the rotation of the piston during operation. This was overcome by threading a spring-loaded rod on to the core of the LVDT, which in turn was maintained in contact with a bracket fixed to the dosator housing. Cables for all transducers were connected to the mercury swivel assembly described earlier.

The force and displacement traces obtained by Mehta and Augsburger confirmed suggestions made earlier. When the dosator enters the powder bed, precompression force develops without piston movement. After the dosator has descended fully, the piston moves downwards, exerting the compression force. Mehta and Augsburger reported that the maximum compression force preceded the point of maximum piston displacement by about 40 ms. They attributed this to the action of the overload spring. However a similar non-coincidence of force and displacement maxima has been noted in the force–displacement curve for tablets and this has been linked to viscoelastic behaviour of the powder particles (Muller, 1996). At the ejection stage, downward movement of the piston results in a rise in force that rapidly falls away as the frictional forces holding the plug in the nozzle are overcome. Though the authors signified their intention to calculate work expenditure during plug formation by calculating areas under the force–displacement curves, few results seem to have been published (Mehta and Augsburger, 1980b). Since forces are so low, the work expended will consequently also be low. The results would be highly dependent on accurate measurement of displacement and the latter would undoubtedly be complicated by the presence of the return spring in the dosator.

Teams of workers other than Augsburger's group have fitted instrumentation to dosator nozzle capsule-filling equipment. For example, Mony et al. (1977) fitted piezoelectric load washers to the ends of the pistons of a Zanasi RV59 machine. With this arrangement, a force can only be detected during the compression and ejection events when the piston is being depressed. Thus precompression, retention and residual forces cannot be studied. These workers investigated the effect of magnesium stearate and talc on compression and ejection forces. A similar study was carried out by Maury et al. (1986) using load washers mounted not on the pistons but on the compression and ejection platens. Again only compression and ejection can be studied. Rowley et al. (1983) attached a load washer to the ejection knob of a Zanasi LZ64. This arrangement can of course only be used to study the ejection event.

An instrumented Zanasi LZ64 was used by Hauer et al. (1993a, 1993b) to examine the formulation variables of a mixture of microcrystalline cellulose (a viscoelastic material) and anhydrous lactose, which is brittle. They found that the better the powder flow, the more variable the fill weight, as the mixture was more difficult to densify. Lubricant concentration was shown to be critical, with magnesium stearate being superior to stearic acid.

Instrumentation studies on dosator tube capsule-filling equipment are summarised in Table 7.1.

Instrumentation of dosing disk capsule-filling machines

The consolidation process on a dosing disk machine is more complex than that of a tablet press or a dosating tube capsule filler. In the latter two, plugs are formed by a single compressive stroke, whereas in the former, the plugs are formed progressively by successive tamps, five in the case of Höfliger & Karg equipment. The aim is to achieve a specified fill weight and this can be obtained by a large number of different combinations of the settings of the tamping pins. If two successive positions are adjusted to the same

depth of penetration, then the full weight gain at the second station depends on the volume of the void left after the preceding tamp.

As reported earlier, the group headed by Professor Augsburger at the University of Maryland had been pioneers in instrumentation of dosator capsule-filling equipment. They commenced with relatively simple instrumentation of a Zanasi AZ64 machine and the progessive complexity of the instrumentation was developed in subsequent years. Augsburger's team then turned their attention to dosing disk machines and a similar pattern developed. In 1983, Shah *et al.* described the instrumentation of a Höfliger & Karg GKF330 machine, fitting strain gauges to the necks of two tamping fingers. One finger was kept at the ejection station and the other could be inserted at any of the five tamping positions. They observed that the fourth tamping position had the greatest influence on fill weight and

Table 7.1   Instrumentation of capsule-filling equipment: dosator tube

Equipment	Component instrumented	Parameters measured	Reference
Zanasi LZ64	Piston	Compression force Retention force Ejection force Drag force	Cole and May, 1975
Zanasi LZ64	Piston	Precompression force Compression force Retention force Ejection force Drag force	Small and Augsburger, 1977
Zanasi LZ64	Piston	Compression force Retention force Ejection force Piston displacement	Mehta and Augsburger, 1980a
Zanasi LZ64	Piston	Precompression force Compression force Retention force Ejection force Drag force	Botzolakis *et al.*, 1982
Zanasi LZ64	Ejection knob	Ejection force	Rowley *et al.*, 1983
Zanasi LZ64	Compression and ejection platens	Compression force Ejection force	Maury *et al.*, 1986
Zanasi LZ64	Piston	Compression force Ejection force	Hauer *et al.*, 1993a
Zanasi RV59	Piston	Compression force Ejection force	Mony *et al.*, 1977

compression force. The second position had the least.

The instrumentation was then developed so that each of the five tamping pistons was equipped with strain gauges and the whole compressive history of a single plug could be examined (Shah *et al.*, 1986). As the tamping finger penetrates the powder bed, it pushes particulate matter into the hole in the disk. On this machine, force rises smoothly to a maximum of about 200 N. Then the tamping finger begins to rise and there is a consequent decrease in force until a plateau is reached. The height of the plateau is dependent on the maximum force, but its duration is constant at about 60 ms. The plateau is caused by a brief halt in the upward movement of the tamping finger, which is due to the intermittent motion of a Höfliger & Karg machine that brings on the next capsule shell to be filled. Thus the plateau is a feature introduced by the design of the machine and has nothing to do with the properties of the plug. Contact between the top surface of the plug and the end of the tamping finger is maintained by the partially relaxed relief spring and hence a force continues to be detected. Once the plateau has been passed, decompression proceeds.

These workers found that the target weight of the plug could be achieved after three tamps, but it was not possible after just two. They also found effective compression began before the piston enters the dosing disk, powder being pushed ahead of the tamping finger. Thus the higher the tamping force, the heavier the plug.

Shah *et al.* also measured the physical strength of the plugs using a three-point flexure test. As compression force was increased, not unexpectedly plug strength increased too. Owing to the repeated application of force during plug formation, it might be expected that the lowest part of the plug would show a progressive increase in strength as it was compressed for a second or third time. This was not the case, provided that the subsequent compressions were at the same force. An increased force led to higher consolidation and hence higher physical strength.

Using this filling equipment, Shah *et al.* examined the plug-forming properties of anhydrous lactose, dicalcium phosphate and microcrystalline cellulose. They found that with the last

named diluent, there was an optimum concentration of magnesium stearate of 0.1%, in that plugs with this level of lubricant were both stronger and heavier than unlubricated plugs for any combination of force and number of tamps. This they attributed to improved powder flow. Magnesium stearate concentrations in excess of 0.1% caused softening of the plugs, ultimately to a lower strength than the unlubricated powder.

In a later paper (Shah *et al.*, 1987), the effects of multiple tamping on plug dissolution were studied, using the same instrumented equipment. In general, increasing the number of tamps resulted in slower dissolution. Higher compression forces accelerated drug release when anhydrous lactose was used as the filler, but the reverse was true with dicalcium phosphate. They made the important observation that provided sufficient disintegrating agent was present (4% croscarmellose was used), altering tamping force and number of tamps had no effect on drug dissolution.

The next logical step was the fitting of displacement transducers so that force and tamping pin position could be monitored simultaneously. This was reported by Cropp, Augsburger and Marshall in 1991. Two LVDTs were fitted to the equipment. One monitored the brass ring to which the tamping pin holder assembly is anchored. The other was attached to a modified pin that rested on the tamping pin head and beneath the overload spring. Thus movement of the overload spring could be detected. The combination of the two LVDTs enabled the penetration of the actual pin into the dosing disk to be determined. Transducer outputs were stored and recorded by computer.

Force–time relationships were obtained as described earlier (Shah *et al.*, 1986) and both ring displacement and pin displacement also showed a pause in upward movement shortly after maximum displacement had been achieved. This confirmed that the plateau in the force–time curve was due to movement of the transport mechanism for the capsule shells. Peak displacement occurred at the same time as peak force.

The authors pointed out that the choice of overload spring could affect the properties of the powder plug. The strength of the spring governs the applied force and consequently plug properties such as strength and dissolution. Plugs made from anhydrous lactose were subjected to a higher peak force during compression than microcrystalline cellulose plugs over a whole range of tamping pin settings. Microcrystalline cellulose has a lower bulk density than anhydrous lactose and so pin displacement is greater for the former.

Force and displacement data were combined to construct force–displacement curves. A force was registered before any tamping displacement was detected. This was because the tamping pin detects a resistance to its movement while travelling through the powder prior to penetrating the dosing disk. After peak penetration is reached, displacement falls to a plateau and then the curve returns to the baseline. Calculation of the work of compaction gave much lower values (a fraction of a joule) than those needed to compress tablets, since though the pins moved several millimetres during compression, the applied force was only a few tens of newtons.

Further development of the same machine was carried out by Davar *et al.* (1997) who fitted each station with instrumented tamping fingers. LVDTs were added to measure tamping pin penetration, pin displacement at peak pressure and the movement of the brass guide block. The relationships between compressing force and plug properties such as length, physical strength and density were investigated.

A totally different approach to instrumentation of tamping disk filling equipment has been described by Podczeck (2000, 2001). In her first paper, Podczeck pointed out that in tablet manufacture, advances in instrumentation have led to feedback mechanisms used to control tablet weight, but such developments in capsule filling had yet to be achieved. One possibility of changing fill weight was by altering the tamping distance of one or more of the tamping pins, but it was shown by Podczeck and Newton (1999) that only modest changes could be achieved in this way. Larger changes in fill weight could be brought about by exchanging the springs inside the tamping fingers. However adjustment of penetrative depth and changing springs can only be carried out when the equipment is stationary and so neither lends itself to a feedback mechanism. For weight adjustment to take place while the machine was running, some form of electrical or electronic control was required.

Podczeck used a Bosch GKF 400S machine. In this equipment, there are five tamping stations each fitted with three tamping pins. In the one tamping block, which was instrumented, the springs were removed and replaced with dashpots and a chamber filled with compressed air. The latter was in contact with a piezoelectric force transducer. By this arrangement, the point at which the springs would be deflected could be altered continuously and hence the volume available in the hole of the dosing disk was also changed. Force rapidly rose to a maximum, was maintained virtually constant for a time and then fell to a plateau level. It then remained constant for a further period before returning to zero. The change of force between maximum and plateau was attributed to deflection of the spring in the tamping fingers, which in this case was simulated by air pressure. The plateau was not detected at tamping forces of less than about 60 N.

In practice, it was found that the pneumatic head could control fill weight, but could only change it in small increments. If larger changes were needed, then the tamping fingers had to be adjusted or the powder bed depth had to be altered. A significant finding in this work was that most of the powder, which ultimately formed the plug, entered the holes of the dosing disk by flow under gravity as the disk rotated. Only a minor portion of the plug came from powder pushed in by the tamping fingers. It follows therefore that on a tamping machine, the ultimate plug weight is largely dependent on the flow properties of the powders rather than the force exerted by the tamping fingers.

In a further study on this apparatus (Podczeck, 2001), the station bearing the instrumented head was varied so that the contribution of each station to plug formation could be assessed. It was found that the plug achieved its final length and density at station four and so the best way to control plug weight would be to position the instrumented head at this station. Podczeck suggested that a feedback device would be achieved with the instrumented head at station four and a non-instrumented pneumatic head at station three. The internal pressure of the latter would be controlled by electrical signals from the former.

An important finding in this work was that for any given plug each successive tamp caused further densification. It will be recalled that Shah *et al.* (1986) reported that the density of each segment of the plug did not increase despite multiple applications of tamping force. Podczeck explained these results with reference to the consolidation mechanisms of the solids involved. Shah *et al.* used lactose and dicalcium phosphate in their studies. Both these undergo consolidation by fracture and since the forces involved in plug formation are well below the yield points of such substances, it is unlikely that these particles would undergo fragmentation. However Podczeck used microcrystalline cellulose and pregelatinised starch. Both of these are ductile materials and are readily deformed by low forces. Hence progressive consolidation can be anticipated.

Instrumentation studies on dosing disk capsule-filling equipment are summarised in Table 7.2.

The simulation of capsule-filling equipment

A disadvantage of conventional tablet presses and capsule-filling equipment is that several hundred grams or even kilograms of particulate material may be needed for them to operate efficiently or even at all. In some circumstances, this quantity may not be available, and even if it were, considerable wastage would be unavoidable.

It was primarily for this reason that tablet press simulators were introduced in the 1970s. These are essentially precisely controlled hydraulic presses fitted with a die and two punches. The die is usually filled manually and individual tablets made and studied. A further advantage of tablet

press simulators is that by adjusting the rate of movement of the punches, the speed of compression as well as the applied force can be controlled. Hence one versatile simulator can, in theory at least, imitate the mode of operation of any tablet press.

Though the advantages of economy and versatility can equally apply to simulation of capsule-filling equipment, there are other considerations that require a different emphasis. As already mentioned, the availability of suitable points of attachment for transducers to capsule-filling machinery is limited without major modification of the equipment. Hence a simulator can be designed so that it has a sufficiency of robust fixing points for the transducers which are not available on the equipment itself. A further feature is that the simulators are usually somewhat cheaper to construct and operate than their tablet-filling counterparts. As mentioned above, tablet press simulators are hydraulic presses and much of their expense arises from the need to move rapidly and precisely large volumes of hydraulic fluid at high pressures. Since capsule-filling simulators require much lower forces than tablet presses, their control systems are less demanding. For example, the simulator described by Britten *et al.* (1995) is operated pneumatically from a commercial compressed air cylinder.

Any simulator must be capable of exerting the forces and reproducing the patterns of component movement of the equipment it is designed to imitate. Since the principles of operation of all tablet presses are essentially the same and equations have been derived to predict punch movement of both rotary and eccentric presses (Armstrong, 1989), a 'universal' simulator for tablet presses is at least theoretically possible. Not so with capsule-filling equipment, since every manufacturer has a different mechanism

Table 7.2   Instrumentation of capsule-filling equipment: dosing disk

Equipment	Component instrumented	Parameters measured	Reference
Hofliger & Karg GKF330	Tamping pins	Tamping force	Shah *et al.*, 1983
Hofliger & Karg GKF330	Tamping pins	Tamping force	Cropp *et al.*, 1991
		Pin movement	
Bosch GKF400S	Tamping pins	Tamping force	Podczeck, 2000

for inducing component movement. Thus movement is simulated either by using isolated parts of the equipment (e.g. Jolliffe *et al.*, 1982), or by measuring component movement on an actual machine with a transducer and incorporating this knowledge into the design of the simulator (e.g. Britten *et al.*, 1995). The latter method of course depends on the transducer being properly sited. Lack of such siting points is one of the reasons for using a simulator in the first place.

Care must also be taken to define what is meant by 'speed'. This is the actual rate of movement by components of the apparatus, e.g. in mm s^{-1} and must not be confused with machine output, i.e. the number of capsules filled in unit time. These two parameters may bear no relation to each other, as demonstrated by Britten *et al.* (1995).

Simulation of dosator tube-filling equipment

An important development in capsule machine instrumentation came with the publication by Jolliffe *et al.* in 1982 of details of the construction of a dosator tube simulator based on an mG2 model G36 machine. The problems of connecting electrical wiring to a rotating component have already been alluded to. In conventional mG2 machines, the filling turret rotates and the powder hopper beneath it is stationary. In this simulator, these roles were reversed so that the turret to which the transducers were connected was stationary and the powder hopper rotated around the dosator. There was no relative movement between the feed tray and the nozzle at the moment when the dosator entered the powder bed. Four semiconductor strain gauges were mounted on the piston in a Wheatstone bridge configuration to measure stress and displacement transducers monitored the vertical movements of the piston and the dosator nozzle. Thus the movement of the whole dosator and the relative movement of dosator and piston could be followed. A force could be applied in two ways. A precompression force was exerted by adjusting the height of the piston in the nozzle. This was found to be particularly useful to consolidate beds of low bulk density. Compression force was exerted by movement of the piston when the

nozzle was in the powder bed and was altered by raising or lowering the compression cam, the precise position of which was recorded by the piston movement transducer.

A considerable body of work carried out on this simulator has been published. Newton and his co-workers were particularly interested in elucidating those factors that contributed to uniformity of plug weight. They found that fine lactose particles gave acceptable uniformity over a wide range of compression settings, whereas the larger the particles, the smaller the range over which satisfactory filling was achieved (Jolliffe and Newton, 1982). Fine cohesive powders gave the best results because they underwent greater volume reduction on compression than coarser particles.

Jolliffe and Newton (1978, 1980) had shown theoretically that for a powder to be retained within the dosator nozzle during transfer from powder bed to capsule shell, a stable arch had to be formed at the outlet of the nozzle. This was related to the flow properties of the powder, in that cohesive powders would require a lower degree of compression for the arch to form. They also surmised that arch formation would depend on the surface characteristics of the inside of the nozzle, which would in turn govern the frictional forces between the nozzle and powder. The surface could be affected either by roughness of the metal or by a coating of powder. They prepared nozzles with a range of surface textures and confirmed that there is an optimum degree of surface roughness needed to ensure powder retention in the nozzle (Jolliffe and Newton, 1983a). These findings were confirmed when they used an mG2 G36 production machine, thereby validating their approach of using a simulator (Jolliffe and Newton, 1983b). They found that fine cohesive powders gave acceptable fill weight uniformity over a wide range of compression settings, but this range was reduced with more free-flowing powders.

A series of papers by Tan and Newton extended the work of this group, using the same simulator, which was now connected to a computer to capture and manipulate data. Using five common capsule diluents, the relationship between uniformity of fill weight and a range of parameters related to powder flow were investigated. Particle size, morphology, bulk density and compressive force were found to be important. They found there was no correlation between uniformity of weight and measures of friction such as angle of internal flow and angle of effective friction (Tan and Newton, 1990a). After each filling cycle, the dosator was weighed, allowing information to be gained on the build-up of powder on the inner surface. It was found that lactose was particularly prone to binding. They found that the texture of the inner wall of the dosator had no significant influence with powders with low binding affinity such as micro-crystalline cellulose and pregelatinised starch (Tan and Newton, 1990b).

In a later paper (Tan and Newton, 1990c), it was found that fill-weight variability also depended on powder bed density. The most uniform weights were achieved when no compressing force was applied during the filling process. As compression was increased, fill weight decreased. This was attributed to coating the wall of the nozzle and loss of powder as particles were forced behind the tip of the piston, which in extreme cases led to the piston jamming in the nozzle.

In a final paper, Tan and Newton (1990d) compared the observed plug densities calculated from plug dimensions with predicted values based on knowledge of powder bed density and piston position. Correlation was poor because of weight variation, which was greatest with fine powders at high compression settings.

Another simulator based on the dosator nozzle principle was constructed by Britten and co-workers (1991, 1995). In this pneumatically driven apparatus which simulated the Macofar MT13-2 machine, there were no rotating components at all. In a conventional Macofar machine, the dosator tubes are plunged into the powder bed, a plug is withdrawn and is then ejected. In this simulator, the dosator mechanism was stationary and the powder brought to it by a powder bowl, which moved in a vertical direction. A precompression force could be exerted, followed by a compression force which was applied by piston movement. Once formed, the dosator ascended out of the powder bed and the plug was ejected by means of the piston. However no attempt was made to eject the plug into an empty capsule shell.

Compression force was measured by semiconductor strain gauges fitted to the dosator piston and arranged in a Wheatstone bridge conformation. An additional development on this apparatus was to fit strain gauges to the outer surface of the dosing funnel in order to measure axial stresses brought about by the presence of the plug in the dosator. The strain gauges were positioned 6 mm from the tip of the dosator. It has been shown that in axial measurements of this type, the positioning of the transducers is critical if meaningful data are to be obtained (Huckle and Summers, 1985). Since direct contact between the piston and the LVDT was not feasible, a small arm, fitted to the piston shank and in contact with the LVDT, was used to determine the position of the piston within the dosator. Vertical movement of the powder bowl was also determined by the LVDT. The output of all transducers was fed into a computer and manipulated on a spreadsheet program.

This simulator could be set to operate in a variety of modes, all at a range of bowl and piston speeds:

- Precompression simulation, when the powder plug was formed solely by the dosator plunging into the powder bed.
- Constant displacement simulation, when an additional tamp is applied to each plug by the dosator piston moving a predetermined distance.
- Constant pressure simulation, when the piston is allowed to travel as far as possible until the resistance of the powder to undergo further consolidation equals the applied compression pressure.

Two-factor, two-level factorial designs were used to study variation in plug weight and density in relation to compression pressure, precompression velocity, compression velocity and ejection velocity using pregelatinised starch and lactose (Britten et al., 1996). It was found possible to form plugs of starch without lubrication, but addition of 1% magnesium stearate was necessary for lactose. The rate of ejection had no effect on plug weight or density. However an increase in the precompression speed caused a fall in plug weight. At higher speeds, powder is pushed ahead of the nozzle rather than entering it and there is also less consolidation. A similar observation was made with tablet compression, especially with pregelatinised starch (Armstrong and Palfrey, 1989).

When the simulator was run in constant pressure mode, the effect of precompression velocity disappears and there is no evidence that higher pressures have a significant effect on plug density. Thus if reproducible and predictable plug weights are required, then a relatively high tamping pressure is indicated. However this may cause an increase in the physical strength of the plugs, which may in turn delay drug release (Mehta and Augsburger, 1981). Hence for any given formulation, an optimum pressure must be sought. Britten *et al.* (1996) noted that no plugs fell out of the dosator tube before active ejection by the piston, despite the radial pressures being as low as 0.01 MPa. It followed that from the point of view of plug retention, high compression pressures are not required, a view also expressed by Tan and Newton (1990d).

A more elaborate study on lactose using the same simulator was carried out by Tattawasart and Armstrong (1997), who studied the effects of lubricant concentration, dosator pressure and dosator piston height on plug properties by means of a three-factor, three-level Box Behnken design followed by multiple regression. Whilst pressure and piston height had significant effects on plug properties, lubricant concentration did not and it was concluded that the lowest concentration of magnesium stearate examined (0.5%) was more than adequate.

Simulation of dosing disk-filling equipment

The earliest attempts to simulate plug formation by tamping were not intended to imitate capsule filling *per se*, but to produce plugs under controlled conditions for dissolution studies. For example, Lerk *et al.* (1979) used a hand-operated press fitted with a plunger and die to produce plugs at a known constant force. This was detected by a load cell fitted to the top of the plunger. A similar approach was used by Ludwig and van Ooteghem (1980).

A device designed by Höfliger & Karg to select the correct dosing disk for a given formulation was used by Jones (1988, 1998) as a simulator. The device had one tamping finger, the force exerted by which could be measured by a load cell. Movement could also be detected, from which plug length could be calculated. Davar *et al.* (1997) used an Instron testing machine for the same purpose and confirmed their results using the instrumented capsule-filling apparatus described earlier.

As stated earlier, the production of powder plugs by dosing disk machines is somewhat analogous to the compression of tablets. In their paper describing pin displacement measurements on dosing disk machines, Cropp *et al.* (1991) pointed out that if tamping pin displacement was known, it should be possible to make a compaction simulator which could mimic the component movement and the low forces involved in plug formation. This development was reported by Heda, Muller and Augsburger in 1999.

Since powder plugs for capsule fills have a greater height to diameter ratio than tablets, it was necessary to use a die which was much deeper than normal. A diameter of 5.71 mm was chosen which was the same diameter as the tamping pin used to prepare a plug for a size 1 capsule. Plugs of height up to 12 mm could be prepared. Anhydrous lactose, microcrystalline cellulose and pregelatinised starch were used and the die was lubricated by hand with a magnesium stearate solution. A feature of tablet press simulators is that they can be operated at a range of punch speeds. Heda *et al.* (1999) studied plug formation at constant punch speeds of 1, 10 and 100 mm s^{-1}. The last speed is slightly faster than that of the tamping pins in a Höfliger & Karg GKF-330 capsule-filling machine and considerably faster than speeds encountered in a Zanasi LZ64 dosator piston machine.

Heda *et al.* (1999) discovered that force transmission through the length of the plug was very dependent on plug length, as measured by the ratio of force detected by the lower punch to that applied by the upper punch. They attributed this to the large difference in packing densities between the two ends of the plug, which leads to poor axial force transmission. Nevertheless, they found that their data could be fitted to the Shaxby–Evans equation (1923), which predicts that force applied by the upper punch decays exponentially towards the lower punch at a rate dependent on plug dimensions and a constant which is substance specific. They also found that the Heckel (1961) and Kawakita and Lüdde (1970–1) equations, which have been successfully applied to the study of tablet compression, apply equally well to the low-force environments of plug formation. Punch speed had no effect on the properties of plugs made from lactose, but with microcrystalline cellulose and pregelatinised starch, peak forces were at a maximum at 10 mm s^{-1}. As plug length decreased, the forces that were also generated, decreased owing to a diminution of the total resistance to compression.

Heda *et al.* (1999) pointed out that though the information gained in this study is more obviously applicable to dosing disk machines, consolidation also occurs in dosator tube equipment and the same low-force powder physics could well apply there.

Development of simulated capsule-filling equipment is summarised in Table 7.3.

Conclusion

Just as the use of the instrumented tablet press has led to meaningful research on the tabletting process, so the fitting of transducers to capsule-filling equipment has resulted in a more profound understanding of the events which lead to plug formation. Jones (2001) has pointed out the belief that 'powder filled capsules are a very simple product that does not need much skill to prepare'. Research using instrumented capsule-filling equipment and simulators has, it is hoped, dispelled this belief. The underlying process of plug formation is, it now appears, similar to that of tablet compression except that the forces involved are much lower. The applied force, the consolidation mechanism of the solid(s) and the presence of other excipients such as lubricants and disintegrants all contribute to the two vital plug properties, uniformity of weight and drug release, just as they do towards tablet properties.

An apparently complicating factor in the study of capsule filling is that there are two types of

Table 7.3 Simulated capsule-filling equipment

Equipment used or simulated	Component instrumented	Parameters measured	Reference
Macofar 13/2	Dosator Powder bed holder Piston shaft Piston tip	Dosator movement Piston movement Compression force Ejection force Radial force at tube tip	Britten *et al.*, 1995
mG2 G36	Dosator piston	Dosator movement Piston movement Compression force Ejection force	Jolliffe *et al.*, 1982
mG2	Dosator piston	Compression force	Veski and Marvola, 1991
Tablet compression simulator	Upper and lower punches	Upper punch movement Lower punch movement Upper punch force Lower punch force	Heda *et al.*, 1999
Instrumented dosing disk	Tamping pin	Pin movement Compression force	Jones, 1998
Instron testing machine	Punch	Punch movement Compression force	Davar *et al.*, 1997

filling mechanism, the dosating piston and the dosing disk. Virtually all published work is based on one or the other of these two types, with hardly any dealing with both. An exception is Heda *et al.* (1998). Furthermore, because of the two modes of action, most texts on capsule filling deal separately with the two types of machine. Consequently reviews such as the present publication are virtually obliged to follow this route. However both involve particles being compressed under a low force in a confined space. It may well be that the underlying mechanisms of plug formation are not so different after all. A comprehensive study, using instrumented dosator and dosing disk machines and the same raw materials would be extremely useful in identifying those formulations and manufacturing variables that are important to both types of capsule-filling equipment and those that are not.

References

Armstrong, N. A. (1989). Time-dependent factors involved in powder compression and tablet manufacture. *Int. J. Pharm.*, 49, 1–13.

Armstrong, N. A. and Palfrey, L. P. (1989). The effect of machine speed on the consolidation of four directly compressible tablet diluents. *J. Pharm. Pharmacol.*, 41, 149–151.

Bolhuis, G. K. and Hölzer, A. W. (1996). Lubricant sensitivity. In Alderborn, G. and Nyström, C. (Eds), *Pharmaceutical Powder Compaction Technology*. New York: Marcel Dekker, 517–560.

Botzolakis, J. E. and Augsburger, L. L. (1984). The role of disintegrants in hard-gelatin capsules. *J. Pharm. Pharmacol.*, 36, 77–84.

Botzolakis, J. E., Small, L. E. and Augsburger, L. L. (1982). Effect of disintegrants on drug dissolution from capsules filled on a dosator-type automatic capsule filling machine. *Int. J. Pharm.*, 12, 341–349.

Brake, E. F. (1951). *Development of Methods for Measuring Pressures During Tablet Manufacture*. MS Thesis, Purdue University.

Britten, J. R. and Barnett, M. I. (1991). Development and validation of a capsule filling machine simulator. *Int. J. Pharm.*, 71, R5–R8.

Britten, J. R., Barnett, M. I. and Armstrong, N. A. (1995). Construction of an intermittent-motion capsule filling machine simulator. *Pharm. Res.*, 12(2), 196–200.

Britten, J. R., Barnett, M. I. and Armstrong, N. A. (1996). Studies on powder plug formation using a simulated capsule filling machine. *J. Pharm. Pharmacol.*, 48, 249–254.

Celik, M. (1992). Overview of compaction data analysis techniques. *Drug Dev. Ind. Pharm.*, 18(6&7), 767–810.

Celik, M. (1996). Overview of tableting technology. Part 1. Tablet presses and instrumentation. *Pharm. Technol.*, 20, 20–39.

Cole, G. C. and May, G. (1972). Instrumentation of a hard shell encapsulation machine. *J. Pharm. Pharmacol.*, 24, 122P.

Cole, G. C. and May, G. (1975). The instrumentation of a Zanasi LZ/64 capsule filling machine. *J. Pharm. Pharmacol.*, 27, 353–358.

Cropp, J. W., Augsburger, L. L. and Marshall, K. (1991). Simultaneous monitoring of tamping force and pin displacement (F-D) on an Hofliger-Karg capsule filling machine. *Int. J. Pharm.*, 71, 127–136.

Davar, N., Shah, R., Pope, D. G. and Augsburger, L. L. (1997). The selection of a dosing disk on a Hofliger-Karg capsule-filling machine. *Pharm. Technol.*, 21, 32–48.

de Blaey, C. J. and Polderman, J. (1970). The quantitative interpretation of force–displacement curves. *Pharm. Weekbl.*, 9, 241–250.

Hauer, B., Remmele, T., Züger, O. and Sucker, H. (1993a). Rational development and optimisation of capsule formulations with an instrumented dosator capsule filling machine. Part 1. Instrumentation and influence of the filling material and the machine parameters (in German). *Pharm. Ind.*, 55(5), 509–515.

Hauer, B., Remmele, T. and Sucker, H. (1993b). Rational development and optimisation of capsule formulations with an instrumented dosator capsule filling machine. Part 2. Fundamentals of the optimisation strategy (in German). *Pharm. Ind.*, 55(8), 780–786.

Heckel, R. W. (1961). Density–pressure relationships in powder compression. *Trans. Metall. Soc. AIME*, 221, 671–675.

Heda, P. K., Muteba, K. and Augsburger, L. L. (1998). Comparison of the formulation requirements of dosator and dosing disc automatic capsule filling machines. *AAPS PharmSci.*, Annual Meeting Abstract 2125.

Heda, P. K., Muller, F. X. and Augsburger, L. L. (1999). Capsule filling machine simulation. 1. Low force powder compaction physics relevant to plug formation. *Pharm. Dev. Technol.*, 4, 209–219.

Higuchi, T., Nelson, E. and Busse, L. W. (1954). The physics of tablet compression. 3. Design and construction of an instrumented tablet machine. *J. Amn. Pharm. Assoc. Sci. Edn.*, 43, 344–348.

Huckle, P. and Summers, M. P. (1985). The use of strain gauges for radial stress measurement during tabletting. *J. Pharm. Pharmacol.*, 56, 722–725.

Jolliffe, I. G. and Newton, J. M. (1978). Powder retention within a capsule dosator nozzle. *J. Pharm. Pharmacol.*, (Suppl) 30, 41P.

Jolliffe, I. G. and Newton, J. M. (1980). The effect of powder coating on capsule filling with a dosator nozzle. *Acta Pharm. Technol.*, 26, 324–326.

Jolliffe, I. G. and Newton, J. M. (1982). An investigation of the relationship between particle size and compression during capsule filling in an mG2 simulator. *J. Pharm. Pharmacol.*, 34, 415–419.

Jolliffe, I. G. and Newton, J. M. (1983a). The effect of dosator wall texture on capsule filling with the mG2 simulator. *J. Pharm. Pharmacol.*, 35, 7–11.

Jolliffe, I. G. and Newton, J. M. (1983b). Capsule filling studies using an mG2 production machine. *J. Pharm. Pharmacol.*, 35, 74–78.

Jolliffe, I. G., Newton, J. M. and Cooper, D. (1982). The design and use of an instrumented mG2 capsule filling machine simulator. *J. Pharm. Pharmacol.*, 34, 230–235.

Jones, B. E. (1988). Powder formulations for capsule filling. *Manuf. Chem.*, 59(7), 28–30, 33.

Jones, B. E. (1998). New thoughts on capsule filling. *S. T. P. Pharma Sci.*, 3, 777–783.

Jones, B. E. (2001). The filling of powders into two-piece hard capsules. *Int. J. Pharm.*, 227, 5–26.

Kawakita, K. and Lüdde, K. H. (1970–1). Some considerations on powder compression equations. *Powder Technol.*, 4, 61–68.

Lerk, C. F., Lagas, M., Lie-A-Huen, L., Broersma, P. and Zuurman, K. (1979). *In vitro* and *in vivo* availability of hydrophilised phenytoin from capsules. *J. Pharm. Sci.*, 68(5), 634–638.

Ludwig, A. and van Ooteghem, M. (1980). Influence of the capsule wall on the availability of compacted drugs packed in hard gelatin capsules. *J. Pharm Belg.*, 35, 351–356.

Maury, M., Heraud, P., Etienne, A., Aumonier, P. and Casahoursat, L. (1986). Measurement of compression during the filling of capsules. *4th International Conference on Pharmaceutical Technology*, Paris, 384–388.

Mehta, A. M. and Augsburger, L. L. (1980a). Simultaneous measurement of force and displacement in an automatic capsule filling machine. *Int. J. Pharm.*, 4, 347–351.

Mehta, A. M. and Augsburger, L. L. (1980b). Quantitative evaluation of force–displacement curves in an automatic capsule-filling operation. *American Association of Pharmaceutical Scientists*, 30th National Meeting, St Louis.

Mehta, A. M. and Augsburger, L. L. (1981). A preliminary study of the effect of slug hardness on drug dissolution from hard gelatin capsules filled on an automatic capsule-filling machine. *Int. J. Pharm.*, 7, 327–334.

Mony, C., Sambeat, C. and Cousins, C. (1977). The measurement of compression during the formulation and filling of capsules. *1st International Conference on Pharmaceutical Technology*, Paris, 98–108.

Muller, F. (1996). Viscoelastic models. In Alderborn, G. and Nyström, C. (Eds). *Pharmaceutical Powder Compaction Technology*. New York: Marcel Dekker, 99–132.

Podczeck, F. (2000). The development of an instrumented tamp-filling capsule machine. I: instrumentation of a Bosch GKF 400S machine and feasibility study. *Eur. J. Pharm. Sci.*, 10, 267–274.

Podczeck, F. (2001). The development of an instrumented tamp-filling capsule machine. II: investigations of plug development and tamping pressure at different filling stations. *Eur. J. Pharm. Sci.*, 12, 515–521.

Podczeck, F. and Newton, J. M. (1999). Powder filling into hard gelatine capsules on a tamp filling machine. *Int. J. Pharm.*, 185, 237–254.

Rowley, D. J., Hendry, R., Ward, M. D. and Timmins, P. (1983). The instrumentation of an automatic capsule filling machine for formulation design studies. *3rd International Conference on Pharmaceutical Technology*, Paris, 287–291.

Shah, K. B., Augsburger, L. L., Small, L. E. and Polli, G. P. (1983). Instrumentation of a dosing disk automatic capsule filling machine. *Pharm. Technol.*, 7(4), 42–54.

Shah, K. B., Augsburger, L. L. and Marshall, K. (1986). An investigation of some factors influencing plug formation and fill weight in a disk-type automatic capsule filling machine. *J. Pharm. Sci.*, 75(4), 291–296.

Shah, K. B., Augsburger, L. L. and Marshall, K. (1987). Multiple tamping effects on drug dissolution from capsules filled on a dosing disk type automatic capsule filling machine. *J. Pharm. Sci.*, 76(8), 639–645.

Shaxby, J. H. and Evans, J. C. (1923). The variation of pressure with depth in columns of powders. *Trans. Faraday Soc.*, 19, 60–72.

Small, L. E. and Augsburger, L. L. (1977). Instrumentation of an automatic capsule-filling machine. *J. Pharm. Sci.*, 66(4), 504–509.

Small, L. E. and Augsburger, L. L. (1978). Aspects of the lubrication requirements for an automatic capsule filling machine. *Drug Dev. Ind. Pharm.*, 4(4), 345–372.

Tan, S. B. and Newton, J. M. (1990a). Powder flowability as an indicator of capsule filling performance. *Int. J. Pharm.*, 61, 145–155.

Tan, S. B. and Newton, J. M. (1990b). Capsule filling performance of powders with dosator nozzles of different wall texture. *Int. J. Pharm.*, 66, 207–211.

Tan, S. B. and Newton, J. M. (1990c). Influence of compression setting ratio on capsule fill weight and weight variability. *Int. J. Pharm.*, 66, 273–282.

Tan, S. B. and Newton, J. M. (1990d). Observed and expected powder plug densities obtained by a capsule dosator nozzle system. *Int. J. Pharm.*, 66, 283–288.

Tattawasart, A. and Armstrong, N. A. (1997). The formation of lactose plugs for hard shell capsule fills. *Pharm. Dev. Technol.*, 2(4), 335–343.

Vesky, P. and Marvola, M. (1991). Design and use of equipment for simulation of plug formation in hard gelatin capsule filling machines. *Acta Pharma. Fenn.*, 100, 19–25.

Watt, P. R. (1988). *Tablet Machine Instrumentation in Pharmaceutics*. Chichester, UK: Ellis Horwood.

Watt, P. R. and Rue, P. J. (1979). The design and construction of a fully instrumented tablet machine. *Proceedings International Conference of Pharmaceutical Technology Production and Manufacture*, Copenhagen.

8

Capsule processing and packing

Walther Pietsch, Kristina Schlauch, Ralf Schmied and Katja Vollmer

Introduction

The capsule fill weight can be checked either during or after the production process. In each case the capsule check-weigher records the capsule weight, compares it to the permissible tolerance values and separates those capsules that fall outside the tolerance range (see Figure 8.1). Systems can be attached to filling machines and these can have a feedback system control in order to optimise the fill weight during continuous production. Weighing systems can be used for check weighing of whole batches after filling has been completed.

After the capsules are filled and their weight has been checked, they are packaged often into

Figure 8.1 Capsule check-weigher, type Bosch KKE 2500: a stand-alone machine with a separated weighing (left) and control (right) area. Capsule in-feed is from the top through the integrated capsule hopper (left upper side).

push-through packages or blisters. Pharmaceutically compliant blister packaging takes place in automated thermoforming machines. This machine type, also known as deep drawing or blister machines, can produce blister packages automatically and dominates the worldwide market for solid pharmaceutical packaging.

The word 'blisters' is used when talking about one entire cut package, mostly containing more than one capsule. Figure 8.2 shows a typical blister package and its handling. Blister packages offer individual product protection against environmental influences such as desiccation, humidity, oxygen, light and microorganisms. They also ensure simple product handling as capsules can be taken specifically and dispensed from the package. Thermoform and aluminium foils are used predominantly as packaging materials. Most common are combinations of thermoform-base webs and aluminium lid webs, as well as aluminium base webs and aluminium lid webs, the so-called 'alu/alu'-blister.

Capsule check-weighing machines

Fundamental requirements

Capsule check-weighers are required to fulfil high demands of measurement accuracy while simultaneously offering high output capacity. Details of some of the machines available are shown in Table 8.1. Capsule check-weighers can either be stand-alone or be connected in-line with a filling machine. The output capacity matches or exceeds the production capacity of the capsule-filling machine to avoid a back up

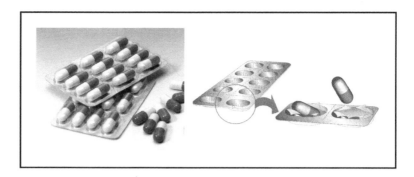

Figure 8.2 Typical blister package and capsule handling. The left picture shows the rectified capsules in the through-push blister with aluminium lid web and a thermoformed base web. The right picture shows the base web of an aluminium blister (alu/alu). The cavities are formed wider than polymer-based films owing to their different forming characteristics.

during a full-scale output. A high-capacity machine can fill at 120 000 to 150 000 capsules per hour and to cope with this output the check-weigher works with multiple tracks and is equipped with multiple weighing cells arranged in parallel. A system with 12 tracks has thus less than 300 ms to complete a weighing cycle and to decide whether the measured weight falls within the predetermined tolerance range. Common systems possess a weighing accuracy of ±2 mg, dependent on the capsules and local installation conditions. Unfavourable environmental and installation conditions such as building oscillations, electrostatic charges, or influences caused by loading arrangements can possibly affect the weighing accuracy significantly. These circumstances need to be taken into consideration during machine assembly and installation.

Dust deposits on top of the weighing cells or the capsules naturally result in flawed measurements. Thus, the application of an upstream capsule polisher is recommended in many cases. Mistakes caused by dust deposits are avoided through regular taring of the weighing cells. This should occur at short and regular intervals, for instance every 5 min, and the process generally runs in the check-weigher's automatic mode.

Table 8.1 Performance overview capsule check-weighers of different manufacturers

	Bosch KKE 2500	IMA Precisa 12/18	Anritsu KW 9001 AP	Anritsu KW 9002 AP
Output capsule/hr	150 000	120 000/200 000	120 000	60 000
Capsule sizes	00, 0el, 0, 1, 2, 3, 4, 5, DB A . . . E	00, 0el, 0, 1, 2, 3, 4, 5, DB A . . . E	00, 0, 1, 2, 3, 4, 5	00, 0, 1, 2, 3, 4, 5
Weighing range/mg	20 . . . 2 000	1 . . . 10 000	20 . . . 1 000	20 . . . 1 000
Weighing accuracy/mg	+/− 2	+/− 2	+/− 2	+/− 2
Feeding hopper capacity/L	20	34/45	100	45
Overall size (H × L × W) mm	1474 × 1051 × 1075 (safety doors closed)	1895 × 1410 × 940	1875 × 893 × 950	1875 × 893 × 950
Weight/kg	900	650	600	500

Source: Robert Bosch GmbH (2002), High-speed capsule check-weigher KKE 2500 S, brochure no. 04/02, Waiblingen 2002; IMA Industria Macchine Automatiche (2002), Precisa, brochure no. 991ZPB2001, Italy 2002; Anritsu Corporation (2000), Capsule Check-weighers, brochure no. K3067-B-1, Japan 2000.

Check-weighers are suited for the processing of all common capsule sizes 00 to 5, as well as special sizes such as DB (double blind) and are independent of the material of which they are composed, gelatin or HPMC (hydroxypropyl methylcellulose). The adjustments required for each capsule size require few specific size-parts. These size-specific parts need to be easily accessible and ideally only one needs to be exchanged in the case of a size change or for cleaning. In an effort to keep downtime short, the size change-over and other adjustment jobs are designed to be made without the use of tools.

Easy access of all operating devices in modern machines is achieved through ergonomic design and supports the size-part changeover as well as all other maintenance and cleaning activities. The display of all measured data and the analysis of the calibration and operating parameters are controlled via an accessible and legible operator panel. Touch-screen monitors currently constitute the best available technology.

Design and functionality

Figure 8.3 shows the functions of an automatic capsule check-weigher. The case panels and control cabinet are composed of stainless steel and the transparent protective covers are usually made from easily cleanable acrylic (polymethyl methacrylic acid, PMMA). The capsules are fed into a capsule hopper either automatically or manually (see Figure 8.4). A check-weigher can be positioned directly behind the capsule-filling machine when fitted with an automatic in-feed device. The minimum and maximum filling level can be controlled via a proximity switch. The conveying performance of both machines needs to be synchronised. If the capsule hopper reaches the maximum fill level, the filling machine stops automatically. In case of the minimum fill level not being reached, one might presume that an operational dysfunction has occurred and the condition of the filling machine should be checked. The larger the capsule hopper's buffer volume, the easier it is to accommodate variations in the production volume and the operation can continue without interruptions.

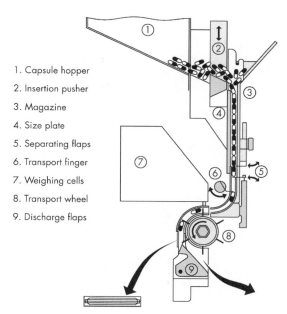

1. Capsule hopper
2. Insertion pusher
3. Magazine
4. Size plate
5. Separating flaps
6. Transport finger
7. Weighing cells
8. Transport wheel
9. Discharge flaps

Figure 8.3 Functions of an automatic capsule check-weigher based on gravity flow. The capsules are fed from the hopper (1) by means of a drive mechanism (2, 3, 5, 6, 8) and a size-dependent guiding plate (4) onto the weighing cell (7). After check weighing, the capsules are classified into correctly and incorrectly filled ones by means of a flap (9).

The perforated base of the capsule hopper gathers loosely adhering powder residues on the capsules into an interception tray. This tray has to be emptied from time to time and this can be done by simply pulling it out of its retainer. For cleaning and maintenance purposes, the capsule hopper should also be dismantled from the machine frame, which can be done without the use of tools.

An insertion pusher, using a swinging motion (see Figure 8.5), guides the capsules into the vertical tracks of the stationary magazine. Most models on the market use non size-specific insertion pushers. Servo-technology is used to achieve gentle motion of the capsules with the minimum of vibration. The capsules are guided to the weighing cells via a stationary vibration-free magazine. Multiple individually adjustable separating flaps accumulate and separate the capsules prior to weighing. To position the capsule on the weighing cell, different transport systems are currently in use, depending upon the construction of the

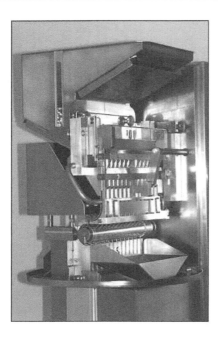

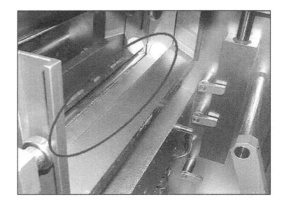

Figure 8.5 Servo-driven size-independent insertion pusher for guiding capsules into the magazine tracks. The capsules are gently inserted into the guide plate (not shown) by a vertical alternating movement.

Figure 8.4 In-feed system with capsule hopper (upper part). The cover for the perforated base is situated underneath the hopper at the right-hand side. The complete hopper unit including the interception tray can be dismantled without the need of tools.

weighing cells. Machines with standing weighing cells either use a continuously rotating drum filled with suction holes to hold the capsules during rotation or a horizontally moving pusher to deliver the capsules onto the standing weighing block. Models with hanging weighing cells use a mechanical transport finger. Some models make use of an optical light barrier, which is not sensitive to dust, to monitor whether all capsules have passed through the lower flap correctly. The transport system either slows down or stops entirely if a capsule is delayed (see Figure 8.6). This ensures that the capsule will not be damaged.

A built-in calibration bar constitutes a definite bottleneck where deformed capsules cannot pass (see Figure 8.7). When the weighing cell registers that a capsule track is blocked a mechanism is immediately activated to eliminate the disturbance. The insertion pusher stops in a position, assuring that no additional capsules will be conveyed into the magazine. Subsequently, those

magazine tracks that are not blocked are emptied to ensure that those capsules will not be lost. The magazine rear wall opens up enabling the blocking capsule to fall into the reject container. The magazine rear wall finally re-closes and the conveyor- and weighing processes restart automatically (see Figure 8.8).

The transport mechanism momentarily stops the capsule on the weighing cell and then releases it. After the weighing process the capsule is transported to a discharge flap where it is directed either into the container for accepted

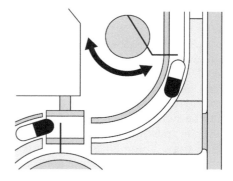

Figure 8.6 Transport finger (right) for smooth conveyance of capsules onto hanging weighing cells. The capsules are pushed through the guide plate onto the weighing cell (left), and after weighing are removed by means of a transport wheel.

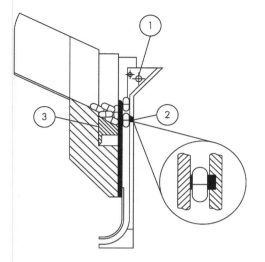

Figure 8.7 Calibration bar to control capsule size and avoid locking of the system. The capsules have to pass a defined 'bottleneck', which is installed at the upper end of the guide plate (2) just behind the transport finger (3). If the capsule is blocked at this point, the rear of the guide plate pivots around (1) and releases it into a reject container.

capsules or is rejected into a separate container. The discharge flaps have to react quickly and reliably and are monitored, for example by proximity switches. Pneumatic cylinders are an appropriate way to drive the discharge flaps. For security reasons, during power cuts or interruptions, the flaps stop in the reject position.

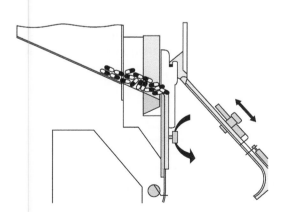

Figure 8.8 Rear wall of the guide plate (magazine) opened to release the blocking capsule.

Depending on manufacturer and model, the transport mechanism guarantees that no faulty capsules can reach the accepted-capsule container.

Calibration

The calibration process is carried out immediately following the machine set-up or during a batch changeover and normally lasts only a few minutes. During the procedure, all the weighing cells are loaded with a control weight representing the upper limit of the measuring range.

Weighing cells

The heart of every check-weigher is its weighing cells. In order to determine the weight of the capsules reliably, they have to meet the highest specifications, particularly:

- precise measurement
- accurate recording of measured values
- robust machine design
- achieve a low measuring tolerance; the state-of-the-art is ±2 mg.

Measuring principle

The weighing cells work according to the principle of an electromagnetic force balance, whereby an adjacent measuring force registers the deflection of a beam. A position control keeps the lever precisely in the measurement position. Minimal load changes are picked up by an optical sensor and compensated for by a PID (proportional-integral-derivative) controller. A suitably placed coil compensates for the deflection of the beam. The necessary current input, which is increased for this operation, is used as a measurement signal (see Figure 8.9).

Construction of a weighing cell

The weighing cells consist of a weighing block, with the weighing pan placed at its end. Bores

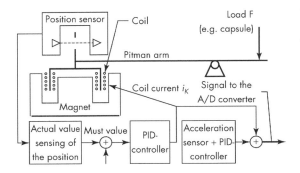

Figure 8.9 The structure and functions of a capsule check-weigher. The load on the right-hand side represents the capsule placed onto the weighing cell. The position sensor on the left detects the displacement of the Pitman arm. This movement is then compensated by an increase in the coil current of the magnet.

are specifically inserted into the weighing block to form integral hinges. Thus, a flexible parallelogram moving a 'Pitman' arm is created in the weighing block. This arm is held in a defined position, controlled in turn by a light sensor. When a load is placed onto the weighing pan, the parallelogram is displaced and the attached arm deflects accordingly. The movement is registered by the light sensor, which sends a corresponding signal to the evaluation unit. The deflection of the arm is then compensated by an increased current flow in the coil. The change in the current is measured and automatically converted into a weight (see Figure 8.10).

Weighing cell alignment

The weighing cells inside the machine can be arranged either in a standing or hanging manner (see Figure 8.11). Hanging weighing cells have the advantage that product deposits cannot accumulate, since any dust shed by the capsules will run off through a drain in the body of the weighing cell. In doing so, a more precise weight accuracy can be achieved. Hanging weighing cells completely surround the capsule. Therefore electrostatic forces have no effect on the result. Devices, such as a tilting hinge, provide for better protection of the hanging weighing cells against possible damage, which could for instance, occur during cleaning.

The weighing process can be divided into several steps:

• Loading period: the time period from the capsule's first contact with the weighing cell to its proper positioning;
• Damping period: the time period required for the voltage time curve to settle;
• Weighing time: the time period required to register and record the measurement value;
• Cycle time: overall time period for one weighing cycle; with multiple balances, for the high-capacity sector, handle up to 4 capsules s^{-1}
• No capsules are placed on the weighing pans during the unloading period.

One additional weighing cell, holding no capsule, serves as a reference cell. It captures the current environmental conditions and interference effects influencing the weighing cells such as translational and rotational

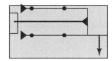

The capsule is pushed onto the weighing pan. The weight of the capsule causes a deflection.

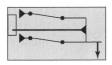

The deflection of the arm is captured by the light barrier.

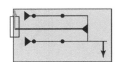

The regulating circuit increases the coil current. As a result, the arm is replaced in its original position.

Figure 8.10 Schematic illustration of the function of the weighing cell (see figure for explanation).

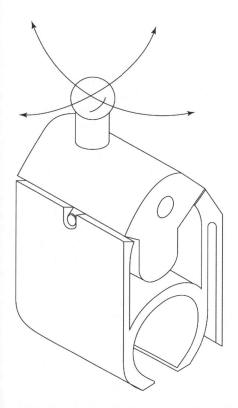

Figure 8.11   Construction of a hanging weighing cell. The pivot connection (upper end) allows the cell to resist overloading and mechanical stress. The capsule support (below the weighing cell) can be released easily since it is only held by a clamp.

accelerations. The signal of the reference cell is subtracted from the measurement signals of the other cells, therefore compensating for these external influences.

The following options exist for storage and release of measurement data:

- standard deviation
- average
- variance
- number of accepted versus rejected capsules
- number of over- versus under-fills
- total number
- target number
- target weight, tare weight
- storage of measuring results for individual batches.

Additional information can be added:

- batch number
- machine calibrations (such as output, motor revolutions, etc)
- product name/type
- capsule size/type.

Thermoforming machines

Fundamental requirements

Pharmaceutical security and production according to good manufacturing practice (GMP) guidelines

From a pharmaceutical safety perspective, the thermoforming machine must, on the one hand, protect the product during the packaging process against external influences and, on the other hand, protect the operator against hazardous materials. The construction and production of modern packaging machines are thus based on cGMP (current good manufacturing practice) guidelines. These principles and guidelines particularly govern product disposal and machine cleaning. Sufficient space needs to be allocated to allow for access and freedom of movement during the cleaning process; bolt heads on smooth surfaces, as well as small corners, hidden cavities and inaccessible areas are not permitted. The production line has to be protected. Modern thermoforming machines are therefore designed using a balcony architecture, which separates the machine operating area from the production area. Heat and dirt particles generated in the machine operating area are contained there and product residues and film vapours are exhausted directly from the production area. Capsules and packaging material remain protected and a clean pharmaceutically conforming packaging process is ensured.

Simple size-part changeovers and packaging machine flexibility

The fault-free packaging of different capsule sizes and quantities into blister packages of varying dimensions is a prerequisite for the economic operation and productivity of a packaging machine. Thermoforming machines therefore possess interchangeable machine parts that can be easily substituted for one another during the changeover to accommodate the many different capsule or blister sizes. The construction and shape of these size-parts are directly linked to the type of product, packaging and the materials used. One can differentiate between complete size-parts and component size-parts, depending on whether it is just the product that is changed, or if it is the blister size as well. Size-parts are used at various stations of the thermoforming machine. In the in-feed area, for instance, they include the guide plate as well as the filling section, immediately followed by the sealing roller. At the forming station, the blower and forming plate are adjusted to the exact capsule cavity measurements. The indexing and deflection rollers that are responsible for the forward conveyance and the guidance of the blister line are counted among the size-parts.

The modern thermoforming machine, an intelligent production unit, must possess a high measure of flexibility in the use of these size-parts. Mechanical adjustment parameters are aligned in a precise and reproducible manner via digital heads. The mounting and disassembly of size-parts are achieved with quick-release clamps and without the use of tools. An entire size-part changeover can therefore be accomplished in less than one hour. The operating and changeover efforts are additionally reduced owing to the application of modern drive and control systems. Servo-drives facilitate quick and even tool-free size-part conversions and simplify the changeover process.

A user-friendly industrial PC with integrated operating panel and touch screen is used to control the machine. The touch screen gives the thermoforming machine operator a menu and self-explanatory symbols to control the machine. The features of the operating panel are multifaceted and particularly comprise the size-part administration, the display and history of performance and consumption data, as well as the analysis of possible machine dysfunction.

Integrated control modules and zero-error concept

Numerous control features in the form of cameras and detection mechanisms control the thermoforming production process, help to eliminate disruptions and therefore implement a zero-error concept. In doing so, the most significant control feature is the scanner of the product blister package. It verifies that the blister cavities have been filled immediately following the capsule insertion into the cavities. As a general rule a black and white or colour camera, comparing the current photograph of the blister track with a previously saved reverence image, is utilised for this purpose. In the case of deviations caused by empty or partially filled blister cavities, the corresponding blister is ejected into a reject gate via a shift register and then replaced with a properly filled blister from a top-up magazine.

The 100% dependability of the inspection result is absolutely crucial for the pharmaceutical security of the operation. Hence, the filled film is completely covered in between the camera check and sealing station, ensuring that no product is lost from the cavities that could negate the inspection result. Several additional controls have been integrated into the thermoforming machine. Controls of air pressure, compressed air, as well as a heat current monitoring system provide a continuous good quality for the forming station. A further noteworthy feature is the early final inspection permitting a timely replacement of a new web reel and enabling continuous production. Many thermoforming machines can be equipped with two web reels as well as a web buffer. When the first reel approaches its end, a light barrier triggers the unwinding of an adequate amount of film into the web buffer. While the second reel is being seamlessly fixed on top of a splicing table, the machine gains access to the web storage and continues production without a machine stoppage. The subsequent recognition of the splice as well as the accurate rejection of the corresponding blister with the aid of a shifting register ensures that only acceptable blisters will be processed further.

Thermoforming machine design and functionality

The production flow of a thermoforming machine encompasses a number of functioning steps that can be found in all types. In the initial step the formation of the receiving pocket cavities for the capsules takes place. The moulded blister cavities inside the base web are then filled with the capsules and subsequently sealed and closed. Embossing and perforating follows, with the blisters being die cut before being packed into folded boxes. The steps are illustrated in Figure 8.12.

Thermoforming machines are available with a range of performances. The maximum machine output depends upon the attributes of the thermoforming film, the capsule composition and size, the product in-feed and the performance characteristics of any successive machines. This could, for instance, be a cartoner that pushes the packaged and stacked blisters into the final package together with any instruction leaflets. Depending upon machine design and layout, the processing of blister packages can occur in either single- or multi-track production. High-capacity machines have an output in the region of 1000

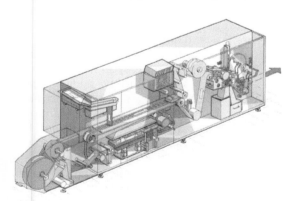

Figure 8.12 Schematic illustration of a thermoforming ('blister packing') machine (BOSCH TLT 1400). The base web is fixed to the reel support (left). After passing the heating and forming station, the formed base web is filled with the capsules (upper left part of the picture). The cavities are then sealed with a foil lid (centre of the machine) and transported to the cutting station. After cutting (right) the blisters are normally transferred to further downstream equipment.

blisters min^{-1}, medium-capacity machines have an output of about 400 blisters min^{-1} and there are smaller machines with outputs of up to 200 blisters min^{-1}. Inside modern thermoforming machines, drive and film transport occurs using manageable and flexible servo-motors, which replace the previously used central drive shaft by a revolution controller, thus facilitating the precise synchronisation of the individual work station drives.

Heating and forming station

The base web is stripped away via the drive reel and subsequently guided into the heating and forming station via deviating and dancing rollers. The film heater is specifically adjustable to each unique film and its performance is reproducible owing to its combination of heating temperature, precise temperature distribution, plate distance and cycle time. In order to ensure a controllable process of the base web, the heating plates lift from it during machine stoppage. Transport rollers control the web prestress and prevent any sagging of the heating plates. The moulding process varies depending on the different base webs. Thermoform films have to be preheated to 120–145°C before the moulding process in order to increase their ductility. Within this temperature range, the film material is readily malleable and mouldable. The heating is made either by contact, convection or radiation and the forming of the film takes place subsequently (see Figure 8.13).

The moulding of polyvinyl chloride (PVC) films is made using compressed air. Intermittent operating forming- and blowing-plates mould each film according to their specified size and shape. To attain an even heat distribution at around 150°C and to prevent crystallisation during machine stoppage, the polypropylene (PP) films are heated using split heating plates. Aluminium foils, however, are moulded cold. Owing to the above-mentioned complex forming properties, plugs (stretch rods) are employed during the cavity moulding for both types of web. For PP, these plugs are aided by compressed air. The plugs preform the cavities and ensure a proper sidewall and base thickness distribution.

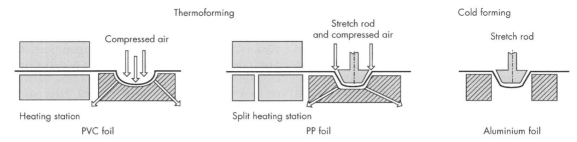

Thermoforming
Compressed air
Heating station
PVC foil

Stretch rod
and compressed air
Split heating station
PP foil

Cold forming
Stretch rod
Aluminium foil

Figure 8.13 Forming process of the base web. Cavities are formed by means of compressed air, which is applied from above and presses the heated PVC foil into the forming cavities (left). The air is exhausted through the bores to the left and right of the cavity. Forming the PP base webs requires the use of a plug or stretch rod, compressed air and a split heating station (middle). Aluminium base webs are formed cold and require only a stretch rod to form the cavities ('plastic forming', right).

Following the forming process, the film line is moved to the filling station. Smooth transition from the intermittent working method of the forming station to the linear movement of the blister line inside the in-feed area is achieved by a dancing roller that compensates for the film run.

Capsule in-feed

Product feeding is automatic via the supply hopper that is attached to the in-feed system. The filling

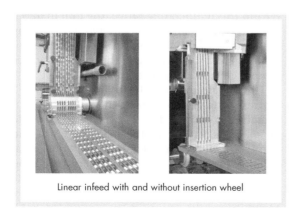

Linear infeed with and without insertion wheel

Figure 8.14 Filling section of a thermoforming machine. The left picture shows the use of an insertion wheel as an in-feed system. The wheel receives the capsules from the feeding track and transports them counter-clockwise into the cavities. Direct in-feed of products into the blister cavities is shown in the right picture.

section can hold various in-feed configurations (see Figure 8.14). Capsules could become jammed during feeding and the machines are designed to avoid this possibility. Capsules can be vertically stacked inside the in-feed chute without wedging. Capsules are therefore predominantly fed using a so-called linear in-feed system. In doing this, the capsules are guided into the vertically positioned dosage chute via an electromagnetic sorting mechanism and subsequently assigned to the individual cavities. One particular advantage of the linear in-feed system is its high degree of cavity filling, which can be adjusted to the individual composition and dimension of the capsules.

If aluminium base webs are used, the linear in-feed system is extended by an insertion wheel. This insertion tool has to be added because aluminium base webs have flattened cavity slopes owing to the material's specific forming properties. The flatter cavity slopes could possibly allow more than one capsule to fall into each cavity. The insertion wheel isolates the capsules and guarantees that there will only be one per cavity. The insertion wheel in-feed is also suitable for filling cavities that are positioned cross-wise or at an angle to the web direction. This feature offers yet another interesting application for the packaging of capsules. Capsules whose halves are of different colours can be specifically aligned and then fed into the cavities. All capsule halves of one colour will thus have the same orientation. The insertion wheel is positioned immediately following the linear in-feed system and the capsules are inserted from the in-feed chute into the respective feeding passage. Depending on whether the capsules drop into the in-feed chute with their cap or body end first, they fall into the insertion wheel to various depths and are aligned in a uniform direction via a mechanical system. A cover prevents the capsules falling out of the insertion wheel during the proceeding movement.

Sealing station

The sealing station melts the hot sealing varnish of the lid web, utilising pressure and heat, thus creating a leak-proof seal with the base web. This occurs at about 180–200°C. The sealing processes can be divided into the plate sealing and continuous roller sealing. The plate sealing can be further divided by the functioning of their plates, one being fixed and the other moving in the direction of the web.

With fixed plate sealing, only the filled blister line moves through the sealing station. The plates are pressed together and the films thus sealed. The film continues to cycle and the next segment can be sealed. High acceleration speeds can affect the capsule in-feed negatively. Moving sealing plates can ameliorate this negative effect by moving alongside the web for a certain blister line segment. However, an issue arises here because considerable energy is required for the movement of the heavy machine parts.

The filled blister line is guided through the sealing station using a positive-fit sealing roller. Inside the sealing station, the lid web is fed directly onto the base web and sealed on top of the filled package. Sealing at the 12 o'clock position prevents the capsules from sticking to the lid web. PP- and paper-laminated lid webs are preheated using a lid web heater for better sealing results. Owing to the short residence time of the web in the actual sealing station and the high sealing pressure caused by the small sealing surface, a best possible sealing seam results. The thermal influence on the product active ingredient is significantly less than during the plate sealing process. In addition, the sealing roller has a higher output in comparison to similar sealing mechanisms and is more efficient in its application.

Embossing, perforating and die cutting station

Following the sealing station, the blister line is forwarded to the embossing station where it is labelled with a date stamp and/or a batch number. Afterwards, it is transported to the perforating station. This step is optional. Here, the blister line is furnished with lengthwise or cross perforations in between the blister cavities. The blister embossing and perforating stations are positioned at a minimal distance from one another to be precisely posted for the servo-driven indexing roller. This is of particular importance for the perforation, since it has to be placed at an exact and predefined position. In a similar fashion to the sealing roller, the embossing and perforating tools can be changed without the use of tools. At the subsequent die cutting station, the blisters are cut out of the endless blister line into their desired size. A low volume of die cutting waste is produced owing to the precise positioning of the blister inside the web section. Empty or partially filled blisters as well as die cutting waste are separated from one another using a rejection flap and are then sorted into separate waste bins. A rotating suction device extracts the blisters from the die cutter and passes them onto the cartoning machine via a positive-fit compact transfer or simply places them on a conveyor belt. The blisters, stacked in specific numbers, are subsequently inserted into their final package along with various inserts.

Packaging materials

Blister package

Thermoformed and aluminium webs are predominantly used as primary packaging materials. In order to optimise their barrier characteristics, the packaging materials are usually constructed as film compounds. The most common film compounds for blisters consist of a PVC base web and an aluminium lid web.

Thermoforming films

The advantage of PVC base webs is primarily their favourable forming properties. The films can be formed starting at 120°C and across a wide temperature range. One further advantage is their relatively low material costs. The film can be additionally coated with a barrier layer, for instance polyvinylidene chloride (PVDC) for improved product protection to reduce oxygen or water vapour permeability. This compound allows for a significant improvement in the barrier qualities of the thermoforming films.

PP films, which are particularly advantageous owing to their easy environmental disposal, can also be processed as base webs. However, forming can only take place at a high temperature of about 150°C over a very narrow range. In order to reach this higher temperature, the heating station has to be precisely adjustable and features a temperature dispersing device. In addition, a higher forming force is necessary for PP films, so that a plug-assist system is utilised for a correct forming process. PP film is also employed for lid webs. If PP is used as the sole packaging material, the blisters are called mono-blisters. Another point to be considered during the production process is the film shrinking behaviour which requires particular attention during the construction of the size-parts for the PP film processing.

Aluminium foils

Next to thermoforming films, aluminium foils can be used also as base and lid webs. Aluminium form packages such as blisters that are used for the lid web as well as the base web, are practically leak-proof against water, water vapour, oils, fats, as well as gases. Owing to their good reflectivity, they are also impermeable to light. Aluminium blisters are well suited for applications in areas with extreme climatic conditions (i.e. the tropics). By means of the use of various combination techniques, the aluminium foil can be processed into a laminate. It becomes childproof when laminated with a layer of paper. A lower mechanical strength can be achieved by lamination with paper or plastic films. During the forming process, attention must be paid to the fact that the paper or plastic films possess low ductility and poor forming characteristics. Hence, the cavity arrangement has to be broader than, for instance, with PVC films. A larger blister layout occurs owing to larger cavities, thus requiring more foil. Additionally, aluminium foil is more expensive than conventional foil. Aluminium foils are formed cold, and similar to the PP-films, plug assists have to be employed in order to handle them.

9

Filling of liquids and semi-solids into hard two-piece capsules

Geoff Rowley

Introduction

Bioactive substances may be formulated as liquids or semi-solid matrices (SSM) for processing and filling into hard capsules, without some of the problems associated with the filling of fine powder preparations. Production machinery is available to fill materials accurately and precisely, and to seal the capsules to prevent leakage of the contents. The filling of liquid-fill formulations offers several possible advantages over powder-filled and liquid-filled soft capsules. It is therefore important to review the developments of liquids and semi-solids in hard capsules, with respect to formulation, processing, filling, drug delivery and product stability.

The advantages of liquid-fill hard gelatin capsules were first demonstrated in a series of papers from the University of Strasbourg (Cuiné *et al.*, 1978a, 1978b, 1978c, 1979a, 1979b). Other workers soon demonstrated the potential advantages in the control of fill weight (François and Jones, 1979) and in drug content uniformity particularly when dealing with low dose bio-actives of relatively high potency (Walker *et al.*, 1980b). Additional advantages include the reduction of airborne particles and the containment of spillage when dealing with toxic or potent materials (Bowtle *et al.*, 1989). In general, fewer excipients are needed for liquid-fill formulations than for solid-fill capsules and tablets. It has been claimed that improvements in the dissolution of poorly water-soluble drugs and also controlled drug delivery can be achieved with relatively simple formulations.

Work published to date has confirmed advantages claimed for this formulation technique (Cole, 1989). It has been shown that stable preparations could be made from liquid and deliquescent unstable materials by preparing them as a semi-solid matrix in commercially available excipients (Doelker *et al.*, 1983). Sustained release formulations have been prepared by using excipients that influence the hydrophilic–lipophilic balance (HLB) of the SSM. Phenylpropanolamine was formulated as an SSM of arachis oil, beeswax and silicon dioxide (François *et al.*, 1982) and Captopril in an SSM consisting of soybean oil and glycerol monostearate (Seta, 1988).

A principal constraint on SSM formulations in hard gelatin capsules is the interaction of certain excipients and/or water with gelatin. The choice of excipients is made in relation to capsule shell integrity and evidence has been presented for the advantages of hard gelatin capsules in comparison to soft gelatin capsules with respect to their stability and product manufacture (Cole, 1989). The use of SSM formulations to improve the bio-availability and the design features and requirements of liquid-fill machinery have been reviewed (François, 1983; Cole, 1989).

Liquid and semi-solid matrix formulation

Rheological considerations

Formulations for liquid-filled hard gelatin capsules may be classified according to their rheological properties during the capsule-filling process (see Table 9.1). They may be either Newtonian liquids, thixotropic gels, both of which are filled at ambient temperatures (Walters *et al.*, 1992a) and thermosoftened matrices, which may exhibit either Newtonian or non-Newtonian behaviour at elevated filling temperatures up to 70°C, but are solid or semi-solid at room temperature, (Hawley *et al.*, 1992).

Where the formulation is a Newtonian liquid, with a viscosity in the range of 0.1–25 Pa s, filling can be achieved with accuracy and precision so that coefficient of variation (CV) of fill-weight values <1% are often achieved. Thus, single components such as dietary supplements, e.g. evening primrose oil, or two-component formulations where the bioactive substance dissolves in the excipient, should fill satisfactorily at ambient temperature, provided that the formulation viscosity is within the capability of the dosing pump of the filling machine.

The scope of liquid-filling can be increased by use of different rheological states, for example, by formulating a thixotropic gel that undergoes shear thinning during filling followed by gel restructuring with an increase in its apparent viscosity in the filled capsule. Alternatively, the viscosity of an SSM can be reduced by increasing the temperature of the filling process so that the formulation is molten at the filling temperature and forms a solid or semi-solid in the capsule on cooling to ambient temperature. The filling process temperature is usually limited up to a maximum of 70°C, in order to avoid thermal damage to the capsule shell and to reduce operational hazards.

If the active substance cannot be formulated as a solution or if the solvent is incompatible with the gelatin shell, then the active substance can be formulated as a dispersion. The essential prerequisites for such a formulation are low hygroscopicity, good dispersion stability to avoid segregation during capsule filling and storage, and also the formation of an SSM with a satisfactory apparent viscosity. The formulation of dispersions for liquid-filling can be approached by two main routes, as summarised in Table 9.1. First, the use of thixotropic gels that undergo shear thinning during the mixing of the disperse phase and filling operation, so that the apparent viscosity permits satisfactory capsule filling. Reformation of the gel structure in the filled capsule causes an increase in apparent viscosity, improved dispersion stability and a reduced tendency to leak from the capsule. Thixotropic formulations have the advantage of being able to be filled at ambient temperatures, whereas the alternative disperse system utilises thermosoftened excipients that are molten during filling, at temperatures up to 70°C and solidify rapidly in the capsule. In these systems, it is essential to have low hygroscopicity, good dispersion stability during filling and solidification and an apparent viscosity that will facilitate filling.

Capsule shell integrity

In addition to formulation rheology, gelatin capsule integrity is a primary consideration in the development of liquid-fill formulations and therefore preformulation information regarding interactions of the active and excipient substances with gelatin is essential for a successful outcome. Data for moisture uptake are required in order to eliminate materials that will change the equilibrium moisture content (EMC) of the capsule outside the 13–16% w/w range, which is required to maintain capsule integrity (see Chapter 4).

The EMC of excipients with the potential for liquid-filling formulations should be determined or obtained from the literature, in order to eliminate the excipients that could change the EMC of the capsule to an unacceptable value. Most active substances are crystalline solids and many excipients that could be used as solvents to formulate a solution, will unfortunately affect gelatin shell integrity. Low molecular weight

Table 9.1 Rheological classification of liquid-fill systems for hard gelatin capsules

Newtonian	non-Newtonian
• Liquid active • Solution of solid active in liquid excipient	• Thixotropic gels with active dissolved or dispersed • Solid dispersions, liquid at filling temperature and solid or semi-solid in capsule

(MW) polyethylene glycols (PEG), e.g. PEG 400 or 600, have good solvent properties for a wide range of active substances, however, the hygroscopicity of these excipients is sufficient to decrease the capsule moisture content and cause gelatin capsule shell brittleness and fracture. Table 9.2 provides data for moisture uptake by gelatin capsules filled with liquid excipients and stored at 18–20°C and 55% relative humidity (RH), along with the resultant effect on capsule brittleness and fracture (Walters *et al.*, 1998; Walters and Rowley, 1999). PEGs 200, 400 and 600 and the triglycerides Imwitor 708 and 908 all had EMCs of >2.5% w/w and the capsules fractured in less than 24 hr. Capsules containing liquid excipients with EMCs of <1% w/w showed no evidence of fracture after 180 days storage. Moisture uptake data are therefore essential preformulation knowledge for the design of liquid-fill formulations for hard gelatin capsules. There is sufficient evidence to suggest that if the EMC of a filled capsule does not change by more than ±2% w/w when stored between 10–65% RH and 25°C, then a gelatin capsule will retain its integrity (Cadé and Madit, 1996).

Near infrared spectroscopy has been investigated (Berntsson *et al.*, 1997) as an at-line process for the determination of moisture content in bulk hard gelatin capsules with an analysis time of 1–2 min. The range of moisture content investigated was 5.6–18% w/w and the root mean square error of prediction was 0.1% w/w. Texture analysis has been used as a non-destructive test to investigate mechanical changes during storage of hard gelatin capsules filled with liquid formulations (Kuentz and Röthlisberger, 2002). Texture analysis can be used to investigate stiffness changes in relation to the quantity of water needed in a formulation to avoid embrittlement on storage. The technique helps to identify hydrophilic or amphiphilic formulations that are compatible with hard gelatin capsules where embrittlement or softening is due to rapid water equilibration processes. This is useful information but the technique is not intended to replace long-term stability tests for deciding the compatibility of formulation and capsule. Several liquid excipients are available with low hygroscopicity, e.g. triglyceride oils, however, the solvent power of these materials is often limited. There is clearly a need for the development of excipients with good solvent properties and detailed investigations into mixed excipients employing one excipient as a cosolvent.

Excipient compatibility

The numerous excipients that are compatible with hard gelatin capsules and can be used to formulate liquid-fill formulations have been classified arbitrarily as lipophilic liquid vehicles, semi-solid lipophilic vehicles/viscosity modifiers and solubilising agents, surfactants, emulsifying agents and absorption enhancers (Cole, 1989). A comprehensive list of compatible excipients is presented in Table 9.3. Excipients that have been found to be incompatible with hard gelatin capsules include ethanol, glycerol, Glycofurol 75, medium chain monoglycerides (Alkoline MCM, Capmul MCM, Imwitor 308), PEG (MW <4000 Da), Pharmasolve, propylene glycol, Span 80 and Transcutol P (Cole, 1989). However, it may be possible to incorporate small amounts of these excipients into liquid-fill formulations without affecting the mechanical properties of the capsule. Therefore, the maximum tolerable concentration of an excipient should be

Table 9.2 Effect of moisture uptake on brittleness and fracture of liquid-filled hard gelatin capsules stored at 55% RH and ambient temperature, 18–20°C (Walters *et al.*, 1998; Walters and Rowley, 1999)

Excipient	Moisture uptake (% w/w)	Capsule splitting time (days)
PEG 200	18.8	<1
PEG 400	13.8	<1
PEG 600	12.2	<1
Imwitor 708	7.3	<1
Imwitor 908	2.5	<1
Imwitor 780K	0.25	>180
Miglyols	Negligible	>180
Poloxamers (liquids)	<1	>180
Poloxamers (solid)	<1	>180
PEGS (solid)	<1	>180

Table 9.3 Classification of excipients compatible with hard gelatin capsules (Cole, 1989)

1 Lipophilic liquids
   (i) Refined oils: arachis, castor, cottonseed, maize, olive, sesame, soybean, sunflower
   (ii) Triglycerides and esters: Alkomeds E and R, Captex 355, Labrafacs CC and PG, Lauroglycol FCC, Miglyols 810, 812, 840, Softisan 645

2 Semi-solid lipophilics and viscosity modifiers
   (i) Hydrogenated oils: Groundnut 36 (arachis), Cutina HR (castor), Sterotex (cottonseed), Softisan 154 (palm), Alkosol 407 (soybean)
   (ii) Miscellaneous: Aerosil (silicon dioxide), cetostearyl, cetyl and stearyl alcohols, Gelucires 33/01, 39/01, 43/01, Compritol 888 ATO (glyceryl behenate), Precirol AT05 (glycerol palmostearate), Softisans 100, 142, 378, 649

3 Solubilisers, surfactants, emulsifiers, absorption enhancers
   Capryol 90, Cremophor RH40, Imwitors 91, 308, 380, 742, 780K, 928, 988, Labrafils M 1944CS, M 2125 CS, Lauroglycol, PEG MW>4000, Plurol Oleique CC 497, poloxamers 124 and 188, Softigens 701 and 767, Tagat TO, Tween 80

---

determined experimentally. Low concentration of impurities, e.g. aldehydes, may be present in or produced during storage by some of the excipients used in liquid-fill formulations. The interaction of a component of the formulation with the gelatin capsule shell can affect drug dissolution (Dey *et al.*, 1993). The effect of such interactions on dissolution can be evaluated by storage of the encapsulated test material at 40°C and 75% RH for periods up to six months (Cole, 1989). After storage, the capsules are emptied, cleaned and then filled with acetaminophen as a dissolution reference material. Comparison of dissolution data for acetaminophen from the stored and reference capsules can provide evidence for interactions between the investigated fill material and the gelatin capsule. If this is a problem, it can be overcome by the use of hydroxypropyl methylcellulose (HPMC) capsules that do not undergo these cross-linking reactions (Nagata *et al.*, 2001).

Pharmaceutical-grade hard capsules are now being produced from HPMC. Their properties are discussed in Chapter 4. In terms of filling liquid and SSM formulations into them they have significantly different properties from gelatin capsules. This is because water plays a different role in the functioning of HPMC capsules. It does not act as a plasticizer for the film and thus when it is removed the shells do not become brittle (Ogura *et al.*, 1998). The interaction on storage of SSM excipients filled into gelatin and HPMC capsules has been studied (Nagata and Tochio, 2002). This showed that there were significantly less problems seen with HPMC capsules compared to gelatin capsules. A range of excipients with different HLB values was filled into the two types of capsule. After a month of storage at 45°C both types of capsule lost water from their shells, however, when a test that simulates the removal of capsules from blisters was applied to them, the gelatin capsules showed signs of brittleness after 1 week and after 4 weeks they were almost all very brittle. The HPMC capsules showed no signs of brittleness in the course of the test. In a further test, the physical compatibility of HPMC capsules with a range with commonly used excipients was tested. Low molecular weight PEG, 400 and 600, distorted the capsules slightly and this was shown to be due to swelling of the shell walls caused by PEG diffusion. PEG 900 did not cause this problem.

In summary, it is essential to investigate moisture uptake propensity of excipients and their compatibility with gelatin capsules during preformulation. Rheological investigations of liquid excipients and trial formulations should also be undertaken at the proposed filling temperature.

Capsule-filling properties

Liquid-fill formulations for hard gelatin capsules can be classified according to their rheological properties, as summarised in Table 9.1. The simplest formulations are Newtonian liquids comprising a liquid excipient and a liquid active or solid active that is soluble in the liquid excipient at the filling temperature. It is preferable to fill the capsules at ambient temperatures in order to avoid unwanted physical or chemical changes in the formulation during filling and storage. Filling can be performed at temperatures up to 70° provided that elevated temperatures do not adversely affect the properties of the filled capsule. The formulation will contain a particulate phase if the active substance does not dissolve in the liquid excipient or melt at the filling temperature and in most cases the rheology will be non-Newtonian. The requirements of the two principal types of non-Newtonian formulations, i.e. thixotropic gels and thermosoftened matrices, are summarised in Table 9.4.

Thixotropic gels

Walker *et al.* (1980b) were amongst the first to demonstrate the successful filling of both thixotropic and thermosoftening formulations. Two thixotropic formulations were made: a Miglyol oil 95% w/w and silicon dioxide (Aerosil 200) 5% w/w mixture and a liquid paraffin 95% w/w, hydrogenated castor oil (Thixcin R) 3% w/w and Aerosil 200 2% w/w mixture. Both were filled satisfactorily on an automatic machine with coefficients of variation of fill weight of 2.8 and 2.0%, respectively.

Thixotropic gels can often be filled under ambient conditions provided that the formulation undergoes a sufficient degree of shear thinning during preparation and filling, so that its apparent viscosity is low enough for the mechanical filling requirements of the machine. A typical formulation would consist of a liquid excipient and gel-forming agent, e.g. a triglyceride oil and silicon dioxide, respectively. The particulate active substance is dispersed uniformly in the gel to produce a formulation that will undergo shear thinning during mixing and filling, but will restructure in the capsule to give adequate dispersion stability. As with all pharmaceutical dispersion formulations, detailed consideration of dispersion properties and dispersion stability are important in relation to drug content uniformity and drug release.

The rheology and properties of thixotropic gels for filling at ambient temperature are affected by the liquid excipient viscosity and by the type and concentration of gel former. If the drug is present as a particulate disperse phase, then particle size and concentration will also need to be considered. The effect of liquid excipient viscosity and type of gel former were investigated for liquid poloxamer/silicon dioxide gels using as a model disperse phase, spray-dried lactose 10% w/w (Saeed *et al.*, 1997, 1998). The viscosity for the liquid poloxamers was in the range 0.17–1.6 Pa s at 25°C and all were Newtonian liquids. Poloxamer gels were prepared using either a hydrophilic (Aerosil 200) or a hydrophobic (Aerosil R974) silicon dioxide. Rheograms showed that the gels were thixotropic and shear thinning, with a higher apparent viscosity for Aerosil 200 gels than for Aerosil R974 gels at

Table 9.4 Principal factors affecting liquid-filling of non-Newtonian systems in hard gelatin capsules

Thixotropic gels	Thermosoftened matrices
• Liquid excipient viscosity	• Molten excipient viscosity at filling temperature
• Type and concentration of gel former	• Effect of dissolved drug on excipient viscosity
• Drug as particulate disperse phase	• Drug as particulate disperse phase

♦ Particle size and concentration
♦ Aeration
♦ Dispersion stability

equivalent silicon dioxide concentrations. The effect of the poloxamer viscosity on filling of 6% w/w Aerosil 200 gels is shown in Table 9.5. All gels, apart from the one prepared with the highest viscosity poloxamer L121, filled with a CV <1.5% for weight of content.

Dispersion stability of the gels was related to the particle size of disperse phase and the viscosity of the continuous phase, as would be predicted by Stokes' law, however, there was a considerable difference in the stability of gels made with hydrophilic and hydrophobic silicon dioxide. In this work, the disperse phase was hydrophilic and produced superior dispersion stability with poloxamer gels prepared with hydrophilic silicon dioxide, than for those made with hydrophobic silicon dioxide.

Thermosoftened systems

Thermosoftened formulations are prepared at an elevated temperature in order to produce a formulation that is sufficiently mobile for satisfactory filling. A typical formulation may be based on a solid excipient, e.g. PEG or poloxamer that will melt below 70°C and in which the active substance will melt, dissolve or disperse. The formulation will be a mobile liquid at the filling temperature and will be either a solution or dispersion of the active substance. When cooled to ambient temperature it will be a solid or a semi-solid dispersion in the capsule. The particle size and polymorphic form of the raw material are unimportant when the active substance dissolves or melts in the SSM at the filling temperature, however, the physical properties of the recrystallised active substance in the capsule may be very influential on drug release. Where the active substance is insoluble or does not melt at the

filling temperature, the dispersion quality and stability must be investigated at both the filling and cooling stages. In these systems, the particle properties of the raw material for SSM dispersions will affect processing and capsule filling as well as drug release from the capsule. The possibility of ageing of solid dispersions prepared by the thermosoftened technique should be considered and will be discussed later in this chapter.

In cases where the active substance has a relatively low melting point or is soluble in the molten polymer, it is possible that the formulation may behave as a Newtonian solution with good filling properties. Ibuprofen, which has a melting point 75°C, formed Newtonian solutions at 70°C with PEG 6000, 10 000, poloxamer (Lutrol F68) and Dynafill (a palmitic acid derivative of polyoxyethylene polyoxypropylene block copolymer) (Hawley et al., 1992). These formulations could be filled with a high concentration of active, e.g. 90% w/w, because it had negligible effect on the viscosity of the molten formulation. An ibuprofen 50% w/w/Lutrol F68 formulation when filled at 70°C using a bench-scale filler (Hibar Systems Ltd) gave a CV of 0.4% with a normal distribution of fill weight.

A comparison of capsule-filling properties in relation to viscosity at 70°C for thermosoftened excipients is shown in Table 9.6 (Hawley et al., 1992). Poloxamer (Lutrol F68), Dynafill and PEG 10 000, with viscosity in the range 1.2–5.5 Pa s, all had excellent filling characteristics with a CV of fill weight <0.5%. Dynasan 114 (myristic acid triglyceride) with viscosity of 0.012 Pa s at 70°C gave a lower mean fill weight than the other excipients but had a CV value of 0.42%. The viscosity of the Dynasan 114 was sufficiently low at 70°C to allow slight leakage from the dosing pump, hence decreasing the mean fill weight. A lower limit of 0.1 Pa s is recommended to prevent

Table 9.5   Effect of viscosity of liquid excipients on capsule fill-weight variation for thixotropic gels of poloxamer and 6% Aerosil A200 at 25°C (Saeed et al., 1997, 1998)

Poloxamer grade	L31	L42	L62	L81	L101	L121
Viscosity/Pa s at 25°C	0.17	0.28	0.42	0.48	0.80	1.60
Mean fill weight/mg	411	413	421	401	416	344
CV/%	0.4	1.3	0.5	1.4	1.4	4.0

Table 9.6 Capsule fill-weight and viscosity of thermosoftened excipients at 70°C (Hawley *et al.*, 1992)

Excipient	Viscosity (Pa s) at 70°C	Mean fill weight (mg)	CV (%)
Dynasan 114	0.012	413	0.42
Poloxamer F68	1.2	504	0.49
Dynafill	1.6	499	0.21
PEG 10000	5.5	521	0.31
PEG 20000	24.0	503	3.06

leakage and splashing of the contents from the capsule body during filling. In addition to prevent leakage it was proposed that the surface tension of the liquid should be in the order of >20–30 mN m^{-1} (François and Jones, 1979).

PEG 20 000 has a viscosity of 24 Pa s at 70°C and filling was characterised by bridging of the molten excipient between capsules. The CV value was 3.06%, whereas a CV of fill weight <1.0% was achieved for the Newtonian and non-Newtonian SSM formulations mentioned earlier. In general, filling problems can be related to formulation viscosity and an approximate range of 0.1–25 Pa s has been proposed for a satisfactory result. Formulations with a viscosity below the lower limit lead to losses through liquid splashing from the capsule during filling, whereas the upper limit is imposed by the limitations of the liquid pumps and dosing mechanisms on currently available filling machines. In addition, formulations with a high viscosity cause problems in transfer from the injector nozzle to the capsule body. For these formulations a machine should be used that inserts the nozzle into the capsule body before pumping commences. This ensures that the liquid enters into the body because viscous ones tend to curl rather than fall in a straight line on leaving the nozzle. Additional problems arise because the liquid stream does not break clearly when the pump finishes its stroke and material is carried over the edge of the capsule bodies as the machine moves the bushings to the next station. In addition, the rheological characteristics may cause incomplete filling and/or bridging of the SSM between the capsules.

In formulations where the active substance does not melt or dissolve in the excipient, the filling properties become dependent upon the disperse phase particle properties and concentration, as well as the viscosity of the molten excipient at the filling temperature. An investigation (Rowley *et al.*, 1998) of disperse systems was undertaken in order to determine the effect of particle size, disperse phase concentration and polymer MW and viscosity on capsule fill-weight variability at 70°C, using a bench-scale filler (Hibar Systems Ltd). Lactose was selected as a model particulate system because it has negligible solubility in the molten PEG 6000, 8000 and 10 000 used in this investigation. The effect of disperse phase concentration on apparent viscosity at a shear rate 12.5 s^{-1} is shown in Figure 9.1. The results comparing the two size fractions of lactose (L and S) with mean particle sizes of 170 µm and 17 µm, respectively, show that there is a critical concentration of the disperse phase above which there is a considerable increase in apparent viscosity. The considerable increase in apparent viscosity correlates with an increase in fill-weight variation and unsatisfactory capsule filling. Whereas lactose (L) produced satisfactory filling at disperse phase concentrations up to 50% w/w in PEG 6000, the smaller particle size lactose (S) in PEG 6000 dispersion showed satisfactory filling only up to 30% w/w. In addition, the maximum disperse phase concentration for satisfactory filling decreased markedly as the polymer MW increased from 6000 to 10 000. The maximum concentration of lactose (S) for satisfactory filling in PEG 8000 formulations was <10% w/w, whereas the same particle size fraction could not be filled satisfactorily at 5% w/w in PEG 10 000.

The principal factors that affect the capsule-filling properties of SSM, where the active substance is present as a particulate disperse phase, are as follows; drug particle size and concentration, liquid excipient viscosity at the filling temperature, apparent viscosity of the SSM at the filling temperature, and dispersion quality and stability. In addition, the incorporation of particulate material into a molten liquid excipient can often cause air entrapment. This problem can be overcome by mixing the disperse phase and molten continuous phase under vacuum (Kattige and Rowley, 1999). Fill-weight variation of a

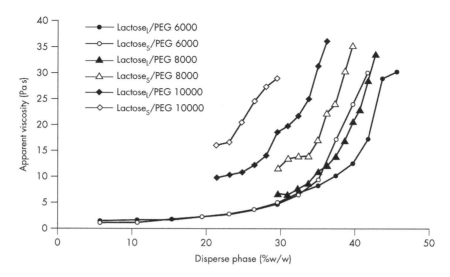

Figure 9.1 Effect of particle size and concentration of disperse phase and polymer molecular weight on apparent viscosity at 70°C and shear rate 12.5 s^{-1}. (Reproduced from Rowley *et al.*, 1998, with permission of publishers and authors.)

molten poloxamer (Synperonic) F88 containing 10% w/w particulate disperse phase without de-aeration was 10.3%, whereas when de-aerated during mixing it had a CV of 1.2%.

The effects of the disperse and continuous phase properties on filling, dispersion stability and drug content uniformity of molten poloxamer formulations have been investigated in detail (Kattige *et al.*, 2000; Kattige, 2001). The effect on the apparent viscosity of an SSM at a shear rate of 12 s^{-1} at 70°C was measured using a bench-scale capsule filler (Hibar Systems Ltd). Two grades of lactose with narrow size ranges, Granulac 230 (mean 23 μm, 10% <2.9 μm, 90% <49 μm) and Sorbolac 400 (mean 15 μm, 10% <3 μm, 90% <36 μm) were used as the disperse phase in two molten poloxamers, Synperonic F68 and F88 in which they have negligible solubility. The SSM were prepared with disperse phase ranging from 0–40% w/w by mixing under vacuum for 60 min at 70°C. The SSM processing and capsule-filling process was evaluated by measuring the capsule-fill weight (Table 9.7) and the disperse phase content uniformity (Table 9.8). The particle size of the lactose had a negligible effect on the increase in apparent viscosity of SSM with low concentrations of disperse phase: <10% w/w for poloxamer

F68 and <7.5% w/w for poloxamer F88 dispersions. The effects of particle size were significant at higher concentrations in poloxamer F68 dispersions and there was an abrupt increase in apparent viscosity between 25 and 30% w/w and 20 and 25% w/w for Granulac 230 and Sorbolac 400, respectively (Figure 9.2). A similar effect of particle size on apparent viscosity was observed with poloxamer F88 dispersions showing an abrupt increase at 25–30% w/w and 15–20% w/w for Granulac 230 and Sorbolac 400, respectively. The viscosity of poloxamers F68 and F88 was 0.92 and 2.8 Pa s, respectively at 70°C, hence SSM made with poloxamer F88 had a greater apparent viscosity than those from poloxamer F68 at equivalent concentrations of the disperse phase. It is advisable in formulation design to use an excipient with a relatively low viscosity at the filling temperature in cases where there is a relatively high proportion of disperse phase. In cases where there is an abrupt increase in apparent viscosity as the concentration of disperse phase increases it is critical to keep below this value in order to achieve good fill-weight uniformity (Table 9.7).

The dispersion stability of the lactose/poloxamer SSM was found to be satisfactory for up to 6 hr storage at 70°C without stirring in the hopper

Table 9.7 Effect of disperse phase particle size and concentration on coefficient of variation for capsule fill-weight (Kattige *et al.*, 2000; Kattige, 2001)

Coefficient of variation (%)		
Disperse phase (%) w/w	Granulac 230 poloxamer F68	Sorbolac 400 poloxamer F68
10	0.44	0.54
20	0.27	0.39
25	0.3	1.1
30	1.3	*
35	3.4	
37.5	*	

* Could not be filled.

of the filling machine. The dispersion quality and stability were evaluated by filling capsules at two-hourly intervals up to 6 hr, followed by measurement of the capsule fill-weight variation and disperse phase content uniformity. The results for dispersions of Granulac 230 (larger particle size fraction) and poloxamer F38 (lowest viscosity) showed good dispersion quality and stability. Initially, the CVs for fill weight and content uniformity were 0.35 and 0.30%, respectively, and 0.18 and 1.40%, respectively, after 6 hr. It was therefore envisaged that the poloxamers F68 and F88 of higher viscosity and with a disperse phase of smaller particle size would produce greater dispersion stability during capsule filling and storage. Shear rate values for the transfer from hopper to pump and from pump to capsule via the nozzle, were calculated to be 12 and 340 s^{-1},

respectively. The formulations that could not be filled had an apparent viscosity >55 Pa s at 12 s^{-1}. An examination of the machine (Hibar Systems Ltd) showed that the pump was unable to suck the SSM from the hopper. Capillary shear flow was used to simulate the high shear transfer stage of SSM from pump to capsule (Kattige, 2001). This showed that the dispersion structure was broken down below 300 s^{-1}, thus providing further evidence that problems with filling occurred during SSM transfer from the hopper to the pump. This work was carried out with the stirrer in the hopper switched off. It was suggested that an SSM with a greater apparent viscosity than 55 Pa s at a 12 s^{-1} shear rate could be filled by using a mixer in the hopper to cause shear thinning.

It is clear from this research that further detailed investigations are required with liquid-fill formulations where the active substance is present as a particulate disperse phase. This work would lead to predictions of filling capabilities and dispersion stability of formulations from the physical properties of the components of the formulation and the rheology of the dispersion. High shear rheological measurements made using, for example, a capillary shear flow rheometer can provide information relevant to the movement of an SSM formulation from the pump into the capsule. This information could be used to predict the filling capability of

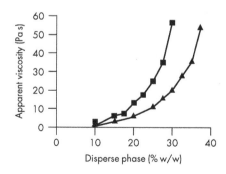

Figure 9.2 The effect of particle size and concentration of disperse phase on apparent viscosity of lactose/poloxamer F68 dispersions at 70°C and shear rate 12 s^{-1}. (▲) Granulac 230, mean size 23 µm; (■) Sorbolac 400, mean size 15 µm. (Data taken from Kattige, 2001, with permission from the author.)

Table 9.8 Disperse phase content uniformity of poloxamer/Granulac 230 liquid-fill formulations (Kattige *et al.*, 2000; Kattige, 2001)

10% w/w Granulac 230 in poloxamer	Mean disperse phase content (mg)	CV(%)
F38	45.1	1.49
F87	44.8	0.60
F68	43.8	1.12
F88	44.8	1.17

machines when dosing pump shear rates are known.

Capsule-filling technology

Two-piece capsules can be filled easily with liquid formulations at all the required levels of demand. Experimental batches can be filled manually using a hypodermic syringe. This process is useful for preliminary investigations into capsule shell integrity and compatibility with excipients, as well as *in vitro* and *in vivo* dissolution and pharmacokinetic studies. There are semi-automatic and automatic filling machines available for all scales of output from bench to production. There are minimal problems in the scale-up to a production scale because increases in output are achieved by using multiples of dosing pumps. McTaggart *et al.* (1984) modified a standard powder fill machine, Zanasi LZ64, for the production of pilot-scale or small production batches. The main difference between liquid- and powder-filling machines is in their speed of output. For an equivalent size machine the output of a liquid-filling machine is about 50–60% that of a powder machine. The reason for this is that the liquid is transferred to the capsule body through a small orifice to ensure uniformity of fill. On a powder-filling machine the same volume of powder passes through an orifice that is only just smaller in diameter than the capsule body and thus it can be done faster.

There are several machines available for laboratory scale production. The Hibar capsule-filling machine (European agents, Hi-Tech Machinery, Hook, UK) is based upon a very accurate pneumatically operated pump, which is widely used in other industries for accurately dispensing of small quantities of liquid. To operate the machine, the capsules are manually separated into cap and body and then the bodies are loaded into a carrier block, which has two rows of 12 holes. The carrier block is then placed on to the filling machine (see Figure 9.3). The machine has a stainless hopper of 1 L capacity, which is heated with an electrical jacket, the temperature of which can be controlled to ±0.1°C. For thermosoftened SSM formulations the hopper and nozzle can be heated indepen-

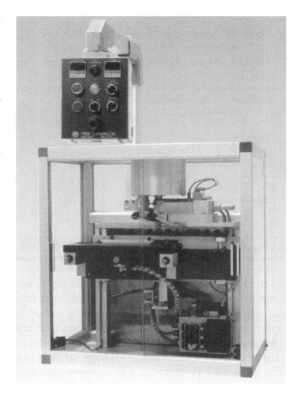

Figure 9.3 Semi-automatic bench-top liquid-filling machine (Hibar Systems Ltd./Hi-Tech Machinery Ltd). (Reproduced with permission from the company.)

dently so that the temperature to which the bulk of the formulation is heated is minimised and the temperature raised in the pump to reduce the viscosity of the material as it passes. There is a variable speed mechanical stirrer to maintain product uniformity. The hopper sits on top of the pneumatic pump (Hibar Systems Ltd, Ontario, Canada). The carrier block is moved along a track that passes under the pump nozzle. The rotary valve dispensing pump sucks liquid from the hopper as the piston inside it is withdrawn. The valve rotates and as the piston returns the liquid is dispensed through the nozzle into a capsule body (see Figure 9.4). The capsule bodies in the carrier block are indexed under the nozzle. The first traverse from left to right fills the bodies in the front of the block and then as it traverses back to the start position the other row is filled. For filling a more viscous formulation the carrier block is raised so the nozzle is inside the bodies

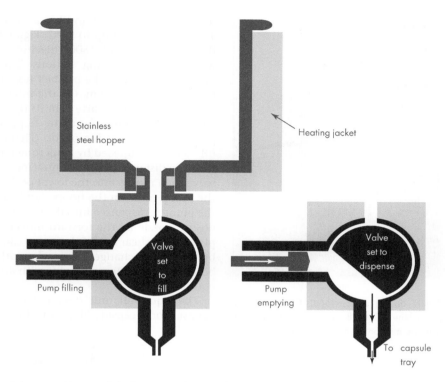

Figure 9.4 Schematic drawing of the hopper and rotary dispensing pump of the semi-automatic bench-top liquid-filling machine (Hibar Systems Ltd./Hi-Tech Machinery Ltd). (Reproduced from Hawley, 1993, with permission from the author.)

before pumping commences. When filling has been completed the caps are manually replaced onto the bodies.

The CFS 1000 machine (Capsugel) is a fully automatic bench-top machine capable of filling capsule sizes 00el to 4 with liquids and semi-solids with viscosity up to 1 Pa s, at an operating speed of 1000 per hour. The hopper can be heated up to 70°C and is fitted with a stirrer to ensure homogeneity during the filling of a batch. The filled capsule bodies are capped and moved automatically to the sealing station which operates automatically using the spray sealing technique described later in this chapter.

The IN-CAP (Bonapace & Company) is a fully automatic bench-top machine for filling solids and/or liquids at a rate of 3000 capsules per hour into sizes 00–4 capsules. The machine can be fitted with both solid and liquid hoppers, the latter has a jacketed heater and mixer. Figure 9.5 shows the detail of the liquid dosing system, which has a micrometer dose adjustment and has

a device for stopping filling if a capsule body is not in position.

Fully automated production-scale capsule machines are available from several manufacturers as shown in Table 9.9. The basic machines for filling powders into hard gelatin two-piece capsules, described in Chapter 6, have been adapted by incorporating liquid-filling and handling devices in place of, or in addition to, the powder, pellet or tablet-filling heads. All sizes of capsules from 00el to 4 can be filled. The volume available for the liquid-filled products is less than for powders, because it is not possible to fill the body completely in order to prevent spillage during filling. The capsule bodies are filled typically to 90% of the total body volume, which represents a fill volume range from 0.83 to 0.19 mL.

Automatic capsule-filling machinery and manufacturers are listed in Table 9.9, which shows that the approximate capsule-filling rates range from 6000 to 100 000 per hour. Most of the

Figure 9.5 Liquid dosing station for IN-CAP bench-top capsule-filling machine (Dott. Bonapace & Company). (Reproduced with permission from the company.)

companies use the same filling mechanism for all rates of filling. Robert Bosch GmbH has three models, GKF 400, 800 and 1500L, with filling rates of about 10 000, 30 000 and 60 000 capsules per hour, respectively. The jacketed hopper is

Table 9.9 Capsule (hard gelatin) liquid-filling rates of fully automatic production-scale machinery

Manufacturer/model	Approximate liquid-filling rate (capsules per hour)
Robert Bosch GmbH	
GKF 400	10 000
GKF 800	30 000
GKF 1500L	60 000
Harro Höfliger GmbH	
KFM	Up to 25 000
IMA Zanasi	8000–85 000
MG2	
Compact	6000–48 000
Futura	6000–48 000
Planeta	up to 100 000
Shionogi Qualicans	
LIOFIL^{Super}80	80 000

fitted with an agitator and has a capacity of 7 L and can be heated up to 90°C. Figure 9.6 shows the interior of a GKF 1500L machine configured with two liquid-dosing stations. The pump is fitted to the base of the product hopper and has its own heating system. On the intake stroke of the pump, the slide valve is in its upper position and liquid is drawn from the hopper into the pump by the retreating piston (Figure 9.7). The fill volume is adjusted by a micrometer that controls the distance of piston movement. The slide valve then moves into the lower position closing the pump off from the hopper and the liquid in the pump is dispensed into the capsule through a dosing nozzle by the forward movement of the piston. The shear rate of the pump can be altered by use of interchangeable nozzles with different bores to allow filling of liquids with different viscosities.

The liquid-fill unit designed to fit all IMA Zanasi E or F capsule-filling machines is shown

Figure 9.6 Liquid-filling stations for Bosch GKF 1500L capsule-filling machine at Encap Drug Delivery. (Reproduced with permission from the company.)

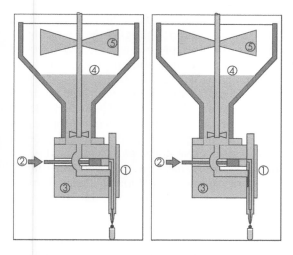

Figure 9.7 Liquid-dosing valve and formulation hopper for Bosch liquid-filling machine. 1, Slide valve with flow valve and fill needle; 2, fill and dosing piston; 3, water-heated pump block; 4, liquid formulation hopper; 5, agitator. (Reproduced with permission from the company.)

the liquid dose from the hopper to the pump and then into the capsule, by the movement of the slide valve and the piston, the stroke of which can be adjusted by a micrometer.

The MG2 range of capsule fillers has a rotary continuous motion, rather than the intermittent mechanism of all the other machines listed in Table 9.9. The liquid formulation is fed through a tube into the rotating liquid container that is fitted with maximum and minimum liquid sensors. Cams are used to open and close the dosing valves that are located at the base of the product container. The sequence of dosing valve rotation for capsule filling is as follows. The dosing cam allows the piston to be lowered to fill the dosing chamber by suction of liquid from the product hopper. Further rotation of the unit allows the opening cam to make contact with the dosing valve and to connect the dosing chamber to the capsule body. The piston rises to expel the liquid dose into the capsule, via a nozzle.

Shionogi Qualicaps manufacture capsule-filling and also the band-sealing machines, described later in this chapter. The LIQFIL^{Super}FS comprises three units: the capsule-filling

in Figure 9.8. This unit is connected into the capsule-filler control unit and consists of a product hopper fitted with heating and agitation facilities, as well as maximum and minimum liquid level sensors. The liquid dosing unit is fitted to its base and the dosing mechanism. Figure 9.9 shows the sequence of movement of

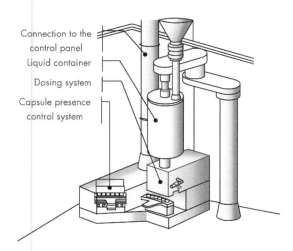

Figure 9.8 Liquid-filling unit for IMA Zanasi capsule-filling machine. (Reproduced with permission from the company.)

Connection to the control panel
Liquid container
Dosing system
Capsule presence control system

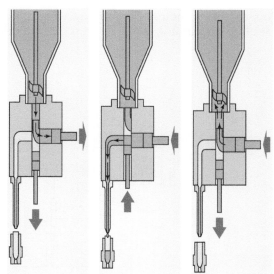

Figure 9.9 Sequence of drawing liquid from the hopper by the slide valve and pushing liquid through a syringe into the capsule. IMA Zanasi liquid-fill capsule machine. (Reproduced with permission from the company.)

machine, a connection unit and the sealing machine and dryer, to give an integrated liquid-filling and sealing process. The 20 L hopper is water jacketed to provide filling temperatures up to 80°C and is fitted with a mixer. The design and operation of the pump for transfer of the SSM to the capsule is shown in Figure 9.10 and fill volume adjustments are easily made, with capability for fine adjustment during operation. The filled and rejoined capsules are automatically conveyed to the connection unit via a pick-up roller, chute and feed roller to the sealing unit and dryer. The operation of the sealing unit is described in the next section.

Sealing of two-piece capsules

One of the reasons that prevented the use of hard gelatin capsules for liquid-filling in the past, was the propensity for leakage of the formulation through the gap between the cap and body of the capsule. Leakage can be minimised by product formulation and eliminated by sealing the two capsule parts together either by coating the junction between cap and base with a layer of gelatin, often termed banding or by using a sealing liquid. Sealing and banding can be made at similar rates as for capsule filling and it is therefore possible to link the filling and sealing/banding machinery into a continuous process, e.g. the LIQFIL^{Super} FS (Shionogi Qualicaps) and the CFS 1000 (Capsugel).

The banding of capsules with a solution of gelatin was first used in the 1950s by Parke, Davis & Company as a means of preventing copying by adding a further colour and making a product with a distinctive appearance. One of the first machine companies interested in the liquid-filling of capsules was Shionogi (Japan Elanco). They built an improved machine that could produce hermetically sealed capsules at high speeds (Jones, 1987) This machine was launched in the USA at the same time as the Tylenol tampering incident. Some packs of Tylenol had apparently been taken off the shelf in a shop, the capsules were opened, contaminated with cyanide, and then closed and replaced onto the shelf. Nobody has ever found out who did this, but from then on tamper-evident closures were sought. Thus, the main use of this machine has been to produce filled capsules with tamper-evident seals. This process produces banded capsules that comply with the tamper-evident packaging requirements in the USA for over-the-counter human drug products (Code of Federal Regulations, 1998).

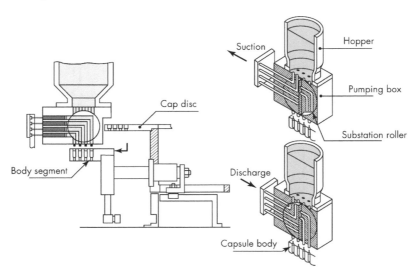

Figure 9.10   Liquid-filling mechanism for LIQFIL^{Super}80 capsule-filling machine, Shionogi Qualicaps Co. Ltd. (Reproduced with permission from the company.)

The banding machine is supplied with filled capsules, which are placed in a storage hopper. The capsules are then passed through a three-drum rectification system that delivers them to slots in slats on a conveyor belt. The movement of the conveyor causes the capsules to spin and they all move to the left until they come in contact with a cap guide. The capsules are now accurately positioned and pass over the sealing units (see Figure 9.11). This consists of a temperature-controlled bath in which the gelatin solution is kept and in which there are two sets of revolving discs. These pick up gelatin solution on their edges and this is transferred on to the capsules when the discs come into contact with them. The banded capsules are now transferred to a continuous-loop conveyor belt in the drying unit. The bands are dried using filtered air at ambient room conditions. This process does not make gelatin capsules brittle. HPMC capsules filled with liquid formulations can be sealed by banding using standard equipment (Tochio and Nagata, 2001). In order to obtain the correct mechanical strength of the seal, HPMC with different MWs was used. It was shown that HPMC (MW 20 000 Da) 16% w/v in an ethanol/water solution gave satisfactory results.

Another method that has been introduced more recently is based on the use of alcohol/water mixtures that are sprayed into the cap/body junction, which has undergone several radical changes since its initial inception (Wittwer, 1985; Cadé *et al.*, 1987). The liquid is rapidly taken into the gap between the cap and body and the capsules are then heated to produce a seal. This has evolved into the LEMS systems used on a production-size sealing machine (Capsugel) and the CFS 1000 bench-scale filling and sealing machine. A solution of equal parts of water and ethanol is sprayed as a fine jet directed at the junction between the cap and body and suction is used to remove excess liquid (see Figure 9.12). A stream of warm air, 40–60°C, is blown across the capsules causing melting and fusion of the two gelatin layers. The capsules are stored overnight at ambient conditions to cool, in order to facilitate the setting and hardening of the gelatin seal.

The production processes for the filling and sealing of capsules can either be carried out as separate unit operations or as one continuous process. The limiting factor usually is the viscosity of capsule contents if they are mobile at ambient conditions. To minimise the chances of leaking for low-viscosity formulations, the capsules need to be sealed as soon after filling as is practically possible.

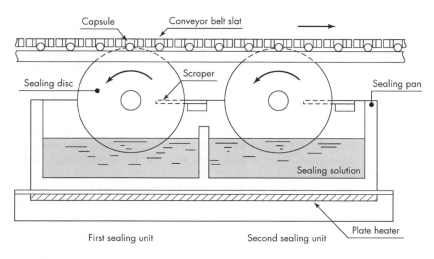

Figure 9.11 Capsule-sealing mechanism, HICAPSEAL 40/100 machines, Shionogi Qualicaps Co. Ltd. (Reproduced with permission from the company.)

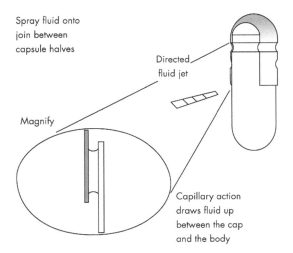

Spray fluid onto join between capsule halves

Directed fluid jet

Magnify

Capillary action draws fluid up between the cap and the body

Figure 9.12 Spray-sealing process, LEMS 30 capsule-sealing machine and CFS 1000 capsule-filling and -sealing machine. (Reproduced with permission from the company.)

Drug delivery

A patent dealing with improvements in pharmaceutical dosage forms (Walker *et al.*, 1980a) claimed that SSM formulations could be used to improve dissolution of active substances with low aqueous solubility and also that there was potential for sustained release from relatively simple formulations. Subsequent research into drug delivery from liquid-fill capsules has provided evidence to support these claims. If the poorly water-soluble active substance is soluble in a liquid hydrophilic excipient, then provided capsule integrity is maintained, there should be minimal problems with bioavailability. Unfortunately, such excipients are not readily available and further research into excipients or mixed excipients with greater solvent capacity for a wide range of drugs, is needed. Despite this problem, thixotropic gels and drug/polymer solid dispersions have been used to improve the dissolution characteristics of actives with low water solubility.

Thermosoftened systems

Thermosoftened formulations solidify in the hard gelatin capsule to form a non-porous crystalline plug, or solid dispersion. The behaviour of the plug in the presence of aqueous media will determine whether the drug release rate is rapid for an immediate release product, or slow for a controlled delivery system. Selection of hydrophilic excipients can produce a plug that will allow rapid penetration of aqueous media, coupled with wetting of the drug crystals and rapid disintegration and/or dissolution of the matrix. Conversely, lipophilic excipients will reduce the rate of penetration of aqueous media into the matrix and the drug release mechanism will be by diffusion from the non-disintegrating matrix and/or surface erosion. It is therefore possible to produce immediate or controlled-release formulations comprising the active and one excipient. However, it may be necessary to use a combination of excipients to produce a matrix with the most appropriate hydrophilic–lipophilic balance properties for the desired drug release rate.

Solid dispersion of drug and water-soluble excipient formed by melting and solidification is a widely researched technique for improvement in drug dissolution (Chiou and Riegelman, 1971; Ford, 1986; Serajuddin, 1999). Hydrophilic polymers with low melting temperatures (<70°C), e.g. PEG and poloxamers are candidates for solid dispersion formulations for liquid-filling into hard gelatin capsules. A great deal of research on drug/PEG solid dispersions was done before the use of hard gelatin capsules for liquid-fill formulations was established. Solid dispersions of griseofulvin with PEG were powdered and the dissolution rates for PEG 4000 or PEG 20 000 dispersions were found to be considerably higher than for unformulated micronised griseofulvin powder (Chiou and Riegelman, 1969). Subsequently, PEG has been used as a solid dispersion to improve the wetting and dissolution of several hydrophobic and poorly water-soluble drugs, however, commercial use has been limited, partly because of manufacturing difficulties encountered with the processing of powdered solid dispersions. The advent of liquid-filling of molten formulations into hard gelatin capsules has overcome the problems of processing solid dispersions since they are filled as a liquid and form a solid dispersion on cooling in the capsule. The temperature of the formation of the melt prior to filling and the cooling rate of the liquid-filled formulation in the capsule, will affect the physical structure of the matrix. The matrix structure, concentration, type and MW of polymer will all affect the dissolution rate of the drug. For example, the effect of PEG molecular weight on the dissolution characteristics of solid dispersions with nortryptyline showed a logarithmic decrease in release rate with increasing carrier MW from 3400 to 20 000 Da (Craig and Newton, 1992).

Thermosoftened ibuprofen formulations made with a variety of excipients were investigated and the results showed that dissolution rate could be controlled by variation of the drug/excipient type and ratio (Smith *et al.*, 1990). Ibuprofen was used as a model drug of low aqueous solubility in a detailed investigation of solid dispersion formulations prepared with hydrophilic excipients PEG 10 000, Lutrol F68 and Dynafill (Hawley, 1993). The results showed that when the carrier was present at the eutectic composition (about 50% w/w) or greater then drug dissolution was complete within 30 min. However, the matrix eroded more slowly in formulations with low carrier concentration, e.g. 10% w/w and drug dissolution occurred over a 6–7 hr period, thus confirming the potential of two-component SSM formulations for sustained release.

The physicochemical properties of both drug and polymer will influence drug release from solid dispersions and the presence of particulate drug can retard the dissolution of the polymer matrix (Corrigan, 1986). Simultaneous measurement of the drug and polymer dissolution rates can provide information on the release mechanisms from capsule formulations (Hawley and Rowley, 1998, 1999). The effect of polymer concentration on the time for 50% dissolution (T_{50}) of both ibuprofen and polymer from Lutrol F68 and PEG 10 000 dispersions is shown in Figures 9.13 and 9.14, respectively. Dissolution of the ibuprofen 100% w/w formulation occurred slowly and was complete in 8 hr, whereas dissolution of PEG 10 000 100% w/w and Lutrol F68 100% w/w was complete after 1 hr, in each case. The dissolution rate of ibuprofen and Lutrol F68 were similar for all dispersion compositions, whereas the dissolution rate of PEG 10 000 was greater than for ibuprofen for all dispersion compositions. The solid dispersions were non-disintegrating and the release mechanisms were

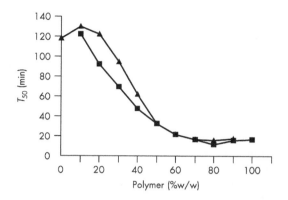

Figure 9.13 T_{50} values for simultaneous dissolution measurements of ibuprofen (▲) and Lutrol F68 (■) from solid dispersion formulations liquid–filled in hard gelatin capsules. (Data taken from Hawley, 1993, with permission from the author.)

further investigated by liquid penetration experiments to investigate the presence of diffusion layers and/or surface erosion. Ibuprofen 0% w/w and 10% w/w in PEG 10 000 formulations showed rapid surface dissolution only and the mechanism of release is by erosion. Ibuprofen 100% w/w and 90% w/w in PEG 10 000 formulations exhibited much slower surface erosion and an absence of diffusion layer within the matrix. Formulations with an intermediate composition, e.g. Ibuprofen 40% w/w and 50% w/w

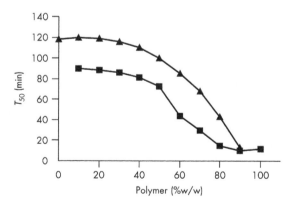

Figure 9.14 T_{50} values for simultaneous dissolution measurements of ibuprofen (▲) and PEG 10 000 (■) from solid dispersion formulations liquid–filled in hard gelatin capsules. (Data taken from Hawley, 1993, with permission from the author.)

in PEG 10 000, showed an advancing penetration front within the matrix and also surface dissolution. In latter cases, the release mechanisms appear to be by diffusion and erosion, whereas liquid penetration experiments for the ibuprofen/Lutrol F68 matrices demonstrated the lack of a penetration front and diffusion layer in the matrix, and hence the release mechanism was predominantly by rapid surface erosion. The simultaneous dissolution tests of ibuprofen and polymer showed that the release of drug and Lutrol F68 proceeded at a similar rate, whereas in the ibuprofen/PEG 10 000 matrices, the rate of dissolution of the polymer was greater than that of the drug. Thus, the evidence from liquid penetration and drug/excipient dissolution indicates that PEG and poloxamer matrices have different drug release mechanisms. Hydrophilic polymers that dissolve rapidly in aqueous media can produce a formulation with rapid dissolution with poorly water-soluble drugs. Hydrophilic polymers that hydrate slowly can form a simple two-component formulation for slow release of the drug by diffusion and/or erosion and this is an area that would merit further research.

Another approach is to use a range of saturated polyglycol glycerides, which contain varying blends of mono-, di- and triglycerides, and mono- and di-esters of fatty acids with PEG and low MW PEGs (Gelucires). These are characterised by their melting point and HLB; e.g. Gelucire 43/01 has a melting point of 43°C and a HLB value of 1. There are grades with melting temperatures ranging from 33 to 64°C and HLB values from 1 to 14. These provide a group of excipients with the potential for liquid-filling in hard capsules with rapid or controlled drug release. Gelucire capsule formulations of ketoprofen with rapid and sustained release properties were investigated *in vivo* (Dennis *et al.*, 1990). A water-miscible, highly ethoxylated excipient produced ketoprofen absorption equivalent in rate and extent to conventional rapid release formulations, whereas a slowly hydrating erodable excipient prolonged the ketoprofen blood concentrations. Gelucire SSM formulations of indometacin filled in to hard gelatin capsules produced sustained drug release in human volunteers (Vial-Bernasconi *et al.*, 1995). The SSM was produced easily, was simple to fill and was shown to have a slow erosion

release mechanism by *in vitro* experiments. Mixed Gelucire systems of 50/13 and 50/02, which have widely different HLB values but similar melting points, were investigated in relation to drug-release mechanisms from thermosoftened formulations that were melted and solidified to form cylindrical moulded tablets (Kopcha *et al.*, 1991). An increase in the concentration of Gelucire 50/02 in the mixed system caused the release rate of theophylline to decrease and the mechanism was predominantly by diffusion in an acidic media (pH 1.2). However, in a basic medium (pH 8.1) the release was controlled by erosion of the matrix and this was attributed to partial hydrolysis of ester-linked matrices.

The Gelucire range of excipients has been shown to give either rapid or sustained release characteristics depending upon their melting points and HLB (Howard and Gould, 1987). Salicylic acid release from mixed Gelucire SSM in hard gelatin capsules was found to be log linearly related to the HLB of a mixed Gelucire system. This work confirmed the potential of relatively simple mixed excipient systems to produce rapid or sustained release formulations from liquid-filled hard two-piece capsules.

The mechanisms by which Gelucires control drug release were investigated with theophylline as the model drug in moulded flat-faced tablets (Sutananta *et al.*, 1995a). Two Gelucires with low hydrophilicity, 43/01 and 54/02, formed a non-eroding matrix with theophylline and the drug-release mechanism was by diffusion. The more hydrophilic Gelucire 50/13 formed an SSM that was easily penetrated by aqueous media, followed by swelling and disintegration of the matrix that was suitable for a rapid drug-release formulation.

Formulations based on ethylcellulose or methacrylate polymers (Eudragit) mixed with polyoxyethylene stearates with different HLB values have been used to develop extended release SSM by control of the hydrophobicity of the matrix and hence decrease the dissolution rate from liquid-filled hard gelatin capsules (Seth, 1992). SSM formulations with ibuprofen as a model were investigated in combination with ethylcellulose or a methacrylate polymer and PEG stearates 300, 400, 1500 and 4000 with HLB of 10, 12, 17 and 19, respectively. The drug release from the SSM formulation could be altered by changing the HLB of the PEG stearate, the type and concentration of polymer and the drug loading. This work led to a patent for Opticaps, an SSM product in hard gelatin capsules for extended release formulations of the Ca-antagonist drugs, nifedipine, verapamil and diltiazem (Mepha Ltd, 1989).

Despite all the research on solid dispersions formulations produced by the thermosoftening technique for liquid-filled capsules, there is still scope to improve the potential of this technology with respect to rapid or sustained drug delivery. The ultimate aim will be to produce formulations from a model based on the physical properties of the bioactive and excipients that consider not only drug delivery, but also capsule integrity, filling and stability.

Thixotropic gels

A thixotropic gel formulation of a glyceride oil, Miglyol and silicon dioxide (Aerosil 200) 5% w/w was shown to have satisfactory filling properties in hard gelatin capsules (Walker *et al.*, 1980b). A rheological investigation of thixotropic gels made from the glyceryl esters of caprylic, capric and succinic acids (Miglyol 829) combined with silicon dioxide (Aerosil 200 and Aerosil R974) to form thixotropic gels was used as part of a detailed investigation into drug release from liquid-filled capsules (Walters, 1994). A wide range of drug-release rates can be achieved by simply altering the concentration of silicon dioxide in thixotropic gels with triglyceride oils (Walters *et al.*, 1992b). A highly water-soluble drug, propantheline bromide, was dispersed in Miglyol 829/Aerosil 200 thixotropic gels and filled into hard gelatin capsules. The drug-release rate was dependent upon the silicon dioxide concentration in the range (0–12% w/w) and the lowest rate was from the formulation with 4% w/w, as shown in Figure 9.15. The marked decrease in release rate between 0–4% w/w Aerosil 200 was attributed to the increased apparent viscosity of the triglyceride gel causing a reduction in drug diffusion from the SSM. The elastic properties of the gel determined by oscillation rheology, increased substantially at Aerosil 200 concentrations >5% w/w and although there was a con-

tinued increase in apparent viscosity, the drug-release rate increased. Thus, at Aerosil 200 concentrations >5% w/w there was a change in the viscoelastic behaviour and the hydrophilic nature of the gel that caused an increase in liquid penetration into the gel with a corresponding increase in drug diffusion from the capsule formulation. The evidence suggests that above a certain concentration of hydrophilic silicon dioxide there is a change in drug release mechanism from that controlled by the apparent viscosity of the gel, to one controlled by liquid penetration in to the gel, via a network of hydrophilic silicon dioxide particles.

The effect of hydrophobic Aerosil R974 on propantheline bromide release from Miglyol 829 gels was markedly different to that for the hydrophilic gel system. Drug release from the Miglyol oil without silicon dioxide was complete within 30 min, whereas it was reduced to 30% and 5% for 1% w/w and 4% w/w Aerosil R974/Miglyol gels after 120 min (Walters, 1994). In contrast to the hydrophilic gels, there was no further effect of increasing the concentration of Aerosil R974 on the release of propantheline bromide. The slow release of water-soluble drugs from Miglyol gels can therefore be achieved by changing the concentration of hydrophilic silicon dioxide or by using hydrophobic silicon dioxide. Aerosil R974 produces a very hydrophobic gel matrix that would give incomplete drug

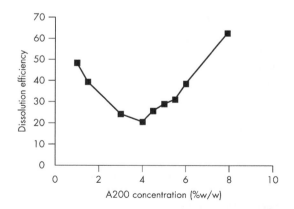

Figure 9.15 Effect of hydrophilic silicon dioxide (Aerosil 200) concentration on dissolution efficiency of propantheline bromide/Miglyol 829 gel fomulations. (Data taken from Walters, 1994, with permission from the author.)

release in aqueous media; however, it could be used to control release when in combination with a hydrophilic silicon dioxide. The effect of mixed hydrophilic and hydrophobic silicon dioxides on drug release from Miglyol 810 gels is shown in Figure 9.16 (Ellison and Rowley, 1999). The total concentration of the mixed silicon dioxide system (Aerosil 200 and R974 grades) was 4% w/w and the release of isoniazid after 150 min was 70% from the gel with 0% hydrophobic silicon dioxide, whereas with 0.02% w/w hydrophobic silicon dioxide the release was reduced to 25%.

Model drugs with different water solubility, e.g. propantheline bromide, isoniazid and tolbutamide, were formulated as similar triglyceride gels by using Miglyol oils 810 and 818 with 4% w/w silicon dioxide. The results showed that the drug-release rate was dependent on drug solubility, silicon dioxide type and the initial viscosity of the Miglyol oil (Ellison *et al.*, 1995). Miglyols 810, 818 and 829 have viscosities of 0.027–0.033, 0.030–0.035 and 0.23 Pa s^{-1} at 20°C and these differences provide a useful range to formulate gels with different filling and drug-release properties where necessary. The *in vitro* studies show that these gel systems are useful for formulating drugs with different physical properties, however, it is important to consider the behaviour of gels

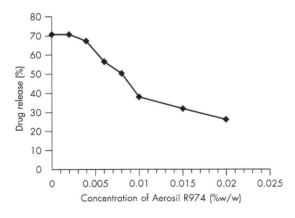

Figure 9.16 Effect of hydrophilic silicon dioxide (Aerosil R974) concentration on drug release (after 150 min) of isoniazid from Miglyol 810 gels containing 4% w/w mixed Aerosil 200 and Aerosil R974. (Data taken from Ellison and Rowley, 1999, with permission from the publisher.)

in the presence of digesting enzymes, e.g. with lipase in the *in vitro* dissolution test media.

Self-emulsifying systems

This type of formulation can be liquid-filled into hard gelatin capsules and is designed to self-emulsify in contact with aqueous media to form a fine dispersion. The selected excipients can often improve the solvent capacity of the system and thereby dissolve hydrophobic and/or poorly water-soluble drugs (Pouton, 1997). This approach has extended the versatility of liquid-fill formulations for hard gelatin capsules and can lead to improved drug delivery and bioavailability. Two problems that need consideration with respect to hard gelatin capsules are the effect of excipients on capsule shell integrity and the physical state of the drug and formulation in the emulsified system in the gastrointestinal tract, e.g. the solvent capacity may be reduced and the drug precipitated.

Self-emulsifying systems have been investigated as formulations that have been hand-filled into hard gelatin or HPMC capsules (Chong-Kook *et al.*, 2001; Kommuru *et al.*, 2001; Nazzal *et al.*, 2002). In each case, the basis of the formulation is a mixture of an oil, a surfactant and a co-surfactant. Table 9.3 presents the excipients that are compatible with hard gelatin capsules and demonstrates the range of excipients that are available for formulating a self-emulsifying drug delivery system (SEDDS) as an SSM for liquid-filling into gelatin capsules. Coenzyme Q_{10} was formulated as a SEDDS using the oils Myvacet 9-45 or Captex-200, the emulsifier Labrafac CM-10 or Labrasol, with lauroglycol as co-surfactant (Kommuru *et al.*, 2001). *In vitro* self-emulsification properties were investigated by determination of droplet size on addition of aqueous media, in order to optimise the formulation. In this case, Myvacet 9-45 (40% w/w), Labrasol (50% w/w) and lauroglycol 10% w/w were used to formulate an SSM that was liquid-filled into hard gelatin capsules and when tested in dogs, showed a two-fold increase in bioavailability, compared to a powder formulation of the coenzyme Q_{10}. A different approach to formulating coenzyme Q_{10} was used to form a self-nanoemulsified system

(SNEDDS) based on mixture of an essential oil, polyoxyl 35 castor oil (Cremophor EL) and medium chain mono- and diglyceride mixture (Capmul MCM-C8), which was filled into HPMC capsules (Nazzal *et al.*, 2002). The release of material was monitored using turbidity measurements and standard dissolution testing. The release of material was in three phases: a lag, a plateau and a pseudolinear. The study showed that the SNEDDS overcame some of the problems, such as low solubility and irreversible precipitation of active drug associated with SEDDS formulations. Biphenyl dimethyl dicarboxylate (BDD) is a poorly water-soluble drug used in the treatment of liver diseases. Formulations comprising an oil, a surfactant and a co-surfactant were investigated as concentrates to form microemulsions, in order to improve bioavailability after oral administration. BDD dissolved in a concentrate prepared from Tween 80, Neobee M-5 and triacetin, was liquid-filled into hard gelatin capsules and gave improved bioavailability after oral administration (Chong-Kook *et al.*, 2001).

Drugs with specific formulation problems

The problem of formulating the very hygroscopic drug vancomycin for oral delivery, which previously had only been available as a dry powder in sealed ampoules for reconstitution prior to administration, was solved using SSM formulation (Bowtle *et al.*, 1988). An SSM formulation was prepared by the thermosoftening technique with PEG 6000 that protected the drug from moisture uptake and produced a stable and effective preparation in a liquid-filled hard gelatin capsule.

A similar approach was used to solve a problem with the hygroscopic and very unpalatable drug mercaptamine hydrochloride used in the treatment of children with cystinosis, a rare hereditary metabolic disorder (Dixon *et al.*, 1994). Previous formulations were found to be unacceptable to the children and compliance was poor, even though the treatment was used to prevent a life-threatening disorder. The SSM

formulation with PEG 6000 protected the drug from moisture uptake and masked the obnoxious taste and smell of mercaptamine hydrochloride that resulted in excellent compliance and the successful treatment of cystinosis in the children.

A drug delivery system (HALO) designed for biphasic drug release was based on two different SSM formulations liquid-filled into a hard gelatin capsule. A problem with this system was the possible mixing of the two formulations with the loss of biphasic release characteristics (Burns *et al.*, 1996). The capsule contained a rapid release phase of propranolol in oleic acid and a sustained release phase of an erodable matrix of the same drug. The mixing of the two phases was prevented by either inserting a hydrophilic Gelucire barrier between the two phases, or by adding a hydrophobic Gelucire to the rapid release phase in order to solidify it at temperatures <37°C.

Liquid-filled formulations can be delivered to the intestines by enteric coating hard capsules, thus providing the opportunity to target actives formulated as liquid-fill preparations to the upper intestine and colon (Nagata and Jones, 2000). A thixotropic gel of peppermint oil has been used for the treatment of irritable bowel syndrome and has been marketed for many years in the form of an enteric-coated hard gelatin capsule (Colpermin). *In vitro* studies of an enteric-coated starch capsule containing a propranolol SSM showed a similar release profile to that of an enteric-coated gelatin capsule with the same formulation (Burns, 1996). Scanning electron micrographs have shown that HPMC capsules have a rougher surface than gelatin capsules and this improves the adhesion of enteric coatings to the HPMC capsule (Cole *et al.*, 2002). Disintegration of HPMC capsules, which had been coated with the polymethacrylate EudragitFS30D, occurred in the distal small intestine and proximal colon, thus providing evidence for the potential of this type of capsule for colonic delivery products (Cole *et al.*, 2002).

Stability issues

The principal stability issues for liquid-fill hard gelatin capsules are the integrity of the capsule shell, as discussed previously, the thermal stability of the SSM during processing and filling at elevated temperatures, and the ageing of SSM structure on storage and its effect on drug dissolution.

Thermosoftened SSM may be held at a processing temperature, e.g. 70°C, for several hours prior to filling into capsules. If filling is not possible on one day, then a cooling, solidification and remelting cycle may be necessary before it can be carried out at a later date. Hence thermosoftened formulations need to have a thermal stability able to withstand such processing conditions. PEGs and poloxamers are stable for several hours at temperatures between 60 and 70°C, which are commonly used for liquid-filling thermosoftened SSM, however, they are unstable on storage at higher temperatures for longer periods. PEGs of MW 4000–10 000 Da and poloxamers have ideal physical properties for liquid-filling into hard capsules by the thermosoftening process, provided that ageing does not adversely affect drug-release properties.

The cooling and solidification of a molten formulation can be investigated by inserting a thermocouple through the wall of the capsule and connecting it to a digital thermometer to record temperature and time during solidification of the formulation in the capsule. Figure 9.17 is a cooling curve for Lutrol F68 showing the characteristic features of a formulation with rapid solidification properties that are required

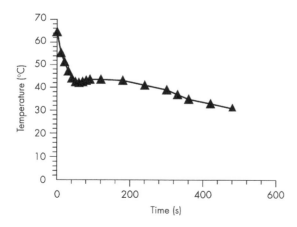

Figure 9.17 Cooling curve for Lutrol F68 immediately after liquid-filling into a hard gelatin capsule. (Data taken from Hawley, 1993, with permission from the author.)

for high-speed liquid-filling in order to prevent leakage from the capsule, prior to sealing. The curve has a rapid cooling phase to a temperature minimum after approximately one minute with an exothermic phase owing to crystallisation indicated by a slight increase in the temperature to a plateau during which complete solidification occurs within 2–4 min. After solidification, the formulation cools slowly to room temperature. Formulations may exhibit excessive supercooling, so that the solidification plateau is not reached during the filling cycle, thereby increasing the risk of leakage prior to sealing.

Liquid-filled capsules of ibuprofen and PEG 10 000 (Hawley and Rowley, 1999) and drug/poloxamer F68 formulations (Utting and Rowley, 1999) showed no change in drug dissolution rate after storage at room temperature for six months. The ageing of SSM of PEGs containing triamterene or temazepam have been investigated in relation to their thermal properties, drug particle size and dissolution (Dordunoo *et al.*, 1991, 1997). The changes in SSM structure that occurred on storage were demonstrated by an increase in the enthalpy and melting temperature of the excipient, crystal growth of the drug and reduced drug dissolution. The change in dissolution was related to MW of PEG in the order 1500 → 2000 → 4000 → 6000 Da.

The ageing effects of glycerides were investigated with respect to thermal and mechanical properties (Sutananta *et al.*, 1994) and it was shown that the effects depended on the composition of the glyceride excipient (Gelucire). Those composed of a mixture of glycerides and PEG esters resulted in changes in thermal properties, whereas Gelucire 55/18 that contains only PEG esters was unaffected. The mechanical strength of Gelucire with mixed components decreased on storage, whereas those with either glycerides or PEG esters showed an increase in strength on storage. The effect of cooling rate of theophylline/Gelucire 50/13 and 55/18 SSM, prepared as moulded tablets, was shown to affect the drug-release mechanism (Sutananta *et al.*, 1995b). A fast cooling rate was achieved by natural cooling to ambient temperature, whereas a slow cooling rate of 10°C per hour was obtained in an oven. It can be seen from the cooling curve (Figure 9.17), that the data from the fast cooling process would be more relevant to liquid-filled capsules, than those obtained at 10°C per hour. The drug release from the moulded tablets was shown to increase on storage after 180 days and therefore similar problems could be anticipated with this type of formulation when liquid-filled into capsules.

Thermosoftened SSM have considerable advantages with respect to processing and drug delivery, however, detailed thermal analysis is essential in order to ascertain stability during manufacture and storage of the capsule product. Changes in SSM structure for thixotropic gels should also be investigated since the reformation of the gel structure in the capsule is a time-dependent process.

Alternatives to gelatin for two-piece capsules

The interaction of the excipients with water, which could potentially be used for SSM formulations, is one of the drawbacks in the use of gelatin capsules. The water that they contain acts as a plasticizer for the film and when it is removed the capsules become brittle. In Chapter 3, alternative materials for manufacturing hard gelatin capsules have been described. These materials and their applications have been reviewed by Bowtle for applications in the field of liquid-filling (2002). Most work to date has been performed on HPMC capsules. It has been shown that HPMC hard capsules do not become brittle when they lose water (Ogura *et al.*, 1998).

Conclusions

Research into liquid-filled hard gelatin capsule formulations has confirmed that the claims and expectations, first proposed in the 1970s, are achievable (Cuiné *et al.*, 1978a, 1978b, 1978c, 1979a, 1979b). These types of product have clear advantages over other oral dosage forms and in-house production can be achieved at filling rates of up to 100 000 capsules per hour. The problem of leakage of contents from the capsule has been solved and machinery is available to seal capsules at a rate that matches the output from the filling machinery. There is sufficient evidence to show that relatively simple SSM formulations can be liquid-filled into hard two-piece capsules designed for either rapid or sustained drug delivery. However, further detailed research is necessary in order to increase the application of SSM technology to a wider range of bioactive substances.

References

Berntsson, O., Zackrisson, G. and Östling, G. (1997). Determination of moisture in hard gelatin capsules using near-infrared spectroscopy: applications to at-line process control of pharmaceutics. *J. Pharm. Biomed. Anal.*, 15, 895–900.

Bowtle, W. J. (2002). Options in materials for liquid-filled capsules. *Pharm. Manuf. Packaging Sources*, 2, 78–81.

Bowtle, W. J., Barker, N. J. and Wodhams, J. (1988). A new approach to vancomycin formulation using filling technology for semi-solid matrix capsules. *Pharm. Technol.*, 12(6), 86–91.

Bowtle, W. J., Burrows, S. L., Holmes, E. D. and Jones, B. E. (1989). SSM formulations and capsules: An improved way to handle toxic compounds. *Pharm. Res.*, 7, PT612.

Burns, S. J. (1996). An *in vitro* assessment of liquid-filled Capill potato starch capsules with biphasic release characteristics. *Int. J. Pharm.*, 134, 223–230.

Burns, S. J., Higginbottom, S., Whelan, I., Bowtle, W. J., Rosenberg, E., Corness, D., Hay, G., Attwood, D. and Barnwell, S. G. (1996). Formulation strategies designed to maintain the biphasic release characteristics of liquid-filled capsules. *Int. J. Pharm.*, 141, 9–16.

Cadé, D. and Madit, N. (1996). Liquid-filling in hard gelatin capsules – preliminary steps. *Bull. Technique Gattefossé*, 89, 15–19.

Cadé, D., Cole, E. T., Mayer, J.-Ph. and Wittwer, F. (1987). Liquid-filled and sealed hard gelatin capsules. *Acta Pharm. Technol.*, 33, 97–100.

Chiou, W. L. and Riegelman, S. (1969). Preparation and dissolution characteristics of several fast release solid dispersions of griseofulvin. *J. Pharm. Sci.*, 58, 1505–1519.

Chiou, W. L. and Riegelman, S. (1971). Pharmaceutical applications of solid dispersions. *J. Pharm. Sci.*, 60, 1281–1302.

Chong-Kook, K., Yeon-Ju, C. and Zhong-Gao, G. (2001). Preparation and evaluation of biphenyl dimethyl dicarboxylate microemulsions for oral delivery. *J. Control. Release*, 70, 149–155.

Code of Federal Regulations (1998). CFR 21, Part 211, Docket No. 92N–0314, Federal Register, Vol. 63, Nov. 4 1998, 59463–59471.

Cole, E. T. (1989). Liquid-fill hard gelatin capsules. *Pharm. Tech. Internat.*, 1(8), 29–33.

Cole, E. T., Scott, R. A., Connow, A. L., Wilding, I. R., Petereit, H. U., Schminke, C., Beckert, T. and Cadé, D. (2002). Enteric coated HPMC capsules designed to achieve intestinal targeting. *Int. J. Pharm.*, 231, 83–95.

Corrigan, O. I. (1986). Retardation of polymeric carrier dissolution by dispersed drug: factors influencing the dissolution of solid dispersions containing PEG. *Drug Dev. Ind. Pharm.*, 12, 1777–1793.

Craig, D. Q. M. and Newton, J. M. (1992). The dissolution of nortriptyline HCl from polyethylene glycol solid dispersions. *Int. J. Pharm.*, 78, 175–182.

Cuiné, A., Mathis, C., Stamm, A. and François, D. (1978a). Mis en gélules de solutions visqueuses de principes actifs. I. Etudes preliminaire – excipients. *Labo-Pharma Probl. Tech.*, 274, 222–227.

Cuiné, A., Mathis, C., Stamm, A. and François, D. (1978b). Mis en gélules de solutions visqueuses de principes actifs. II. Etude rhéologique d'excipients gras. *Labo-Pharma Probl. Tech.*, 276, 421–429.

Cuiné, A., Mathis, C., Stamm, A. and François, D. (1978c). Das Einbringen viskoser Lösungen von Aktivstoffen in Hartgelatinekapseln. *Pharm. Ind.*, 40, 654–657.

Cuiné, A., Mathis, C., Stamm, A. and François, D. (1979a). Mis en gélules de solutions (ou suspensions) visqueuses de principes actifs. III. Formulation de princips actifs. *Labo-Pharma Probl. Tech.*, 292, 863–868.

Cuiné, A., Mathis, C., Stamm, A. and François, D. (1979b). Mis en gélules de solutions (ou suspensions) visqueuses de principes actifs. IV. Dissolution *in vitro*. *Labo-Pharma Probl. Tech.*, 292, 863–868.

Dennis, A. B., Farr, S. J., Kellaway, I. W., Taylor, G. and Davison, R. (1990). *In vivo* evaluation of rapid release and sustained release Gelucire capsule formulations. *Int. J. Pharm.*, 65, 85–100.

Dey, M., Enever, R., Kraml, M., Prue, D. G., Smith, D. and Weierstall, R. (1993). The dissolution and bioavailability of etodolac from capsules exposed to conditions of high relative humidity and temperatures. *Pharm. Res.*, 10, 1295–1300.

Dixon, F. K., Rowley, G., Sharkey, I. M., Hurst, J. and Coulthard, M. G. (1994). Liquid-fill hard gelatin capsule and pellet formulations of the hygroscopic drug cysteamine hydrochloride for the treatment of cystinosis in children. *Proceedings Pharmaceutical Technology Conference (Strasbourg)*, 13(1), 260–270.

Doelker, C., Doelker, E., Buri, P., Glas, B. and Waginaire, L. (1983). Incorporation des principes actifs liquides, delequescents ou instables dans des excipients pour gélules. *Proceedings 3rd International Conference on Pharmaceutical Technology*, APGI, Paris, May 31–June 2 (1983, pp.V: 73–81.

Dordunoo, S. K., Ford, J. L. and Rubinstein, M. H. (1991). Preformulation studies on solid dispersions containing triamterene or temazepam in polyethylene glycols or Gelucire 44/14 for liquid-filling of hard gelatin capsules. *Drug Dev. Ind. Pharm.*, 17, 1685–1713.

Dordunoo, S. K., Ford, J. L. and Rubinstein, M. H. (1997). Physical stability of solid dispersions containing triamterene or temazepam in polyethylene glycols. *J. Pharm. Pharmacol.*, 49, 390–396.

Ellison, M. J. H. (1997). *An Investigation of the Physicochemical and Rheological Properties of Drug–Gel Formulations in Relation to Drug Release Mechanisms*. PhD Thesis, University of Sunderland (UK).

Ellison, M. J. H. and Rowley, G. (1999). The effect of mixed hydrophilic and hydrophobic silicon dioxide on drug release from semi-solid matrices in hard gelatin capsules. *J. Pharm. Pharmacol.*, 51, 292.

Ellison, M. J. H., Rowley, G., Walters, P. A. and Barnes, D. M. H. (1995). Release of drugs from thixotropic triglyceride gel formulations liquid-filled into hard gelatin capsules. *Proceedings Pharmaceutical Technology Conference (Barcelona)*, 14(2), 91–98.

Ford, J. L. (1986). The current status of solid dispersions. *Pharm. Acta. Helv.*, 61, 69–88.

François, D. (1983), Flüssige und pastöse Füllgüter in Hartgelatinekapseln. In: Fahrig, W. and Hofer, U. (Eds.), *Die Kapsel, Grundlagen, Technologie und Biopharmazie einer modernen Arzneiform*. Stuttgart: Wissenschaftliche Verlagsgesellschaft, 112–126.

François, D. and Jones, B. E. (1979). The making of the hard capsule with the soft centre. *Manuf. Chem. Aerosol News*, 50(3), 37–41.

François, D., Dennet, A., Waugh, A. and Woodgate, T. (1982). The *in vitro* and *in vivo* availability of phenyl-propanolamine from oil/paste formulations in hard gelatin capsules. *Pharm. Ind.*, 44, 86–89.

Hawley, A. R. (1993). *Investigation of Thermosoftened Solid Dispersions Formulations of Ibuprofen for Hard Gelatin Capsules*. PhD Thesis, University of Sunderland (UK).

Hawley, A. R. and Rowley, G. (1998). Release mechanisms for drug polymer formulations in hard gelatin capsules. *Eur. J. Pharm. Sci.*, 6, S1, 11.

Hawley, A. R. and Rowley, G. (1999). Diffusion and erosion release mechanisms for drug/polymer formulations in hard gelatin capsules. *Proceedings Pharmaceutical Technology Conference (Utrecht)*, 18(2), 38–45.

Hawley, A. R., Rowley, G., Lough, W. J. and Chatham, S. M. (1992). Physical and chemical characterisation of thermosoftened bases for molten-filled hard gelatin capsule formulation. *Drug Dev. Ind. Pharm.*, 18(16), 1719–1739.

Howard, J. R. and Gould, P.L. (1987). Drug release of thermosetting fatty vehicles filled into hard gelatin capsules. *Drug Dev. Ind. Pharm.*, 13, 1031–1045.

Jones, B. E. (1987). Practical advances in two-piece gelatin capsules. *Manuf. Chem.*, 58(1), 27–31.

Kattige, A. (2001). *The Effect of Particulate Phase on Rheology and Capsule Filling of Poloxamer Dispersions*. PhD Thesis. University of Sunderland (UK).

Kattige, A. and Rowley, G. (1999). The effects of particulate dispersion on aeration and filling of semi-solid matrices in hard gelatin capsules. *J. Pharm. Pharmacol.*, 51, S-285.

Kattige, A., Rowley, G. and Walters, P. A. (2000). Particulate/polymer dispersions. The influence of rheological behaviour on liquid-filling characteristics for hard gelatin capsules. *Proceedings Pharmaceutical Technology Conference (Stresa-Baveno)*, 19(2), 147–155.

Kommuru, T. R., Gurley, B., Khan, M. A. and Reddy, I. K. (2001). Self-emulsifying drug delivery systems (SEDDS) of coenzyme Q_{10}: formulation development and bioavailability assessment. *Int. J. Pharm.*, 212, 233–246.

Kopcha, M., Lordi, N. G. and Tojo, K. J. (1991). Evaluation of release from selected thermosoftening vehicles. *J. Pharm. Pharmacol.*, 43, 382–387.

Kuentz, M. and Röthlisberger, D. (2002). Determination of the optimal amount of water in liquid-fill masses for hard gelatin capsules by means of texture analy-sis and experimental design. *Int. J. Pharm.*, 236, 145–152.

McTaggart, C., Wood, R., Bedford, K. and Walker, S. E. (1984). The evaluation of an automatic system for filling liquids into hard gelatin capsules. *J. Pharm. Pharmacol.*, 36, 119–121.

Mepha Ltd, Aesch/BL, Basel (1989). *Medicament with a delayed release of active ingredient*. US Patent 4 795 643.

Nagata, S. and Jones, B. E. (2000). Hard two-piece capsules and the control of drug delivery. *Eur. Pharm. Rev.*, 5(2), 41–46.

Nagata, S. and Tochio, S. (2002). Compatibility between liquid and semi-solid excipients and HPMC capsules. *3rd CRS International Meeting*, Seoul, July.

Nagata, S., Tochio, S., Sakuma, S. and Suzuki, Y. (2001). Dissolution profile of drugs filled into HPMC capsules and gelatin capsules. *AAPS PharmSci.*, 3(3), Poster.

Nazzal, S., Smalyukh, I. I., Lavrentovich, O. D. and Khan, M. A. (2002). Preparation and *in vitro* characterization of eutectic based semi-solid self-nanoemulsified drug delivery system (SNEDDS) of ubiquinone: mechanism and progress of emulsion formation. *Int. J. Pharm.*, 235, 247–265.

Ogura, T., Furuya, Y. and Matsuura, S. (1998). HPMC capsules – an alternative to gelatin. *Pharm. Technol. Europe*, 10(11), 32–42.

Pouton, C. W. (1997). Formulation of self-emulsifying systems. *Adv. Drug Delivery Rev.*, 25, 47–58.

Rowley, G., Hawley, A. R., Dobson, C. L. and Chatham, S. M. (1998). Rheology and filling characteristics of particulate dispersions in polymer melt formulations for liquid-fill hard gelatin capsules. *Drug Dev. Ind. Pharm.*, 24, 605–611.

Saeed, T., Rowley, G. and Walters, P. A. (1997). Rheological characteristics of poloxamers and poloxamer/silicon dioxide gels in relation to liquid-filling of hard gelatin capsules. *Proceedings Pharmaceutical Technology Conference (Athens)*, 16(2), 217–224.

Saeed, T., Rowley, G., Walters, P. A. and Bowtle, W. (1998). The effect of disperse phase particle size and continuous phase viscosity on the filling and dispersion stability of thixotropic gels for hard gelatin capsules. *Proceedings Pharmaceutical Technology Conference (Dublin)*, 17(1), 98–106.

Serajuddin, A. T. M. (1999). Solid dispersions of poorly water-soluble drugs: Early promises, subsequent problems, recent breakthroughs. *J. Pharm. Sci.*, 88, 1058–1066.

Seta, Y. (1988). Design of captopril sustained-release preparations and their biopharmaceutical properties. *Int. J. Pharm.*, 41, 263–269.

Seth, P. (1992). The development of extended release semi-solid matrix (SSM) hard gelatin capsules. *Proceedings Capsugel Symposium on Hard Gelatin Capsules: Controlled release formulations in hard gelatin capsules.* London, 28–31.

Smith, A., Lampard, J. F., Carruthers, K. M. and Regan, P. (1990). The filling of molten ibuprofen into hard gelatin capsules. *Int. J. Pharm.*, 59, 115–119.

Sutananta, W., Craig, D. Q. M. and Newton, J. M. (1994). The effects of ageing on the thermal behaviour and mechanical properties of pharmaceutical glycerides. *Int. J. Pharm.*, 111, 51–62.

Sutananta, W., Craig, D. Q. M. and Newton, J. M. (1995a). An evaluation of the mechanisms of drug release from glyceride bases. *J. Pharm. Pharmacol.*, 47, 182–187.

Sutananta, W., Craig, D. Q. M. and Newton, J. M. (1995b). An investigation into the effects of preparation conditions and storage on the rate of drug release from pharmaceutical glyceride base. *J. Pharm. Pharmacol.*, 47, 355–359.

Tochio, S. and Nagata, S. (2001). Study on optimal sealing method of HPMC capsules. *AAPS PharmSci.*, 3(No. 3), Poster.

Utting, A. and Rowley, G. (1999). Drug/poloxamer solid dispersion formulations for liquid-fill hard gelatin capsules. *J. Pharm. Pharmacol.*, 51(suppl.), 187.

Vial-Bernasconi, A.-C., Buri, P., Doelker, E., Beyssac, E., Duchaix, E. and Aiche, J.-M. (1995). *In vivo* evaluation of an indomethacin monolithic, extended zero-order release hard gelatin capsule formulation based on saturated polyglycolysed glycerides. *Pharm. Acta. Helv.*, 70, 307–313.

Walker, S. E., Bedford, K. and Eaves. T. (1980a). *Improvements in and relating to pharmaceutical preparations in solid unit dosage form.* British Patent 157 222 630.

Walker, S. E., Ganley, J. A., Bedford, K. and Eaves, T. (1980b). The filling of molten and thixotropic formulations into hard gelatin capsules. *J. Pharm. Pharmacol.*, 32, 389–393.

Walters, P. A. (1994). *A Rheological Approach to the Formulation and Drug Release of Thixotropic Gels or Hard Gelatin Capsules.* PhD Thesis, University of Sunderland (UK).

Walters, P. A. and Rowley, G. (1999). Moisture uptake of excipients for liquid-filling into hard gelatin capsules. *Proceedings Pharmaceutical Technology Conference (Utrecht)*, 18(1), 97–101.

Walters, P. A., Rowley, G., Pearson, J. T. and Taylor, C. J. (1992a). Formulation and physical properties of thixotropic gels for hard gelatin capsules. *Drug Dev. Ind. Pharm.*, 18, 1613–1631.

Walters, P. A., Rowley, G., Pearson, J. T. and Taylor, C. J. (1992b). *In vitro* drug release control by colloidal silicon dioxide in a semi-solid matrix hard gelatin capsule. *Proceedings International Symposium on Controlled Release of Bioactive Matererials (Amsterdam)*, 18, 65.

Walters, P. A., Rowley, G. and Ellison, M. J. H. (1998). Moisture uptake for vehicles for liquid-filling into hard gelatin capsules. *Eur. J. Pharm. Sci.*, 6, S1, 11.

Wittwer, F. (1985). New developments in hermetic sealing of hard gelatin capsules. *Pharm. Manuf.*, 2, 24–27.

10

Technology to manufacture soft capsules

Fridrun Podczeck

Introduction

Soft capsules are single unit dosage forms consisting of a flexible shell and normally a semi-liquid or liquid filling. They can be administered via various routes, although the oral route is the most common one. Soft capsules can contain fills up to 20 mL in volume (see Table 10.1) and their shape varies between round and oval, for oral administration, to torpedo or tube form, when used as a suppository or a container for topic formulations. In most instances, soft capsules will contain the active in the fill. However, for oral administration by sucking, the active drug can also be contained in the outer shell (Gelsolets), which is in these cases extremely thick.

One main difference between hard and soft capsules lies in the manufacturing sequence. Soft capsules are produced in one step, i.e. the container and fill are made and combined on one and the same process line, whereas for hard capsules, the shells are produced separately by different manufacturers. While the pharmaceutical and other industries are mainly concerned with the correct storage of the empty shells and compatibility of the shells with the fills, for the development and manufacture of hard capsules, the manufacturers of soft capsules must have a detailed knowledge about the physical properties of both shells and fill. These are discussed in detail in Chapter 11, while this chapter will only introduce some technical aspects of the manufacture of soft capsules.

Manufacturing methods

Globex method

This method is principally different from all other methods that are or have been used to manufacture soft capsules. The process is outlined in Figure 10.1. The lipophilic fill and the shell formulation are stored in separate containers,

Table 10.1 Typical shapes, fill volumes and applications of soft capsules

Shape	Typical fill volume (mL)	Application	Example
Round	0.05–6 (mainly small volumes)	Orally typically 0.15–0.3 mL	Orally, to be swallowed (Adalat, Targretin)
Oval	0.05–7 (mainly larger volumes)	Orally typically 0.1–0.5 mL	Orally, to be bitten (Nitrangin Isis, Gepan)
Oblong	0.1–20	Orally typically 0.3–0.8 mL	Orally, to be sucked (Mebucalets F Gelsolets, Orofar Gelsolets)
Torpedo-shaped	0.1–20	Rectally Vaginally	Emesan E, Klismacort Polygynax, Fossyol
Tube-like	0.1–30 (mainly larger volumes)	Topically (content only)	

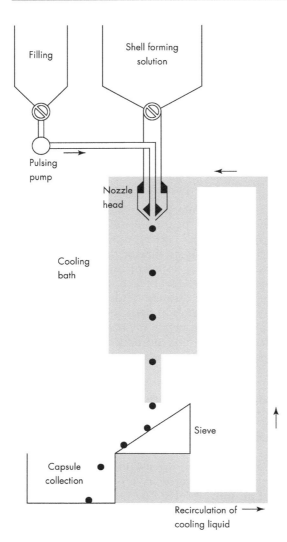

Figure 10.1 Process outline of the Globex method.

surface tension. In an attempt to reduce its surface tension, droplets are formed, which enclose the normally lipophilic fill. As a result, seamless round soft capsules are formed. The end of the double capillary is immersed in a cooling bath containing a liquid, which is not miscible with the shell material and kept at about 4°C. Liquid paraffin is the most common liquid used. The cooling bath ensures an immediate sol–gel transformation of the shell material and hence the formation of a flexible, yet firm and robust outer film. The soft capsules are finally collected, washed with organic solvents to remove the residues of the cooling bath liquid and gently dried at a relative humidity of 20% in infrared tunnels.

One advantage of the Globex method is the production of seamless capsules, which are tamper-evident and free of contamination or entrapped air, and the risk of leakage is minimal. For the Globex Mark III encapsulator, a maximum variability in weight of 1.5% is quoted by the manufacturer (ITS Machinery Development Ltd, UK). The process is less expensive as there is no waste of shell material and reduced maintenance costs because there are few moving parts. The process can run for hours unattended and product changeover times are fairly short. There are no requirements of the production facilities other than prescribed by GMP (good manufacturing practice), i.e. even in temperate climates no air conditioning is required. However, the restrictions imposed by the physical principle of droplet formation, i.e. the enforced round capsule shape and the requirements on optimisation of the surface tension between shell and filling material are often considered to be a main disadvantage of the process. The production rates vary with capsule size and fill material and range between 10 000 and 40 000 capsules per hour. This is considerably slower than the rotary die method (see below).

Rotary die method

The rotary die method was developed in 1930 by Robert Pauli Scherer, who founded the R. P. Scherer Corporation in 1933 and transformed the production of soft capsules (then soft gelatin

both kept warm, i.e. in the liquid/sol state. The fill and the shell material are pumped through a concentric double capillary, with the fill in the inner one. The fill is pumped in pulses to give fill volumes between 20 and 600 mg. Owing to the pulsing action, the shell material, which runs continuously through the outer capillary, is constricted after each pulse, resulting in the separation of individual dose units. The formation of the capsules depends on the interfacial tension between fill and shell material. The shell material is usually hydrophilic and possesses a higher capsules or 'softgels') into a commercial process that is widely used today in the health food and nutritional supplements manufacturing industry and to some extent in the production of medicines.

The R. P. Scherer Corporation is now part of 'Cardinal Health', which in 2002 owned 15 manufacturing facilities in 11 countries, employing about 3600 people (Woodward Heritage, 2002). That the manufacture of soft capsules by this method has not been taken up to a larger extent by the pharmaceutical industry is likely to be the result of R. P. Scherer's policy of withholding information about the development and composition of shell and fill formulations. In fact, for a long time R. P. Scherer insisted on developing the formulations themselves, which is in conflict with the policy of the pharmaceutical industry not to disclose information about their active substances to third parties. In connection with this, contract manufacture would have been required, which was also not normally sought by many pharmaceutical companies in the past. However, since the patents expired, other contract manufacturers have begun to offer the development and manufacture of soft capsules (e.g. Banner Pharmacaps Inc., Eurocaps, Ltd, Swiss Caps AG). In addition, Swiss Caps AG produces soft gelatin encapsulation machinery to enable pharmaceutical and healthcare companies to produce soft capsules in their own facilities. To date, they offer three models (SGM 1000, 1010 and 2000), which vary in the die roll dimensions and the number of baskets in the drying units. The output increases from SGM 1000 up to the 2000 machine. The company has also modified the casting system (see below) to enable the production of soft capsules from gelatin-free raw materials. The so-called VegaGels are mainly made from thermoplastic potato starch, plasticizers (sorbitol, maltitol or glycerol) and anti-caking agent (glyceryl monostearate) and water. A set of further optional ingredients is available (Swiss Caps, 2002).

To produce soft capsules by the rotary die method, the shell material is converted into the liquid state in jacketed tanks. For gelatin, the usual temperature is between 60 and 80°C and the dissolving of the gelatin and mixing with plasticizer and other additives is performed under vacuum to prevent the formation of air bubbles. The tank is connected to the encapsulation machinery via transfer pipes. The outline of the encapsulation machinery is shown in Figure 10.2.

Two separate spreader boxes, situated to the left and right of the encapsulation unit, above cooling drums, are fed with the shell-forming material from the tanks. With the help of a casting drum rotating parallel to a precisely controlled gap between spreader box and cooling drum, shell (e.g. gelatin) ribbons are formed. These are about 150 mm wide and normally about 800 µm thick. The thickness can be varied for special applications, for example, if particularly thin (capsules to be opened by biting) or thick shells (Gelsolets) are needed. Owing to the casting onto the cooling drums, the shell material passes from the sol into the gel state and is characterised by an elastic, rubbery consistency. In the VegaGels system, the spreader boxes are replaced by an extrusion system in order to accommodate the properties of the thermoplastic potato starch mixtures. After the ribbons have been formed, they are lifted off the cooling drums and are lubricated on either side with liquid paraffin or vegetable oil with the aid of felt rollers. The ribbons are then guided over counter-rotating rolls containing sharp-edged dies. Each die cavity has the size and shape of half of the capsule to be formed. The temperature of the die rolls is kept between 36 and 39°C so that the shell ribbons behave highly elastically, yet do not melt.

The die rolls are pressed together with a defined pressure and, as a result, the two passing ribbons can be sealed and cut along the slightly raised rims of the dies. Initially, only the lower part and sides of the capsules are sealed and the two ribbons form sack-like structures. The sacks are filled via a wedge with fill material. The wedge itself can be heated to ensure an equilibrium in temperature between shell ribbons and fill and to improve sealing. The amount of fill injected into each sack is controlled by a dosing pump with a precision typical for liquid dosing systems, i.e. 1–3%. The amount of fill injected results in a widening of the sack-like structures to accommodate the volume so that there is no room for air to be entrapped. As the shell ribbons proceed to

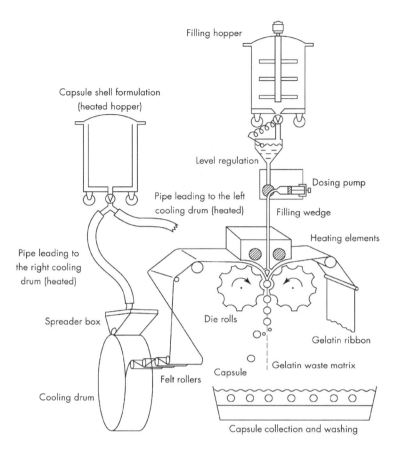

Figure 10.2   Outline of the rotary die method (modified from Bauer, 1983, with permission).

pass between the die rolls, the capsules are now fully sealed and a continuous seam will have been formed. Capsules can be under- or overfilled by this method, whereby overfilling, i.e. stretching of the sacks to fill completely the die cavity is preferred. Underfilling can result in uneven shape of the capsules.

The wedge is constructed so that the fill material is delivered sideways into the forming capsule (Figure 10.3) to avoid contamination of the upper seam, which could result in leakage. The bores in the wedge are fairly narrow and hence suspension fills should contain only fine particles to prevent blockage.

After the shell-forming ribbons have passed through the die rolls, the capsules are separated from the remaining matrix, which, owing to the outer film of lubricant, cannot be recycled to

form ribbons. Instead, the matrix has to be treated as waste. The capsules have to be washed with organic liquids to remove the lubricant from their outer surface. They are then placed into rotating baskets and predried at a temperature between 20–30°C until they are more or less stable in shape. During this step, about 50% of the water content of the shell is removed. The final drying takes place in tunnel dryers at a relative humidity of 20%. This may take several days or even up to two weeks, depending on the formulation of the shell and the fill.

One major advantage of the rotary die method is that the capsules produced can have all kinds of shape or sizes and they can be made with different colours for each half. The formation of the capsules is not governed by surface tension phenomena and hence there are no restrictions

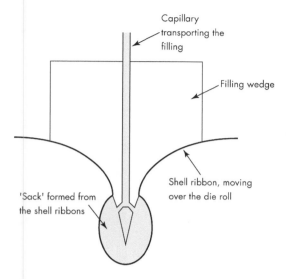

Capillary transporting the filling

Filling wedge

Shell ribbon, moving over the die roll

'Sack' formed from the shell ribbons

Figure 10.3 Wedge construction (the material is emptied to the sides well above the tip of the wedge; the shell ribbons 'wipe' the wedge tip clean and hence contamination of the sealing area by the tip is prevented).

on the fills other than compatibility/incompatibility with the shell. Hence, a wider variety of fills than is possible in the Globex method can be encapsulated. In addition to the prevention of air entrapment, there is the possibility of degassing with nitrogen during filling, which can enhance the shelf life of oxygen-sensitive medicaments. Up to 100 000 capsules per hour can be produced. However, the cost advantage gained by a higher output might be reduced because of the large amount of shell waste material produced and the considerably longer drying times when compared to the Globex method. There is also a large risk of leakage through imperfect seams and the machinery requires more maintenance. Thus idle times for maintenance and changeover times are clearly longer.

Compared with hard capsules, the flexibility in choosing fill materials is reduced in soft capsules. The filling of solid dosage forms such as pellets or tablets, either singly or in combinations, is currently impossible. However, Banner Pharmacaps have a patent for what they describe as a method that 'enrobes standard tablets into gelatin-enrobed caplets' (Banner, 2002). The method (Soflet Gelcaps) provides individual tablets with a coat of gelatin, which can be two-coloured. It is an aesthetic rather than a modified-release measure for oral solid dosage forms but satisfies a need in the American market for such over-the-counter tablet products.

Other methods

Other methods such as the Norton, Colton, Upjohn and Accogel encapsulation processes are nowadays rarely employed. A detailed description of these processes can be found in Bauer (1983).

References

Banner (2002). Electronic information material. http://www.banpharm.com/ (accessed 11 February 2003).

Bauer, K. H. (1983). Die Herstellung von Hart- und Weichgelatinekapseln. In: Fahrig, W. and Hofer, U. (Eds.), *Die Kapsel: Grundlagen, Technologie und Biopharmazie einer modernen Arzneiform*. Stuttgart: Wissenschaftliche Verlagsgesellschaft, 58–82.

SwissCaps (2002). Electronic information material. http://www.swisscaps.com/data/gif/VegaGelsTech.pdf (accessed 11 February 2003).

Woodward Heritage (2002). Historical Sites. http://www.woodwardheritage.com/ (accessed 11 February 2003).

11

Formulation and physical properties of soft capsules

Gabriele Reich

Introduction

Soft capsules are a single-unit solid dosage form, consisting of a liquid or semi-solid fill enveloped by a one-piece hermetically sealed elastic outer shell. They are formed, filled and sealed in one continuous operation, preferably by the rotary die process. Depending on the polymer forming the shell, they can be subdivided into two categories, namely soft gelatin capsules or 'softgels' and non-gelatin soft capsules. The majority of soft capsules are made from gelatin owing to its unique physical properties that make it an ideal excipient for the rotary die process. Soft capsules based on plant-derived and/or synthetic non-gelatin alternatives have, however, been patented and a few prototype products have recently entered the market. Formulation and physical properties of both soft capsule categories will be discussed.

Soft gelatin capsules

General aspects

Originally developed in the 19th century to mask unpleasant taste and odour of drug substances, soft gelatin capsules are used in many applications, for pharmaceutical and health and nutrition products, cosmetic applications and even recreational products such as paint balls.

In the pharmaceutical field soft gelatin capsules are increasingly being chosen for *strategic reasons* (line extension), *technological issues* (high content uniformity of low-dose drugs), *safety aspects* (reduced operator and environmental contamination with highly potent or cytotoxic compounds) and *consumer preference* (easy to swallow). The most interesting advances have recently been made in the area of developing liquid and semi-solid formulations in a soft gelatin capsule to address *particular bioperformance issues*, namely increased bioavailability and decreased plasma variability by improved solubility and absorption-enhancing techniques.

The proper design for a specific soft gelatin capsule formulation requires the appropriate selection of shell and fill composition and the optimisation of the two to allow for the efficient production of a chemically and physically stable product with the desired biopharmaceutical properties.

Composition of the capsule shell

The shell of a soft gelatin capsule is composed of gelatin, a plasticizer or a combination of plasticizers and water. In addition, it may contain preservatives, colouring and opacifying agents, flavourings and sweeteners, possibly sugars to impart chewable characteristics to the shell, gastroresistant substances and in special cases even active compounds.

The water serves as a solvent to make a molten gelatin mass with a pourable viscosity at 60–70°C. The ratio by weight of water to dry gelatin (W/G) can vary from 0.7 to 1.3, depending on the viscosity of the gelatin being used. After capsule formation, most of the water is removed by drying, leading to finished capsules with a moisture content of 4–10%.

Gelatin

The gelatins used for pharmaceutical or health and nutrition soft capsule products are described by the official pharmacopoeias such as USP (United States Pharmacopoeia), PhEur (European Pharmacopoeia) etc., or approved by local authorities, with additional physicochemical specifications (Babel, 2000). The specifications and controls for gelatins are discussed in Chapter 2.

For soft capsule production, the pharmacopoeial specifications generally represent the minimum requirements. Capsule manufacturers' specifications are more detailed and stringent with respect to the performance-related physicochemical properties of the gelatins (Reich and Babel, 2001). This is due to the fact that these parameters are critical for economic soft capsule production by the rotary die process and for the quality of the final product. Gelatin types and grades that are adequate for continuous commercial soft capsule production require the ability to set at a fast rate to ribbons of defined thickness and reproducible microstructure and to produce films with a mechanical strength and elasticity sufficient to survive all the manipulations on the encapsulation machine, i.e. to allow the wet films to be easily removed from the drums, to stretch during filling, to be sealed at temperatures below the melting point of the film and to be dried quickly under ambient conditions to a specified capsule strength. Moreover, the dissolution characteristics of the resulting capsules have to fulfil the pharmacopoeial requirements.

Considering these aspects, the technologically relevant gelatin parameters are gel strength, viscosity at 60°C and $6\frac{2}{3}$% w/w concentration in water, viscosity breakdown (the impact of temperature and time on the degradation of gelatin), melting point, setting point, setting time, particle size and molecular weight distribution. A perfect soft capsule gelatin should have the following specifications:

- Gel strength: 150–200 Bloom, depending on the gelatin type;
- Viscosity (60°C/$6\frac{2}{3}$% w/w in water): 2.8–4.5 mPa s, depending on the gelatin type;
- Well-controlled degree of viscosity breakdown;
- Well-defined particle size to allow fast dissolution and deaeration of the molten mass, even at high gelatin concentrations;
- A broad molecular weight distribution to provide a fast setting and the fusion temperature being well below the melting temperature of the plasticized wet film.

The main gelatin types and grades used for the manufacture of soft capsules are listed in Table 11.1 together with their physicochemical specifications.

The proper choice of the gelatin type and grade is related to technological issues, consumer preference and pricing. For pharmaceutical or health and nutrition products, medium bloom limed bone (LB) gelatins, or blends of limed bone and pigskin (LB/PS) or limed bone, pigskin and limed hide gelatin (LB/LH/PS) are commonly used, with a certain preference for LB gelatin in the United States and for blended gelatins in

Table 11.1   Physicochemical properties of pharmacopoeial-grade soft capsule gelatins

Gelatin	Raw material	Type	Bloom (g) (10°C; $6\frac{2}{3}$% w/w)	Viscosity (mPa s) (60°C; $6\frac{2}{3}$% w/w)
160 LB (= limed bone)	Bovine/porcine bone	B	155–185	3.4–4.2
160 LH (= limed hide)	Bovine hide	B	150–170	3.5–4.2
160 LB/LH	Blend of bovine/porcine bone and bovine hide	B	150–170	3.5–4.2
200 AB (= acid bone)	Bovine bone	A	180–210	2.7–3.2
200 PS (= pigskin)	Pigskin	A	190–210	2.5–3.1
160 PS/LB/LH	Blend of pigskin, bovine/ porcine bone and bovine hide	A/B	145–175	2.7–3.3

Europe. Low-viscosity, high-bloom gelatins such as a 200 Bloom pigskin (PS) or acid bone (AB) gelatin are often used for the encapsulation of hygroscopic formulations and/or water-sensitive drugs, where standard gelatin formulations have to be modified to contain less water and dry faster, thus improving the product stability during capsule manufacturing. Mixtures of low (<100 Bloom) and medium Bloom (>150 Bloom) gelatins have been proposed for the formulation of chewable soft capsules (Overholt, 2001) to achieve the desired mouthfeel and solubility of the shells, a low stickiness for improved machin-ability and sufficient integrity for stable fill encapsulation.

In addition to the pharmacopoeial grade gelatin types listed in Table 11.1, succinylated pigskin gelatin (Bloom: 190–210 g; viscosity: 3.3–4.1 mPa s) is often used for products with reactive fill ingredients, such as aldehydes, to prevent cross-linking of the shell. Gelatins derived from poultry, fish or other sources have recently been proposed in the patent literature as alternatives to gelatin of bovine and porcine origin. Poultry and fish gelatins have recently been approved by PhEur.

From a technological point of view, poultry gelatin is a potential alternative to the con-ventional soft capsule gelatins derived from bovine and porcine origin, since its physico-chemical properties are comparable to those of pigskin gelatins. In practice, it has not gained high commercial interest yet because its avail-ability is limited.

The use of cold- or warm-water fish gelatins for soft capsule production is limited by the fact that their gelling, setting and drying properties are more or less different to those of mammalian gelatins. Owing to their low degree of proline hydroxylation, cold-water fish gelatins lack the gelling and setting attributes that are required to allow their use in the conventional rotary die process. Although addition of a setting system, such as carrageenan, has been described to enable the adaptation of specific and desired gelling properties (Scott *et al.*, 1997), this approach has not yet reached commercial status. Only warm-water fish gelatins with a somewhat higher degree of proline hydroxylation, and thus an intrinsic gelling and setting ability sufficient for conventional soft capsule production, have been used for a small number of products. Acceptable soft capsules can be produced by adjusting the formulation and process parameters, such as the production speed in accordance to the reduced setting rates, the mechanical properties and the drying characteristics of this gelatin type.

The use of plant-derived genetically engineered gelatins for soft capsule production is not practi-cable. This is mainly due to technological issues, supply problems, high costs and, for pharma-ceutical products, the regulatory issues. Only small amounts of gelatins, with gelling and setting properties and mechanical characteristics different to mammalian gelatins, are available at a multiple of the price of conventional gelatins.

Plasticizers

The formation of a soft capsule requires the use of a non-volatile plasticizer in addition to water to guarantee the mechanical stability, i.e. the elas-ticity of the capsule shells during and after the drying process. The additional plasticizer has to counterbalance the stresses that are induced in the shrinking capsule shells, as the plasticizing effect of water in the shells decreases upon drying.

Practically, only a few plasticizers are in use, namely polyalcohols, which are approved by the official pharmacopoeias or by local regulatory authorities. Glycerol (85% and 98% w/w), special grades of non-crystallising aqueous sorbitol and sorbitol/sorbitan solutions and combinations of these are the most used. In addition, propylene glycol and low molecular weight polyethylene glycol (PEG 200) have been used.

The type and concentration of plasticizer(s) in the shell is related to the composition of the fill, i.e. the possible interactions with the fill, the capsule size and shape, the end use of the product and the anticipated storage conditions. The ratio by weight of dry plasticizer to dry gelatin (P/G) determines the strength of the shell and usually varies between 0.3 and 1.0. The choice of the proper shell formula with respect to the gelatin/ plasticizer combination is crucial to the physical stability of the capsules, during manufacture and on storage. A rational shell design therefore requires analytical tools that allow the perform-ance-related test parameters to be assessed.

Differential scanning calorimetry (DSC) and dynamic mechanical thermal analysis (DMTA) have been reported as suitable methods to monitor phase transitions and elastic moduli indicating molecular gelatin/plasticizer interactions and their effect on shell elasticity, i.e. to evaluate plasticizer effectivity and compatibility (Reich, 1983, 1994).

An ideal plasticizer should interact with the gelatin molecules in such a way as to reduce effectively the glass transition temperature (T_g) of the gelatin shell without inhibiting the formation of crystallites that stabilise the three-dimensional gel network structure. In addition, if present in concentrations higher than saturation, it should be physically embedded in the sol phase of the gel network to avoid bleeding out (Reich, 1994).

Glycerol, the most frequently used soft gelatin capsule plasticizer, combines these advantages of a high plasticizer effectivity, a sufficient compatibility and a low volatility with the ability to interact specifically with the gelatin allowing for the formation of a stable thermoreversible gel network. Its plasticizing capability is mainly resulting from *direct* interactions with the gelatin and only slightly from its hygroscopicity allowing for an additional indirect moisturising effect (Reich, 1994).

Sorbitol, on the other hand, is an *indirect* plasticizer, mainly acting as a moisturising agent with water being the effective plasticizer. Compared to glycerol, its direct plasticizing capability is very much reduced, as indicated by a minor reduction of the gelatin glass transition temperature. Gradual differences of various grades of non-crystallising sorbitol solutions in their plasticizing capability and their compatibility with gelatin are the result of differences in the amount of by-products, namely hydrogenated oligosaccharides and sorbitol anhydrides, i.e. sorbitans (Reich, 1996). Only sorbitol grades with a high amount of sorbitans, such as Anidrisorb, can effectively replace glycerol owing to a certain direct plasticizing effect. On the other hand, hydrogenated oligosaccharides such as maltitol in combination with glycerol are very effective additives for the formulation of chewable soft gelatin capsules, since they augment the taste and chewability and assist in the rapid dissolution of the shell upon chewing, thus improving the mouthfeel (Berry *et al.*, 1988; Montes and Steele, 1999).

Regarding plasticizing capability, propylene glycol is superior to sorbitol/sorbitan blends and even to glycerol. However, owing to its high solvent power for gelatin, it has a slightly negative effect on the formation of the gel structure that has to be compensated for by adjusting the manufacturing parameters at the encapsulation machine (Reich, 1994). Liquid polyethylene glycols can only be used in combination with glycerol or propylene glycol, since their compatibility with gelatin is limited.

Other additives

In addition to gelatin, the plasticizer(s) and water, optional components in the capsule shell are limited in their use. For economic reasons, the addition of active ingredients in the shell is usually not recommended and limited to inexpensive compounds in chewable capsules. The use of water-insoluble polymers to impart sustained-release characteristics to the capsules has failed owing to their limited compatibility with the gelatin mass (Reich, 1983). Formulations with gastroresistant enteric soluble polymers are under development.

Colouring and opacifying agents are used frequently to give the shells the desired colour and a proper finish, i.e. to allow the shell to protect the fill from light and to mask the unpleasant look of the fill. As a general rule, the colour of the capsule shell is chosen to be darker than the colour of the fill. Before a colour is chosen, mixtures should be checked to ascertain that fading or darkening of the capsule shells does not occur on storage, as a result of reactions between the colouring agent and other components of the shell or fill.

Fill compositions

Soft gelatin capsules can be used to dispense active compounds that are formulated as a liquid or semi-solid solution, suspension or microemulsion preconcentrate. The formulation of the fill is individually developed to fulfil the following requirements:

- to optimise the chemical stability of the active compound
- to improve bioavailability of the active compound
- to allow for an efficient and safe filling process
- to achieve a physically stable capsule product.

Final product stability is related to shell compatibility and will be discussed later.

For a soft gelatin capsule-filling operation, the technologically important factors to bear in mind are temperature, viscosity and surface activity of the fill material and, in the case of suspensions, the particle size of the suspended drug. Liquids or combinations of liquids for encapsulation must possess a viscosity sufficient to be precisely dosed by displacement pumps at a temperature of 35°C or below and may not show stringing to allow for a clean break from the dosing nozzle. The temperature specification is necessary owing to the sealing conditions, which are usually in the range of 37 to 40°C. Owing to certain tolerances of the encapsulation equipment, suspended solids should have a particle size below 200 µm to ensure maximum homogeneity of the suspension. Moreover, the surface-active properties of the fill, whether a solution or a suspension, may not interfere with the formation of the seals.

Interestingly, soft gelatin capsule fill formulations have changed over time from basic lipophilic to hydrophilic solutions or suspensions and recently to more complex self-emulsifying systems. The reason for these developments is that new chemical entities (NCEs) present increasing biopharmaceutical formulation demands.

Basic lipophilic solutions or suspensions have been the traditional and most frequently used soft gelatin capsule fill formulations in the past. They have been applied successfully to formulate oily and lipophilic low melting point drugs such as the vitamins A, D and E, drugs with unpleasant taste and/or odour such as the vitamins of the B group or herbal extracts, drugs with critical stability, i.e. oxygen- and light-sensitive drugs and low-dose or highly potent drugs. The vehicles used for this purpose are lipophilic liquids and semi-solids, and the optional use of surfactants. The lipophilic liquids are refined speciality oils such as soya bean oil, castor oil etc. and/or medium chain triglycerides (MCT). The semi-solids, acting as viscosity

modifier for the liquid oils, are hydrogenated speciality oils or waxes, such as hydrogenated castor oil or bees wax. Surfactants such as lecithin may be present to improve the dispersion of suspended drug particles, thus improving content uniformity. Antioxidants are usually added to stabilise oxygen-sensitive drugs. Moreover, impregnation of solid polymer particles with the drug or drug coating prior to suspension in the oil formulation has been reported as a successful means to improve the content uniformity of low-dose suspended drugs and further increase chemical stability of sensitive drugs. Examples are vitamin B12 (Sanc *et al.*, 2000) and retinol (Rinaldi *et al.*, 1999).

Hydrophilic soft gelatin capsule fill formulations are based on polyethylene glycols (PEGs). Low molecular weight polyethylene glycols are usually used for liquid solutions, with PEG 400 and PEG 600 being the most frequently used. For the formulation of semi-solid solutions and suspensions, the low molecular weight polyethylene glycols (PEG 300–600) are mixed with high molecular weight solid polyethylene glycols, such as PEG 4000–10 000, to increase the viscosity.

PEG-based formulations are often chosen to address bioavailability concerns, i.e. to improve the solubility of poorly soluble drugs, or to dispense low-dose and/or high-potency drugs. Digoxin (Gardella and Kesler, 1977; Ghirardi *et al.*, 1977), nifedipin (Radivojevich *et al.*, 1983), temazepam (Brox, 1983) and ibuprofen (Gullapalli, 2001) are active compounds that have been successfully formulated as PEG solutions in soft gelatin capsules.

Complex self-emulsifying lipophilic systems and microemulsion preconcentrates are additional approaches that have gained increasing interest as soft gelatin capsule fill formulations to increase the bioavailability and/or reduce the plasma variability of poorly soluble and/or poorly absorbed drugs (Charman *et al.*, 1992; Amemiya *et al.*, 1998, 1999). These systems are typically composed of a lipophilic solvent and surfactant(s), and optional use of co-solvent(s) and/or co-surfactant(s), and may exert solubilising and absorption-enhancing effects. On contact with aqueous gastrointestinal fluids, these formulations spontaneously produce an emulsion with an average droplet size of less than 100 nm, thus improving drug delivery.

Active compounds that have been successfully formulated as a microemulsion preconcentrate in soft gelatin capsules are ciclosporin and the protease inhibitor ritonavir. A patent has also been filed for ibuprofen (Rouffer, 2001). Examples of microemulsion pre-concentrate soft gelatin capsule fill formulations are given in Table 11.2, indicating the use of hydrophilic co-solvents such as ethanol and propylene glycol.

Formulation strategies

Soft gelatin capsule formulation strategies have to consider the specific shell/fill interactions that may occur during manufacture, drying and on storage and control their rate and extent to achieve a stable product.

Two major types of interactions have to be distinguished:

• Chemical reactions of fill components with the gelatin and the plasticizer
• Physical interactions, i.e. migration of fill components in or through the shell and vice versa.

Cross-linking of gelatin leading to solubility problems of the shell is a well-known problem associated with the encapsulation of drugs containing reactive groups such as the aldehyde group. It can be successfully reduced by the use of succinylated gelatin, an approach that is often used for health and nutrition products, and in some countries even for pharmaceutical products. Esterification and transesterification of drugs with polyols present another unwanted chemical reaction that may occur. Since glycerol is more reactive than other polyols, glycerol-free shell formulations and/or the addition of polyvinyl pyrrolidone to the fill (Gullapalli, 2001) are preferred to reduce this problem.

The rate and extent of physical shell/fill interactions depend strongly on the qualitative and quantitative composition of both, the shell and the fill. As a general rule, the water content of the fill should not exceed a critical value of about 5%.

Fill formulations simply composed of a lipophilic drug in a lipophilic oily vehicle do not interact with the hydrophilic gelatin capsule shell at any time, i.e. either during production or on storage. The proper choice of the shell composition therefore only depends on the stability of the active ingredient, the capsule size, shape and end use and the anticipated storage conditions. For very soft capsules and those stored at ambient conditions, glycerol is the plasticizer of choice. For more rigid soft gelatin capsules and those intended to be used in hot and humid climates, glycerol/sorbitol blends are preferred. The latter is also valid for soft capsules containing oxygen-sensitive compounds in the fill (Hom et al., 1975; Meinzer, 1988). In any case, the P/G ratio is adjusted to the size and intended use of the capsules. To obtain light protection, the shell

Table 11.2   Examples of microemulsion preconcentrate compositions for soft gelatin capsules

Active ingredient	Fill excipients
Ciclosporin	Ethanol
	Propylene glycol
	Mono-, di-, triglycerides from corn (maize) oil
	Polyoxyethylene (40) hydrogenated castor oil
	DL-alpha-tocopherol
Ritonavir	Ethanol
	Propylene glycol
	Polyoxyethylene (35) castor oil
	Polysorbate 80
	Polyoxyethylene/glyceryl mono-, di-, tri-alcanoate (C8–C18)
	Medium chain triglycerides
	Citric acid

can be formulated with pigments such as titanium dioxide and/or iron oxides.

Compared to lipophilic solutions, fill compositions with hydrophilic components are more challenging to encapsulate, since they are prone to interact with the shell (Armstrong *et al.*, 1984, 1985, 1986). The most critical period for diffusional exchanges between shell and fill is the manufacturing process, since the moisture content of the initial shells before drying is around 40% and the equilibrium moisture level is only reached after several days. Thus, during manufacture and drying, hydrophilic components of the fill may migrate rapidly into the shell and vice versa, thereby changing the initial composition of both, the shell and the fill (Serajuddin *et al.*, 1986). On storage, these processes may continue until equilibrium is reached. As a result, the capsule shells can become brittle or tacky and the fill formulation may be deteriorated, either shortly after production or on storage. To guarantee the stability of the final product, the initial composition of shell and fill has to be designed in such a way as to minimise exchange processes. Several approaches to demonstrate the proof of this concept will be discussed as follows.

Hydrophilic and/or hygroscopic drug particles suspended in an oily vehicle may attract and retain water from the shell and/or migrate themselves into the shell. This can lead to stability problems such as hydrolysis or oxidation of the active ingredient, to assay failure and/or shell discoloration. To overcome these problems, the following solutions have been proposed:

- Use of high-Bloom, low-viscosity pigskin or acid bone gelatin to reduce the initial water content in the capsule shell and accelerate the drying process;
- Replacement of glycerol by glycerol/sorbitol or sorbitol/sorbitan blends to minimise diffusion of glycerol-soluble active ingredients into the shell;
- Coating of drug particles to inhibit the browning reaction between active ingredients, such as ascorbic acid and gelatin (Oppenheim and Truong, 2002).

Considerable difficulties have been encountered with the design of physically stable and durable soft capsules containing liquid polyethylene glycols (PEG 300–600) as the fill vehicle. This is owing to the fact that polyethylene glycols have a high affinity for water, glycerol and even gelatin, i.e. they have a high tendency to attract water and glycerol from the shell and may migrate to a certain extent into the shell. As a result of these processes, capsules may become brittle shortly after production or on storage, especially when exposed to cold temperatures (Shah *et al.*, 1992). Several approaches have been reported in the patent literature to provide PEG-containing soft capsules, in which the optimum shell strength and elasticity and the desired constitution of the fill, adjusted after production, remain unchanged on storage (Brox, 1983, 1988).

EP 0 121 321 (Brox, 1983) describes the combined use of glycerol and a sorbitol/sorbitan blend, namely Anidrisorb 85/70, as shell plasticizers. At the same time the addition of minor amounts of glycerol and/or propylene glycol to the liquid PEG fill is proposed. The combination of these two strategies prevents capsule shell embrittlement, since exchange processes between shell and fill are reduced to a minimum. The tendency of PEG to migrate into the shell is significantly reduced owing to the fact that PEG is less soluble in the sorbitol/sorbitan blend than in glycerol. On the other hand, the excess of plasticizer in the fill prevents the glycerol from migrating from the shell into the fill (Shah *et al.*, 1992; Reich, 1996). US 4 744 988 (Brox, 1988), an extended version of EP 0 121 321, recommends the selection of PEG 600 with a higher molecular weight and a lower hygroscopicity compared to PEG 400 as an additional means of reducing shell/fill interactions and improving capsule shell elasticity.

Microemulsion preconcentrates, comprising hydrophilic co-solvents such as propylene glycol and/or ethanol, in addition to oil(s) and surfactant(s), are another type of fill composition with challenging demands on the soft gelatin formulation concept. The hydrophilic co-solvents are prone to migrate into the shell, with propylene glycol softening the shell and ethanol volatilising through the shell, thereby upsetting the fill formulation in such a way as to change its solubilising and/or emulsifying properties.

The problems associated with *propylene glycol* may be solved by adjusting the shell formulation in such a way as to reduce the tendency of propylene glycol to migrate, during production and on storage, by using it as a plasticizer component in the shell and adjusting the manufacturing conditions at the drums to reduce tackiness of the ribbons (Brox *et al.*, 1993; Woo, 1997). An additional benefit of this approach is, that the amount of water required for dissolving and melting the gelatin may be reduced owing to the lower viscosity of propylene glycol compared to glycerol and sorbitol solutions, thus reducing the overall water exchange between shell and fill.

The problems associated with the use of a volatile solvent such as *ethanol* are more difficult to solve. To prevent volatilisation of ethanol, the finished capsules have to be enclosed in a solvent-tight packaging material such as an aluminium blister. Moreover, replacement of glycerol by higher polyols such as xylitol, sorbitol, sorbitol/sorbitan blends and/or hydrogenated starch hydrolysates has been reported as an effective means of reducing the rate and extent of ethanol diffusion into the shell (Reich, 1996; Moreton and Armstrong, 1998). In certain cases, however, both approaches may not be sufficient to prevent fill deterioration, since ethanol diffusion cannot be fully prevented. Thus, for a microemulsion preconcentrate formulation that is very sensitive to the co-solvent concentration, the only way to overcome the problem at present, is the use of a co-solvent other than ethanol, that is not volatile and does not show any diffusion into the capsule shell. For ciclosporin microemulsion preconcentrate soft capsules, such approaches have been filed in two patents, namely a European Patent Application (Woo, 1995) describing the use of dimethylisosorbide and a US Patent Application (Shin *et al.*, 2000) that describes the use of a microemulsion preconcentrate containing a lipophilic instead of a hydrophilic co-solvent.

Post-treatments and coatings

Soft gelatin capsules may be post-treated after production or coated to improve product stability, to modify the dissolution rate and to enable enteric capsules to be produced. Several patents have been filed describing the use of protective coatings to overcome the stability problems of soft

capsules arising from the hygroscopic nature and heat sensitivity of the soft capsule shell. However, most of these attempts have failed in practice, since coating of soft capsules is not an easy task. The low surface roughness of soft capsule shells and the intrinsic insolubility of the shell components in organic solvents means that coatings applied as an organic solution usually do not adhere properly to the capsules, resulting in onion-like coatings of layers peeling off immediately after drying or on storage. Aqueous coatings, on the other hand, may result in capsule swelling, softening and/or sticking together, since water is acting as a plasticizer for the gelatin capsule shells. To balance the two extremes, emulsion-based formulations or solutions in a mixture of water and alcohol have been recommended (Osterwald *et al.*, 1982). The technological approach of choice for soft capsules to be coated is using the fluidised-bed air-suspension technique.

Capsules with modified dissolution characteristics, such as gastroresistant enteric soft gelatin capsules, have been described in the scientific and patent literature and can be achieved by adding gastroresistant, enteric-soluble polymers to the gelatin mass prior to capsule formation, or by aldehyde post-treatment or enteric coating of the dried capsules. All three attempts have their specific difficulties. For soft gelatin capsules produced by the rotary die process, the last two approaches are in practical use.

Aldehyde post-treatment of soft gelatin capsules has been known for many years as a popular means to reduce their dissolution rate, i.e. the capsules take a long time to dissolve and have left the stomach before this occurs. Formaldehyde has been described to cross-link effectively soft capsules to render them gastroresistant. Since safety questions have been raised about the presence of trace amounts of formaldehyde in foods and pharmaceuticals, the use of aldehydes without health concerns such as aldoses have been claimed in a patent (Fischer, 1986) and are actually used. The major disadvantage of any aldehyde treatment of soft gelatin capsules is that cross-linking can continue on storage. Alternatively, soft gelatin capsules may be coated with a gastroresistant, enteric-soluble polymer. Owing to the aforementioned difficulties associated with organic and aqueous soft capsule coating, a protective subcoat is usually applied as an alcoholic solution prior to the application of the gastroresistant, enteric polymer layer (Virgilio and Matthews, 1989).

Non-gelatin soft capsules

Traditionally, gelatin has been used almost exclusively as shell-forming material of soft capsules. This is due to its legal status and its unique physicochemical properties, namely its oxygen impermeability and the combination of film-forming capability and thermoreversible sol/gel formation, that favour its use for the industrial soft capsule production especially in the rotary die process.

Despite these great advantages, which have been described in detail in the section above on 'Soft gelatin capsules', gelatin has several drawbacks that limit its use for soft capsules:

- The animal source of gelatin can be a problem for certain consumers such as vegetarians or vegans and religious or ethnic groups (Jews, Muslims, Hindus, etc.) who observe dietary laws that forbid the use of certain animal products.
- Since unmodified gelatin is prone to crosslinking when in contact with aldehydes, solubility problems might be expected with certain fill formulations.
- Transparent low-colour capsules are difficult to produce owing to the effect of the intrinsic Maillard reaction on gelatin colour.
- The temperature and moisture sensitivity of gelatin-based soft capsules is an issue that complicates the use of soft gelatin capsules in very hot and humid regions and requires special packaging and storage conditions to ensure product stability.
- For low-price health and nutrition products, pricing of commercially available gelatin might be an additional problem.

To address these concerns, there has been a great interest in the soft capsule industry in looking for gelatin substitutes. Indeed, several concepts based on synthetic polymers and/or plant-derived hydrocolloids have been described

in the patent literature. However, only few have gained commercial interest. This is due to the fact that a change in the capsule shell polymer material requires more than just overcoming the aforementioned shortcomings of gelatin. It requires both legal approval and machinability, i.e. either to mimic most of the physicochemical gelatin characteristics that are important for rotary die soft capsule production with some adjustments of the production equipment for the new material characteristics or to use a completely redesigned machinery.

To date, three non-gelatin soft capsule concepts with different process adjustments have reached prototype status: two are based on plant-derived hydrocolloids (Draper *et al.*, 1999; Menard *et al.*, 1999), the third is based on a synthetic polymer (Brown, 1996).

WO 0 103 677 (Draper *et al.*, 1999) describes the use of a combination of iota carrageenan (12–24% w/w of dry shell) and modified starch, namely hydroxypropyl starch (30–60% w/w of dry shell), as a gelatin substitute. Both components are accepted as food additives with E numbers, thus allowing their use in health and nutrition products. Hydroxypropyl starch is also approved as a pharmaceutical excipient. The combination of the two hydrocolloids leads to a synergistic interaction that produces a gel network, which is suitable for soft capsule production using the rotary die process. It can be formulated with conventional plasticizers such as glycerol, sorbitol, etc. (10–60% w/w of dry shell) and water to form a molten mass that can be extruded to set within less than 20 s producing mechanically strong, elastic films on temperature-controlled casting drums. Sealing may be performed at temperatures between 25 and 80°C, by a fusion process comparable to the one observed with soft gelatin capsules. After drying, mechanically strong and highly elastic products can be achieved.

Prototype capsules with lipophilic fill formulations are shiny with a high appearance stability on storage. The capsule shells do not show crosslinking and exhibit a greater mechanical stability than gelatin shells when exposed to elevated humidity and temperature, i.e. even under hot and humid storage conditions they may not become sticky. Formulation approaches with hydrophilic fills are expected to be as challenging as for soft gelatin capsules. Oxygen permeability is comparable to gelatin-based shells. The dissolution mechanism is completely different to the one of a soft gelatin capsule. On contact with an enzyme-free aqueous medium at 37°C, the capsule shell only swells, at a rate and to an extent depending on the type and concentration of electrolytes present. The capsule content may be released when the shell bursts at its point of lowest resistance, i.e. at the seams. Under *in vivo* conditions, capsule shell dissolution may be induced by enzymatic degradation.

WO 0 137 817 (Menard *et al.*, 1999) describes the formation of soft capsules from a potato starch (45–80% w/w), with a specific molecular weight distribution and amylopectin content, together with a conventional plasticizer such as glycerol (>12% w/w), a glidant and a disintegrant. Soft capsule production may be performed with a rotary die machine with nearly water-free formulations that are processed by hot melt extrusion. A narrow production window and the use of a high molecular weight amorphous starch with a high amylopectin content (>50% w/w) are necessary for the formation of acceptable capsules.

From the regulatory point of view, starch-based soft capsules are a low-price alternative to soft gelatin capsules, appropriate for pharmaceutical and health and nutrition products. Moisture sensitivity and fill compatibility of the capsule shells are comparable to soft gelatin capsules, with the exception that cross-linking is not a problem. Oxygen permeability is expected to be a little higher compared to soft gelatin capsules (Reich, unpublished results). Shell dissolution requires enzymatic degradation by amylases; on contact with amylase-free aqueous media at 37°C, the capsules release their content only by swelling-induced disintegration. The addition of calcium carbonate is one option to enhance capsule disintegration further.

The visual appearance, the seam quality, and the long-term stability of the finished product of the prototype starch capsules cannot compete with soft gelatin capsules. This is due to the structural rearrangements within the capsule shells associated with the tendency of starch to retrograde on storage, in some instances leading to a

subsequent plasticizer syneresis (Reich, unpublished results).

WO 9 735 537 (Brown, 1996) describes the preferable use of polyvinyl alcohol (PVA) and optional use of some other materials, all being film-forming polymers that lack the gelling properties that are necessary for soft capsule production using the conventional rotary die process. The invention therefore provides the use of pre-formed rolls of nearly water-free plasticized films that may be fed to a rotary die encapsulation unit for soft capsule production. To render the film material more flexible and to assist seam formation at temperatures depending on the film composition, the films are partially spray-solvated prior to encapsulation.

PVA films according to this invention may be composed of 70–75% w/w PVA, 10–15% w/w glycerol and 5–10% w/w starch, with a sealing temperature of 140–180°C, depending on the degree of solvation. PVA as an optional gelatin substitute has the advantage of being less hygroscopic, thus leading to soft capsule shells that are less sensitive to moisture than soft gelatin capsule shells. Moreover, the capsules are readily water soluble with no cross-linking tendency. However, prototype capsules lack the shiny and smooth surface appearance and the seam quality of conventional soft gelatin capsules. In addition, the regulatory issues and the formulation of hydrophilic fills are problems that have to be solved.

To summarise, it may be concluded that none of the gelatin-free soft capsule concepts is fully developed yet. Nevertheless, soft capsules based on plant-derived or synthetic polymers are an interesting line extension to soft gelatin capsules with the potential to gain a market share for certain niche products.

Analytical approaches for soft capsule testing

Finished soft capsule products, either gelatin- or non-gelatin-based, are routinely specified by strength values, drug content, dissolution properties and in some cases by their water content. Moreover, they are checked for long-term stability under ICH (International Conference on

Harmonisation) conditions. The official tests for soft capsules are discussed in Chapter 13.

Drug content is determined by either HPLC (high performance liquid chromatography) or any other appropriate QC (quality control) method. Dissolution properties are checked according to the pharmacopoeial requirements, with a two-tier test being approved for products with reduced solubility. Water content may be assessed by Karl Fischer titration.

Strength measurements ('hardness' measurements) are performed using a commercially available tester (Barreiss Hardness Tester) in which the capsules are compressed to a certain extent between a measuring detector and a slowly moving plate. The counter force exerted by the capsules is displayed in newtons. Under these test conditions, an optimum strength range is specified by the manufacturer for each product. Strength values above or below this specified range are indicative of insufficient flexibility or softening, respectively.

Recently, the use of conventional or modulated DSC has been proposed as an additional analytical tool to determine performance-related microstructural features of soft capsule shells, such as the glass transition temperature (T_g), the melting temperature (T_m) and the melting enthalpy (H_m) (Reich, 1994, 1995, 1996; Nazzal and Wang, 2001). T_g, T_m and H_m are important parameters for monitoring process- and storage-induced structural changes within a capsule shell and have been successfully applied to the design and optimisation of soft capsules, i.e. to evaluate plasticizer effectivity and compatibility, and shell/ fill interactions.

Near infrared (NIR) spectroscopy is another modern analytical technique with great potential for soft capsule specification and efficient stability testing (Reich, 2000). NIR is a fast and non-invasive spectroscopic method that allows for simultaneous evaluation of soft capsule identity, drug and water content (Buice et al., 1995) and in some cases film coat thickness of the finished product directly after production. In addition, storage-induced chemical and physical changes in both the shell and the fill can be assessed, thus allowing for an early and non-destructive detection of stability problems such as moisture changes and shell/fill interactions leading to

cross-linking, softening or hardening of the capsule shell (Gold et al., 1997; Reich, 2000).

References

Amemiya, T., Mizuno, S., Yuasa, H. and Watanabe, J. (1998). Development of emulsion type new vehicle for soft gelatin capsule. I. Selection of surfactants for development of new vehicle and its physicochemical properties. *Chem. Pharm. Bull.*, 46, 309–313.

Amemiya, T., Mizuno, S., Yuasa, H. and Watanabe, J. (1999). Emulsion type new vehicle for soft gelatin capsule available for pre-clinical and clinical trials: effects of PEG 6000 and PVP K30 on physicochemical stability of new vehicle. *Chem. Pharm. Bull.*, 47, 492–497.

Armstrong, N. A., James, K. C. and Pugh, K. L. (1984). Drug migration into soft gelatine capsule shells and its effect on in-vitro availability. *J. Pharm. Pharmacol.*, 36, 361–365.

Armstrong, N. A., James, K. C., Collett, D. and Thomas, M. (1985). Solute migration from oily solutions into glycerol–gelatin mixtures. *Drug Dev. Ind. Pharm.*, 11, 1859–1868.

Armstrong, N. A., Gebre-Mariam, T. and James, K. C. (1986). An apparatus for investigating drug migration into gelatin capsule shells. *Int. J. Pharm.*, 34, 125–129.

Babel, W. (2000). Gelatine in pharmaceutical applications. *Pharmaceut. Manuf. Packing Sourcer*, 6, 63–66.

Berry, I. R., Borkan, L. and Shah, D. (1988). Chewable, edible soft gelatin capsule. European Patent Application 0 374 359.

Brown, M. D. (1996). Improvements in or relating to encapsulation. International Patent Application WO 9 735 537.

Brox, W. (1983). Pharmaceutical compositions. EP 0 121 321.

Brox, W. (1988). Soft gelatin capsules and methods for their production. US Patent 4 744 988.

Brox, W., Zande, H. and Meinzer, A. (1993). Soft gelatin capsule manufacture. EP 0 649 651.

Buice Jr., R. G., Gold, T. B., Lodder, R. A. and Digenis, G. A. (1995). Determination of moisture in intact gelatin capsules by near-infrared spectrophotometry. *Pharm. Res.*, 12, 161–163.

Charman S. A., Charman, W. N., Rogge, M. C., Wilson, T. D., Dutko, F. J. and Pouton, C. W. (1992). Self-emulsifying drug delivery systems: formulation and biopharmaceutical evaluation of an investigational lipophilic compound. *Pharm. Res.*, 9, 87–93.

Draper, P. R., Tanner, K. E., Getz, J. J., Burnett, S. and Youngblood, E. (1999). Film forming compositions comprising modified starches and iota-carrageenan and methods for manufacturing soft capsules using same. International Patent Application WO 0 103 677.

Fischer, G. (1986). Gelatin capsules for the controlled release of the active agent, and process for their preparation. EP 0 240 581.

Gardella, L. A. and Kesler, H. (1977). Gelatin-encapsulated digoxin solutions and method of preparing the same. US Patent 4 002 718.

Ghirardi, P., Catenazzo, G., Mantero, O., Merotti, G. C. and Marzo, A. (1977). Bioavailability of digoxin in a new soluble pharmaceutical formulation in capsules. *J. Pharm. Sci.*, 66, 267–269.

Gold, Th. B., Buice Jr., R. G., Lodder, R. A. and Digenis, G. A. (1997). Determination of extent of cross-linking in hard gelatin capsules by near-infrared spectrophotometry. *Pharm. Res.*, 14, 1046–1050.

Gullapalli, R. P. (2001). Ibuprofen-containing softgels. US Patent 6 251 426.

Hom F. S., Veresh, S. A. and Ebert, W.R. (1975). Soft gelatin capsules. II. Oxygen permeability study of capsule shells. *J. Pharm. Sci.*, 64, 851–857.

Meinzer, A. (1988). *Studies on Oxygen Permeability of Soft Gelatin Capsules*. PhD Thesis, Freiburg i.Br., Germany.

Menard, R., Tomka, I., Engel, W. D. and Brocker, E. (1999). Process to manufacture starch-containing shaped bodies, mass containing homogenized starch and device to manufacture soft capsules. International Patent Application WO 0 137 817.

Montes, R. and Steele, D. R. (1999). Chewable softgel oral hygiene product. US Patent 5 948 388.

Moreton, R. C. and Armstrong, N. A. (1998). The effect of film composition on the diffusion of ethanol through soft gelatin films. *Int. J. Pharm.*, 161, 123–131.

Nazzal, S. and Wang, Y. (2001). Characterization of soft gelatin capsules by thermal analysis. *Int. J. Pharm.*, 230, 35–45.

Oppenheim, R. C. and Truong, H. C. (2002). Discoloration-resistant vitamin composition. US Patent Application 2002 004 069.

Osterwald, H., Bauer, K. H., Jäger, K. F. and Reich, G. (1982), Verfahren und Lösungen zum Beschichten von Gelatinekapseln. DE 3 243 331.

Overholt, S. M. (2001). Chewable soft capsules. US Patent 6 258 380.

Radivojevich, F., Joss, H. and Dvorsky, S. (1983). Therapeutic coronary composition in the form of soft gelatine capsules. EP 0 143 857.

Reich, G. (1983). *Soft Gelatin Capsules: Effect of Gelatin Gel Structure on Thermal, Mechanical and Biopharmaceutical Properties*. PhD Thesis, Freiburg i.Br., Germany.

Reich, G. (1994). Action and optimization of plasticizers in soft gelatin capsules. *Pharm. Ind.*, 56, 915–920.

Reich, G. (1995). Effect of drying conditions on structure and properties of gelatin capsules: studies with gelatin films. *Pharm. Ind.*, 57, 63–67.

Reich, G. (1996). Effect of sorbitol specification on structure and properties of soft gelatin capsules. *Pharm. Ind.*, 58, 941–946.

Reich, G. (2000). Monitoring structural changes in hard and soft gelatin films and capsules using NIR transmission and reflectance spectroscopy. *Proceedings 3rd World Meeting APV/APGI*, Berlin, 3/6 April 2000, 507–508.

Reich, G. and Babel, W. (2001). Performance of different types of pharmaceutical grade soft capsule gelatins. *AAPS Annual Meeting/Pharmaceutical Science Supplement*, W4323.

Rinaldi, M. A., Tutschek, P. C. and Saxena, S. J. (1999). Softgel formulation containing retinol. US Patent 5 891 470.

Rouffer, M. T. (2001). Self-emulsifying ibuprofen solution and soft gelatin capsule for use therewith. US Patent 6 221 391.

Sanc, Y., Ito, H. and Enomoto, I. (2000). Soft capsule including vitamin B12. Japan Patent 2000 095 676.

Scott, R., He, X. and Cade, D. (1997). Gelatine compositions. WO 9 933 924.

Serajuddin, A.T., Sheen, P.C. and Augustine, M.A. (1986). Water migration from soft gelatin capsule shell to fill material and its effect on drug solubility. *J. Pharm. Sci.*, 75, 62–64.

Shah, N. H., Stiel, D., Infeld, M. H., Railkar, A. S., Malick, A. W. and Patrawala, M. (1992). Elasticity of soft gelatin capsules containing polyethylene glycol 400 – quantitation and resolution. *Pharm. Technol.*, 3, 126–131.

Shin, H. J., Choi, N. H., Kim, J. W., Yang, S. G. and Hong, C. I. (2000). Cyclosporin-containing microemulsion preconcentrate composition. US Patent 6 063 762.

Virgilic, G. and Matthews, J. W. (1989). Process for coating gelatin capsules. US Patent 4 816 259.

Woo, J. S. (1995). Cyclosporin soft capsule composition. EP 0 650 721.

Woo, J. S. (1997). Cyclosporin-containing soft capsule preparations. International Patent WO 9 748 410.

12

Drug release from capsules

J Michael Newton

Concepts of bioavailability

After administration of an oral preparation, the active substance is delivered to the site of pharmacological activity by a complex process which involves solution of the drug in the gastrointestinal (GI) fluid, absorption, usually by passive diffusion across the membrane of the GI wall into the capillary blood supply and distribution via the portal circulation to the systemic circulation and the site of action. There are some instances, e.g. inflammatory bowel disease, where a local action may be required but this is not the usual situation with most drugs. In relation to absorption, drugs have been classified into four different types by Amidon *et al.* (1995), i.e. those which are readily soluble and which readily permeate the intestinal mucosal wall, those which are insoluble but have a high permeability, those which are highly soluble and have a low permeability and those which are insoluble and have a low permeability. Thus as it was thought at one time, the dissolution of the drug at the site of absorption is not the only rate-limiting factor involved in the distribution process. Over the years it has been established that physicochemical characteristics of the drug and its formulation into the dosage form has a marked influence on the dissolution and permeation processes and hence on the pharmacological performance of the drug. The rate at which and the extent to which the active ingredient is delivered to the circulation is used to quantify these processes and is referred to as the bioavailability.

To evaluate the performance of a formulated product, the bioavailability can be assessed by measuring the concentration of the drug in plasma or serum over a period of time after administration. The variation in the plasma concentration with time gives an indication of the amount and rate of absorption of the drug. Increasing the rate of absorption will decrease the time taken to reach peak concentration T_{max}, but it may not change the total area under the plasma concentration–time curve (AUC (area under the curve), total amount absorbed) or the peak concentration C_{max}. Changing the extent of the absorption could involve change in both the peak height and the total area under the curve. Alternatively, the rate and extent of urinary excretion of the drug or its metabolites can be measured. There are now guidelines for the evaluation of the bioavailability of a drug and the comparison of different products (CDER, 2000).

Another approach to the assessment of bioavailability, which is less time-consuming and costly, is to use *in vitro* tests, which attempt to simulate the *in vivo* performance of the dosage form. Such tests are based on the ability of the dosage form to disintegrate in a fluid under given conditions (disintegration test), or upon the amount of drug that is released into solution in a specified fluid under given conditions (dissolution test). These tests are used in official standards for certain preparations in the British Pharmacopoeia (BP) and the United States Pharmacopoeia (USP). Dissolution tests provide reproducible conditions for the solution process and can indicate the way in which formulation variables influence the solution rate process. However, some caution is necessary in the interpretation of *in vitro* tests in terms of *in vivo* performance, as they do not always correlate. Again, there are guidelines (CDER, 1997) about where these tests can be applied and monographs which define the procedures to be adopted.

Details of the test procedures in the standardisation of the quality of capsules are discussed in Chapter 13.

No single test procedure has been devised which can simulate accurately what happens to a dosage form after administration, because of the variations in the physicochemical properties of the drugs and the way individuals vary in their responses. The apparent simplicity of capsule formulations as a blend of powders, which will be readily available for dissolution, promotes the concept that hard gelatin capsules are a readily bioavailable oral dosage form. Thus in pharmaceutical textbooks written when the area of biopharmaceutics was established, an order of drug release solution → suspension → capsule → tablet was presented by Levy (1970) and some 30 years later, York (2002) states 'in most cases the drug is released from a capsule faster than from a tablet'. Is there any evidence to support this hypothesis?

Evidence that capsule formulations may be subject to problems of bioavailability may be obtained from the literature when comparisons of the plasma concentration–time curve with that of other preparations containing a drug have been reported. These may be other commercial products, such as capsules or tablets, which have been found to be clinically acceptable. Alternatively, an intravenous injection or other type of preparation such as a solution or a suspension may be used as a reference. Comparison with an intravenous injection will give a measure of the absolute bioavailability because it eliminates the absorptive phase. Comparison with other types of preparation only indicates the relative bioavailability of the formulation. A solution of the drug is considered to be the most useful oral reference preparation as it eliminates the dissolution phase. Hence, comparison with a solution should indicate whether dissolution is the rate-limiting factor. Intravenous injections and solutions may not be feasible for insoluble drugs, yet these are the ones that are more likely to give rise to bioavailability problems. Solutions in non-aqueous polar solvents or in oils may offer an alternative as reference formulations for such drugs.

A drug, which is clinically effective over a wide range of blood concentrations, would provide products that may not differ significantly in bioavailability. However, if there is a narrow range

between the minimum therapeutic concentration and the minimum toxic concentration, then changes in bioavailability may have serious clinical consequences. For example, phenytoin has dose-dependent elimination kinetics and a narrow therapeutic range and even small differences in bioavailability may be hazardous. The earliest reports that there could be problems with capsule products (Eadie, 1968; Martin, 1968; Rail, 1968; Tyrer *et al.*, 1970) indicated that patients had shown the toxic effects of phenytoin overdosage when the capsule formulation of an established brand was changed. In this particular example, the difficulty was shown to be due to the substitution of lactose for calcium sulfate dihydrate as the excipient. Since then, numerous studies of this drug have shown that different formulations may not be bioequivalent. Neuvonen (1979) has reviewed the bioavailability of phenytoin.

Administration of hard gelatin capsule products

There have been reports in the literature, which suggest, that after administration, solid dosage forms, including capsules, can stick in the oesophagus, resulting in damage to the mucosa (Hey *et al.*, 1982). They found that the incidences of lodging were associated with taking the dosage form with a small volume of fluid when patients were in the supine position. This lead to reports of *in vitro* studies to develop methods of evaluation of the tendency of drug products to adhere to the oesophagus (Marvola *et al.*, 1982, 1983) and their application (Al-Dujuaili *et al.*, 1986; Honkanen *et al.*, 2002). Bailey *et al.* (1987) clearly established by gamma scintigraphy studies, that the lodging of size 00 capsules in the oesophagus was related to the volume of liquid taken with the capsule. When this was only 15 mL, the incidence of lodging was 61% as opposed to 17% when the capsule was taken with 120 mL. In those cases where lodging did occur, the volunteers were over 70 years old and the retention was probably associated with reduced amplitude of contractions in this group. The authors suggested that the maximum size of capsules should be restricted to a size 0 and that patients should take the capsule with a large volume of liquid (the volume however was not specified). In view of these findings, it is surprising that in a gamma scintigraphy study by Perkins *et al.* (1994) of the oesophageal transit of capsules and tablets in the elderly, only 50 mL of liquid was used when the dosage forms were administered. Hence the findings that the tablets had a more rapid oesophageal transit (mean 4.3 s, range 1.0–14.0) than capsules (mean 20.9 s, range 1.5–174.3 s) may not be representative of the situation if sufficient fluid had been used.

Another factor to be considered in assessing the comparison of formulations is the physiology of the GI tract and its involvement in drug absorption. The rate of subsequent transit of the dosage form through the GI tract could influence the absorption of a drug. For example, Stewart *et al.* (1979) compared an experimental capsule formulation with a solution of riboflavin as a standard, in healthy volunteers. The higher urinary excretion observed after administration of capsule formulations could be associated with the rapid transit of the solution past the limited area of the intestine capable of absorbing riboflavin. Similarly, Hansford *et al.* (1980) found that a capsule formulation containing micronised griseofulvin gave rather variable blood levels when it was administered to rabbits. However, these levels were considerably higher than those obtained with a solution of the drug in PEG (polyethylene glycol) 400, which was presumably due to the rapid passage of the liquid form through the GI tract.

Further evidence for the importance of GI transit is illustrated by the work of Sun *et al.* (1996), who explained the variability of the absorption of a hypolipidaemic agent in healthy volunteers in terms of variable gastric emptying of a capsule formulation, determined by gamma scintigraphy. Thus, as with most other oral dosage forms, food influences the transit of capsules and this could have consequences for the bioavailability of a capsule product. An example of such an effect is the gamma scintigraphy study of capsules of avitriptan (details of formulation not given) carried out by Marathe *et al.* (1998). These workers concluded that the qualitative appearance of the plasma–concentration profiles

for the drug could be related to the manner in which the drug emptied from the stomach. Inter-subject variability in C_{max} and AUC could be explained by the inter-subject variability in gastric emptying in both the fed and fasted state. The value of T_{max} was significantly delayed by food. Further evidence for the variability in transit of an intact capsule through the GI tract is provided by the work of Mojaverian *et al.* (1989) who tracked the movement of a Heidelberg capsule labelled with Indium-111 by gamma scintigraphy. This capsule contains a pH-sensitive radiofrequency (1.94 MHz) radio transmitter in an inert, indigestible shell, which was approximately the same size as a size 1 capsule and therefore allows monitoring of GI pH and position in the GI tract. For six volunteers it was only possible to record a mean value for the gastric emptying time (3.5 ± 0.8 hr). The variation in the time the capsule spent at the ileo-caecal junction exceeded the total observation time (9 hr) in three volunteers, which prevented clear values for this time being obtained. This also prevented values being found for the small bowel and the mouth to caecum transit times. Two typical pH–transit time curves for individual volunteers are presented in Figures 12.1 and 12.2. These results also clearly have implications for the administration of capsules with pH-sensitive coatings. The importance of gastric pH on the release of drugs whose solubility is pH-dependent is illustrated by the work of Ogata *et al.* (1986) who studied the bioavailability of commercial capsules of cinnarizine in 12 healthy volunteers who could be divided into those with low ($n = 8$) and high ($n = 4$) acidity. The solubility of cinnarizine ranges from 2.2 μg mL^{-1} at pH 6.7 to 1549.5 μg mL^{-1} at pH 1.1. They selected the two extremes from 32 products, which showed a wide range of dissolution performance in dissolution media of different pH values. The dependence of dissolution on the pH of the dissolution media of the two products is shown in Figure 12.3. The serum level–time curves for the two products, shown in Figure 12.4, clearly illustrate that the performance of both types of capsule is strongly influenced by the gastric acidity of the subjects in the group. There was no significant difference in the pharmacokinetic parameters obtained from the two products in

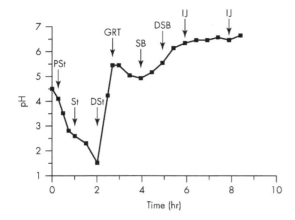

Figure 12.1 Gastrointestinal pH versus time profile measured by radiotelemetry in a healthy volunteer (subject 6). The intraluminal location of the radiotelemetry capsule is shown by the arrows at the appropriate time point: PSt, proximal stomach; ST, stomach; DSt, distal stomach; GRT, gastric residence time; SB, small bowel; DSB, distal small bowel; IJ, ileocaecal junction. This subject showed a 2.8 hr GRT and >2.3 hr delay at the ileocaecal junction. Transition to the large bowel was not observed in this subject within the 9.0 hr study period. (Reproduced from Mojaverian et al., 1989, with permission from the publishers.)

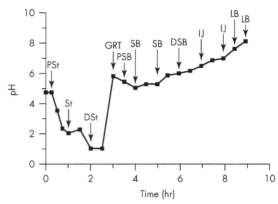

Figure 12.2 Gastrointestinal pH versus time profile measured via the radiotelemetry Heidelberg capsule in a healthy volunteer (subject 5). The arrows at each time point (abbreviations as per Figure 12.1, PSB, proximal small bowel; LB, large bowel) show the luminal location of the telemetry capsule. This subject showed a GRT of 3.0 hr followed by an ileocaecal lag time of 1.0 hr. Passage through the ileocaecal junction was observed in this subject 8.0 hr after administration of the Heidelberg capsule. Small bowel transit time was about 5.0 hr as measured by external gamma scintigraphy. (Reproduced from Mojaverian et al., 1989, with permission from the publishers.)

the group with the high acidity. In the case of the low-acidity group, however, while there was no significant difference in the value of T_{max}, the values of both C_{max} and the AUC_{0-8h} did differ significantly. Thus depending on the gastric acidity of the volunteers, different assessments of the bioavailability of the two products are obtained.

The administration of capsules is usually by the oral route but there are reports in which hard gelatin capsules have been formulated for rectal use, e.g. Hagenlocher et al. (1986). A study by Eerikainen et al. (1996) described the potential of rectal administration of hard gelatin capsules. After identifying the correct procedure for insertion of the size 0 hard gelatin capsule, they used ibuprofen as the model drug and found that the rectally administered capsule was bioequivalent to an orally administered capsule in healthy volunteers. It had a lower value for C_{max} and a longer T_{max} indicating a slower rate of absorption by the rectal route.

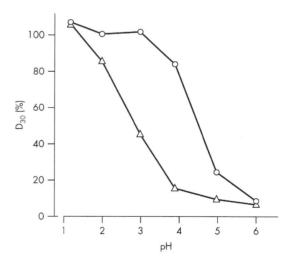

Figure 12.3 Dissolution rate versus pH profile for cinnarizine capsules, (O) capsule A, (△) capsule B. (Reproduced from Ogata et al., 1986, with permission of the publishers.)

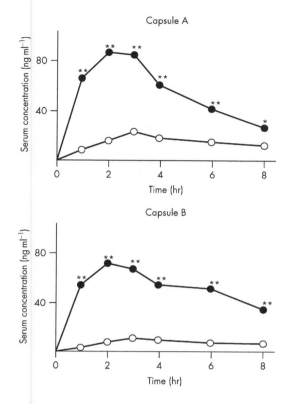

Figure 12.4 Mean serum levels of cinnarizine administered as a capsule to subject groups having high gastric acidity (●) and low gastric acidity (○). * and ** represent significant differences from the serum concentration of the low gastric acidity group at the same sampling time with $P < 0.05$ and $P < 0.01$, respectively. (Reproduced from Ogata *et al.*, 1986, with permission of the publishers.)

Bioavailability of hard gelatin capsule products

Consideration of published papers may provide an answer to the question of whether hard gelatin capsules do present problems in terms of ensuring bioequivalence. Unfortunately, individual papers taken in isolation may provide conflicting evidence. For example, Tannenbaum *et al.* (1968) concluded that a commercial capsule formulation of triamterene and hydrochlorthiazide was less effectively absorbed than an experimentally formulated tablet. Randolph *et al.* (1985), however, found that a capsule formulation

of the same two drugs was bioequivalent to an aqueous solution, suggesting that the capsule formulation of the earlier study was probably not optimised.

In a study to establish the performance of a new tablet product, randomised cross-over design of 40 subjects allocated randomly to a tablet formulation of the drug carbocysteine, a capsule formulation and a syrup were used as standards (Bron, 1988). All three preparations were found to be bioequivalent. Chaulet *et al.* (2002) found that a capsule containing a mixture of chloroquine and proguanil was equivalent to the same two drugs administered as separate tablet formulations. A capsule formulation of lithium carbonate was found by Terao *et al.* (1994) to be bioequivalent to two marketed tablet preparations. Sechaud *et al.* (2002) in a single-centre, open-label, two-treatment randomised cross-over design of 40 subjects, established that a new 320 mg tablet formulation of the drug valsartan was bioequivalent to two 160 mg capsules of a product which was already on the market. An example of bioequivalence of a generic and marketed capsule formulation of the drug fluconazole is provided by the study of Ribeiro *et al.* (2000) and the drug mefenamic acid by Rawashdeh *et al.* (1997). Similarly Chouchane *et al.* (1995) found that two preparations of rifampicin were bioequivalent, as did Kim *et al.* (2000) with the drug cycloserine. Capsules containing caffeine (with details of formulation) were used as a reference preparation in an *in vivo* study in healthy human volunteers, which evaluated the performance of chewing gum preparations (Kamimori *et al.*, 2002). The chewing gum formulations had a shorter value for the $T_{max,}$ no doubt owing to buccal absorption, but a lower bioavailability. A capsule and an intravenous formulation of mycophenolate mofetil, an ester prodrug of the immunosupressant mycophenolic acid, were studied in healthy volunteers by Bullingham *et al.* (1996) and found to be bioequivalent. In a comparison of a capsule formulation of amoxicillin with that of a tablet, Molinaro *et al.* (1997) showed that the two commercial preparations were bioequivalent.

Since many published papers, especially in the medical literature, lack details of formulation, it is generally impossible to make an accurate assessment of which formulation factors are important. It is also important to realise that reports of capsule formulations being less bioavailable than other preparations could be due to the drug being poorly absorbed by the oral route for various reasons. For example, Sasahara *et al.* (1980) concluded that the relatively low absolute bioavailability of levodopa in capsules is probably due to first pass metabolism. Similarly, the absolute bioavailability of the antitumour drug, idarubicin in patients, was reported to be about 30% (Lu *et al.*, 1986; Gillies *et al.*, 1987; Smith *et al.*, 1987; Stewart *et al.*, 1991), again probably owing to first pass metabolism. On the other hand, the high water solubility and extensive ionisation have been suggested as the cause of the poor oral absorption of the bisphosphonates (Lin, 1996). Thus the very low absolute bioavailability (about 2%) of two dose levels of capsules was related to the properties of the drug rather than the capsule formulation (Villikka *et al.*, 2002). This conclusion is supported by the paper of Lapham *et al.* (1996), where a tablet formulation of the drug was found to provide only 52% of the bioavailability of a capsule formulation. Failure to provide the drug etoposide solubilised with a co-solvent, which is the formulation of the commercially available product, was suggested as the reason for a 'home made' powder capsule formulation having an absolute bioavailability of 33% rather than the usual value of 70–80%, for the commercial tablet product (Jonkman-de Vries *et al.*, 1996).

An alternative source of difference between the various preparations of the same drug is the use of non-optimum formulations in the comparison. Wagner *et al.* (1966) provided clear evidence that a capsule formulation of indoxole was inferior to emulsion, soft gelatin capsule and suspension formulations. Whether this was the best hard gelatin capsule formulation cannot be judged from the evidence available in the paper. Thus comparison of capsules with other dosage forms must be judged with caution.

Comparison of different capsule formulations of the same drug should provide better evidence that formulation needs to be considered to ensure comparable bioavailability of capsule formulations, but even here caution must be used in interpretation of some of the papers that appear in the literature. In several instances conclusions are drawn from poor *in vivo* experimental data, e.g. Brice and Hammer (1969), where serum levels produced by commercial formulations of oxytetracycline were compared at only four time intervals, providing insufficient data for a reasonable pharmacokinetic analysis. The evidence of the papers contained in the literature, while not being unequivocal, does lead to the conclusion that it is necessary to consider the formulation to ensure an adequate *in vivo* performance of certain drugs.

Formulation and the release of drugs from hard gelatin capsules

The release of drugs from hard gelatin capsules can be influenced by the formulation, hence it is important to consider the ways in which maximisation and/or consistency of drug release can be achieved. For this purpose, it is necessary to understand the mechanism of absorption of the drug to be formulated and especially to know which stage is the rate-controlling step in the process. For instance, if it is lack of drug permeability through the membrane of the GI tract rather than solubility which is the problem, then it will be necessary to tackle this problem with the formulation rather than that of solubility. Much time can be wasted in attempting to improve the dissolution of the formulation if this is only a minor aspect of the absorption process. In the present account, it will be assumed that the drug is in the correct crystalline form; there will be no consideration of the influence of crystal structure or the existence of solvates or hydrates.

An important feature of formulation is to ensure that the capsule contains the correct, uniform dose of the drug. The formulation will only exist as a single-component system if the drug completely and reproducibly fills the capsule volume. Small dose levels of drug require prior blending with an inert diluent. Similarly, larger doses can be blended with a diluent if, by this addition, a greater reproducibility of bulk volume can be achieved. High-speed filling machines, which operate by pre-forming a plug of powder prior to transferring it to the capsule, often involve a degree of consolidation of the powder bed and friction between this plug and metal makes necessary the addition of a powder/metal lubricant. Glidants may also be needed to improve powder flow and ensure reproducible bulk volume. Details of the methods of ensuring reproducible drug content are presented in Chapters 5 and 6.

The work involved in studying formulation variables by *in vivo* techniques is costly and time-consuming hence many studies involving formulation factors use *in vitro* testing, particularly dissolution. This makes two important assumptions: that dissolution is the rate-controlling step in the absorption process and that the particular *in vitro* dissolution test reflects the *in vivo* performance of the formulation. Unfortunately, these two restrictions are only rarely assessed. Any deductions made from dissolution results should be used only as guidelines to formulation, not as absolute values. An illustration of the type of problems that can arise with the sole reliance on dissolution testing was the case of phenytoin capsule formulations. Here owing to cross-linking of the decomposition product and the gelatin shell on storage, there was a failure of the product in the dissolution test (Rubino *et al.*, 1985). Similar findings were reported for capsules containing acetaminophen filled into hard gelatin capsules into which cross-linking had been induced (Meyer *et al.*, 2000). Stressed capsules, which failed the USP dissolution test in simulated gastric fluid (SGF), were found to be bioequivalent when tested *in vivo*. If pepsin was added to the dissolution medium, then the unstressed and stressed capsule passed the dissolution test. There had been no evidence of failure of cross-linked capsules to disintegrate *in vivo* when studied by gamma scintigraphy (Brown *et al.*, 1998). The lack of *in vivo* effect of capsules, which have been subjected to cross-linking induced stress, has also been confirmed by the combined pharmacokinetic and scintigraphy study of capsules containing amoxicillin as a marker (Digenis *et al.*, 2000). As discussed in Chapter 13, this leads to a change in the choice of dissolution media in the USP when testing capsule formulations *in vitro*.

In order to ensure adequate bioavailability when formulating hard gelatin capsules, it is necessary to consider various factors. These include the solubility, particle size and wettability of the drug, together with the combination of possible additives, the filling process to be used and the requirement to produce a granulation. An essential feature is to ensure that the capsule disintegrates both *in vitro* and *in vivo*. Whilst the application of the *in vitro* test may not be a conclusive indicator of a bioavailable capsule formulation, a capsule that does not disintegrate, is very unlikely to be effective. These factors will now be considered.

Drug solubility

Unless other factors dominate, the rate-controlling step in the absorption process is the rate at which the drug is transferred from the solid state into solution. For a wide range of compounds, the intrinsic rate of dissolution is directly proportional to the solubility (Hamlin *et al.*, 1965). Hence the lower the drug solubility, the lower will be the rate of dissolution and so absorption. The combined effects of drug solubility and of additives within the capsule on the dissolution rate of a range of drugs were studied by making measurements at two levels (20 and 80%) of three diluents and in the presence or absence of magnesium stearate and sodium lauryl sulfate (Newton and Razzo, 1977a). There was a strong indication that the rate was proportional to the logarithm of the drug solubility (Figure 12.5). Thus one can anticipate problems when presenting drugs with low water solubility in capsules. Blending of simple additives does not always overcome the formulation problems.

Particle size

The standard method of increasing the rate of solution of a drug is to increase the surface area in contact with the solvent by reducing the particle size. However, the effectiveness of this method will depend on the contact between liquid and solid. When nitrofurantoin, a relatively insoluble substance, was administered to rats in a hard gelatin capsule, the proportion of the dose excreted in the urine increased as the

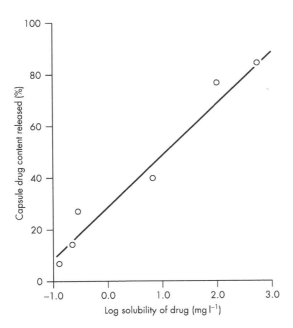

Figure 12.5 Influence of drug solubility on the *in vitro* release of drug from capsules (Newton and Razzo, 1977a, with permission from the publishers).

particle size was decreased (see Figure 12.6) (Paul *et al.*, 1967). Mason *et al.* (1987a) established that by using a larger particle size of this drug, a sustained release absorption could be achieved.

Three production lots of capsules of this drug based on this formulation principle were found to be reproducible in their bioavailability (Mason *et al.*, 1987b). Capsules containing sulfafurazole of mean particle size 1.7, 39 and 95 μm, were tested in dogs by Fincher *et al.* (1965); the peak blood concentration increased with decreasing particle size. Results such as these for relatively water-insoluble drugs make it appear that reduction of the particle size of the drug should solve the bioavailability problems of capsules. However, the opposite effect can occur. Capsules of ethinamate, a drug with a solubility of 1 in 400, were tested for dissolution *in vitro*, using various particle sizes packed to give different porosities. For equivalent packing densities as judged by porosity, a greater drug release was obtained with the largest particle size fraction (Figure 12.7) (Newton and Rowley, 1970). This was because the powder bed with smaller particle sizes was less permeable to liquid. Similar effects were obtained with acetylsalicylic acid (Newton and Bader, 1980).

When four different particle size fractions of acetylsalicylic acid (≦6, <10, <50 and <100 μm) were administered in capsules to dogs, there was no significant difference in absorption, but when the powder was more closely classified to give better-defined size ranges, namely between 6 and 12 μm and between 60 and 100 μm, higher blood

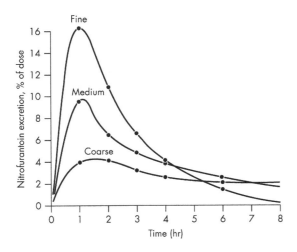

Figure 12.6 Effect of particle size of orally administered nitrofurantoin on urinary excretion in rats (Paul *et al.*, 1967, with permission from the publishers).

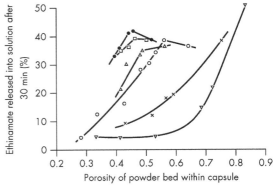

Figure 12.7 Percentage of ethinamate released into solution after 30 min from capsules containing different particle size fractions packed to give different porosities. (●) 251–420 μm; (○) 125–152 μm; (□) 177–251 μm; (✕) 66–76 μm; (△) 152–177 μm; (▽) 8.3 μm (Newton and Rowley, 1970, with permission from the publishers).

concentrations resulted from the administration of the capsules containing the larger particle size fraction (Ljungberg and Otto, 1970). Ridolfo *et al.* (1979) established that for capsules containing particles of 67 μm or 640 μm mean equivalent diameter of a relatively water-insoluble drug, benoxaprofen, the capsules containing the smaller size particles dissolved more rapidly and gave a higher *in vivo* bioavailability, judged by plasma concentration–time curves and urinary excretion. The capsules, however, contained nearly twice as much starch as drug. This could ensure water penetration between the fine particles and thus adequate contact with their larger surface area. Confirmation of this result was reported from the same laboratories (Wolen *et al.*, 1979). Again, the drug was incorporated into a large quantity of starch within the capsules. Aoyagi *et al.* (1990) compared the *in vitro* and *in vivo* bioavailability of a branded capsule formulation containing fine particle size indometacin and had a rapid dissolution, with an experimental formulation containing the same dose of the drug with 125–177 μm size fraction, mixed with seven times its weight of lactose. The branded capsule had the more rapid dissolution (see Figure 12.8), and a higher value of C_{max}, a smaller T_{max} and a shorter mean residence time, yet comparable values for the total amount of drug absorbed, as measured by area under the serum level–time curves (AUC) in the fasted state (see Figure 12.9) (open symbols). In the fed state, the only parameter that differed statistically was the mean residence time (see Figure 12.9) (closed symbols). The authors stressed the need to consider the influence of food when considering the formulation of drugs whose solubility was dependent on pH. It is also worth noting that the authors added a considerable excess of lactose to their experimental preparation and the branded product had been formulated to utilise the fine particle size of the drug.

Further evidence that it is not just the particle size that is the controlling factor in drug release from capsule formulations is provided by the work of Kubo *et al.* (1996). Working with a sparingly water-soluble drug, TA-7552 (1-(3,4-dimethoxy-phenyl)-2,3-bis (methoxycarbonyl)-4-hydroxy-6,7,8-trimethoxynaphthalene), a hypo cholesterolaemic agent, these workers found that both wet milling and air jet milling could reduce the original particle size of 39 μm to about 1 and 3 μm, respectively. When capsules containing these powders were administered to dogs, the bioavailability was 14% for the powder from the fluid energy mill and 50% for the wet ground powder, when compared to a lecithin solution of the drug. When the drug was dry ground in a ball

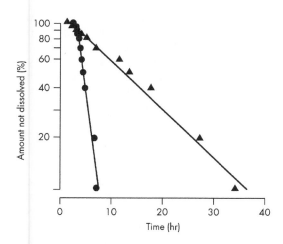

Figure 12.8 Dissolution of indometacin from preparations A (branded capsule ▲) and D (experimental formulation ●) plotted in a first order fashion. The values are the means of three determinations by the JP XI paddle method at 120 rpm in 900 mL pH 7.2 medium. (Reproduced from Aoyagi *et al.*, 1990, with permission from the publishers.)

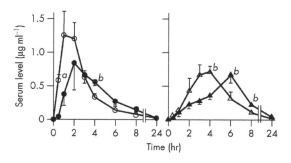

Figure 12.9 Mean serum levels of indometacin after oral administration of 25 mg of preparations A (left hand figure) and D (right hand figure) to humans (*n* = 6) having normal gastric acidity in the fasting (open symbols) and nonfasting states (closed symbols). Vertical lines show standard error. (a) *P* < 0.01, (b) *P* < 0.05. (Reproduced from Aoyagi *et al.*, 1990, with permission from the publishers.)

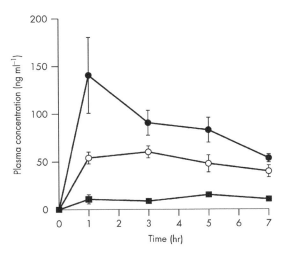

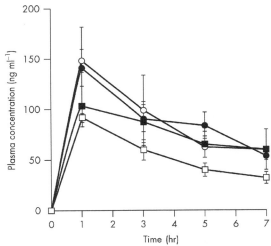

Figure 12.10   Plasma concentration profiles of TA-7552 and its glucuronide after oral administration of micronised powder and lecithin solution of TA-7552 at a dose of 50 mg in beagle dogs. (●), lecithin solution; (○), micronised powder (Eiger mill); (■) micronised powder (jet mill). Each point represents the mean of three or four dogs. (Reproduced from Kubo *et al.*, 1996, with permission from the publishers.)

Figure 12.11   Plasma concentration profiles of TA-7552 and its glucuronide after oral administration of co-ground powders and lecithin solution of TA-7552 at a dose of 50 mg in beagle dogs. (●), lecithin solution; (○), co-ground with D-mannitol at a weight ratio of 1:9; (■) co-ground with D mannitol at a weight ratio of 1:3; (□) co-ground powder with equal amount of D-mannitol by weight. Each point represents the mean and standard error from three or four dogs. (Reproduced from Kubo *et al.*, 1996, with permission from the publishers.)

mill with various ratios of mannitol, a ratio of 1 part drug to 9 parts mannitol was bioequivalent to the lecithin formulation (see Figure 12.10). While a decrease in the particle size of the drug increased the relative bioavailability of the capsules from both the single ground powder and the co-ground powder, the extent of the improvement was clearly improved by the co-grinding process as is illustrated by the results in Figure 12.11. The improvement was not associated with a change in the crystallinity of the drug, as there were no changes observed in the X-ray diffraction patterns or the DTA (differential thermal analysis) curves. This also suggests that the systems should be stable on storage, as the improvement in performance was not associated with the formation of an unstable amorphous form of the drug. As the best powder-filled capsule was able to match the bioavailability of the solution of drug in lecithin, it would appear that the drug would fall into the drug classification of 'low solubility–high permeability'. No details of the capsule sizes used or how the formulations were actually packed into the capsule were given. One must assume therefore, that those capsules that

contain the additional mannitol, are either larger or packed more tightly than those that contain the drug alone. It does appear, however, that the benefits of the co-grinding with a hard hydrophilic crystal, such as mannitol, are real effects and well worth consideration if a drug has low water solubility and high permeability.

Wetting

As discussed in the previous section, decreasing the particle size of a drug, although it increases the surface area for dissolution, does not necessarily increase the dissolution rate because there may be a reduction in the contact between the liquid and the solid. This is particularly likely if the liquid does not wet the solid. Whether a liquid will spread over the surface of a solid is determined by the relative values of the attraction of the molecules of the liquid for those of the surface and of those of the liquid molecules for each other. If the former exceed the latter, spreading of the liquid over the solid surface will occur. The ease of wetting is best expressed by the contact angle between the edge of the liquid meniscus and the solid surface. A zero value implies ready and complete wetting of the solid by the liquid while a value of 180° would correspond to absolute non-wetting. Water-insoluble drugs usually have a low affinity for aqueous fluids and hence have high values of the contact angle, indicative of non-wetting (Lerk *et al.*, 1977). Similarly the addition of the hydrophobic lubricant, magnesium stearate, can also prevent wetting of powders and hence retard dissolution.

Wetting, as indicated by the liquid penetration test of Studebaker and Snow (1955) and drug release as indicated by dissolution have been shown to correlate inversely. Interesting behaviour of magnesium stearate is shown in Figure 12.12 (Samyn and Jung, 1970). However, for more complex formulations, systems giving poor wetting do not necessarily give poor dissolution (see Figure 12.13) (Rowley and Newton, 1970). Sodium lauryl sulfate dissolves with swelling and this may retard penetration in the wetting test, yet aid disruption of the capsule in the dissolution test. Most surfactants tested were found to produce only small increases in the dissolution

rate of the drug ethinamate in capsule formulations (Newton, 1972). An alternative method of increasing the wettability of a capsule formulation is to incorporate a hydrophilic material such as starch or lactose. Such additives may of course act by disintegration as well as wetting and their effectiveness is not reliably disclosed by dissolution techniques (Newton, 1972). Large proportions (up to 80%) of the diluent may be needed to be effective and do not always guarantee

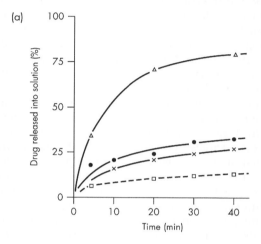

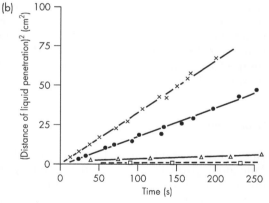

Figure 12.13 (a) Dissolution of four different capsule formulations. (b) Liquid penetration test for the same formulations. (□) drug with 0.5% magnesium stearate; (✕) drug with 0.5% magnesium stearate, 1% sodium lauryl sulfate and 5% lactose; (●) drug with 0.5% magnesium stearate, 1% sodium lauryl sulfate and 20% lactose; (○) drug with 1.0% magnesium stearate, 1% sodium lauryl sulfate and 50% lactose (Newton and Rowley, 1970, with permission from the publishers).

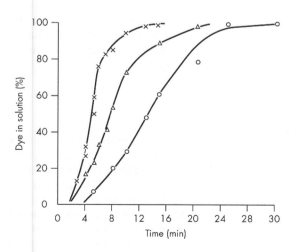

Figure 12.12 Dissolution of dye from capsules containing lactose (✕), and with 2% (△) or 5% (○) magnesium stearate (Samyn and Jung, 1970, with permission from the publishers).

complete drug release, especially of highly insoluble drugs (Newton and Razzo, 1974).

The inability of physical mixing to ensure wetting and drug release was confirmed by mixing hexobarbitone with hydroxymethyl or hydroxyethyl cellulose, which failed to reduce the contact angle or increase dissolution rate (Lerk *et al.*, 1978). However, by intimate mixing of the drug with a solution of the hydrophilic material, followed by drying to give mini-granules, a reduction in the contact angle and a marked improvement in dissolution were achieved. The contact angles, although reduced, remained relatively high, so that the improved dissolution should perhaps be attributed to the structure introduced by granulation. Treating griseofulvin with hydroxypropyl cellulose, by a similar process, improved the drug release, as assessed by dissolution and urinary excretion of the major metabolite (Fell *et al.*, 1978). As no measurements of contact angles were reported, it could again be the granulation as well as the wetting process, which is involved in improving drug release. The same process has been shown by Lerk *et al.* (1979) to improve the rate of absorption, relative to pure drug, when phenytoin capsule formulations were administered to human volunteers. The authors were able to show that the process improved water uptake into the powder plug, employing penetration tests. *In vitro* dissolution tests were found to reflect the improved liquid penetration, irrespective of the presence of surfactant in the dissolution fluid.

Granulation

Granulation provides a method of assembling individual particles in a regular manner, as opposed to the random aggregation, which occurs under the influence of the natural interparticulate forces between small particles. It is known that granulation can increase the dissolution rate of fine particles (Finholt *et al.*, 1968). The technique was found to be applicable to capsule formulations (Newton and Rowley, 1970). In this case a solvent is the only binder, so there is no change in contact angle; only the particle arrangement is altered. The improvement in dissolution rate is related to the permeability of the structure produced (Figure 12.14).

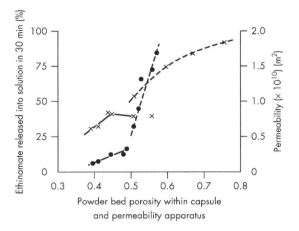

Figure 12.14 Dissolution of drug from capsules containing crystals and granules of micronised ethinamate, packed to give different porosities, compared with the liquid permeability of beds of the same samples. (——) crystals, 251–420 μm; (- - - -) granules, 251–420 μm; (✗) dissolution; (●) liquid permeability (Newton and Rowley, 1970, with permission from the publishers).

Granulation appears to be an extremely beneficial process in capsule formulation; it should improve flow, aid mixing of low-dose drugs and uniformity of bulk density in addition to assisting drug release. It is however not always considered owing to the increase in processing cost, plus the concept that having made a granulation, why not compress it to form a tablet? The granulation required to aid capsule performance may not always be of the same nature as that for compressing into a tablet.

Disintegration

Disintegration of the capsule into primary particles is a necessary requirement, especially for high-dose drugs. The deaggregation performance and serum levels of four chloramphenicol capsule formulations have been shown to be related (Aguiar *et al.*, 1968). Incorporation of twice the drug weight of sodium bicarbonate into tetracycline capsules ensured disruption of the capsule and improved dissolution (Nelson, 1960) but the incorporation of an equal weight of citric acid did not affect the drug release at all (O'Reilly and Nelson, 1961). Improved drug release, as indicated by dissolution testing, was claimed to follow the incorporation of sodium starch glycolate (Primojel, a tablet disintegrant) into capsule formulations, but as no disintegration test results are reported, improved drug release may also be a function of improved wetting as well as of disintegration (Newton and Rowley, 1975). This is also true of a comparison of capsule disintegrants by dissolution testing (Ryder and Thomas, 1977). A ranking order of starch glycolate (Primojel) > Nymcel > no disintegrant was established and the results also suggested that maize starch is not an effective disintegrant.

To assess the *in vivo* disintegration of capsules, Eckert (1967) used capsules filled with sodium bicarbonate and determined the disintegration time by the change in pH of the gastric juice, measured by a radio-endoscope. Casey *et al.* (1976) reported the use of external scintigraphy, using a gamma camera to estimate the dispersion of a gamma-emitting isotope (Tc^{99m}) contained in capsule formulations. They found that a more rapid disintegration occurred when the capsule contents were soluble (6 min) than when they were insoluble (30–40 min). Using the same technique, Hunter *et al.* (1980) found that the condition of the subjects was important. Capsules containing resin beads labelled with Tc^{99m} provided little dispersion when administered to fasting subjects, whereas after food the dispersion was more rapid and reflected *in vitro* disintegration times. A profile scanning technique was used by Alpsten *et al.* (1979) to compare the *in vivo* and *in vitro* disintegration of granule formulations of aspirin labelled with ^{53}Cr. Subjects were scanned with a movable detector whilst sitting in a well-defined position in an armchair within a low-activity laboratory. The disintegration times were considerably longer *in vivo* for both formulations than those obtained from *in vitro* disintegration tests. Tuleu *et al.* (2000) demonstrated that it was possible to detect *in vivo* disintegration of a capsule in humans by placing a 1 mm pellet, which had been impregnated with Tc^{99m}, at each end of a size 0 capsule and observe by gamma scintigraphy the change in separation of the pellets when the capsule disintegrated.

Combination of additives

The presentation of drugs in hard gelatin capsules is usually considered to be a relatively simple matter, but it is seldom possible to provide formulations, which meet the needs of filling machines without resorting to a multicomponent system. The added components, however, do not always result in a simple additive effect on drug release (Newton *et al.*, 1971a, 1971b; Newton and Razzo, 1974). The effect of changing the quantity of additives has been modelled for a single drug by fitting a second-degree equation to relate capsule composition to dissolution (Newton and Razzo, 1977b). Such a relationship is the simplest model that allows the interactions between pairs of factors to be represented. The results also include the influence of formulation combinations on the filling performance, allowing the feasibility of filling a particular combination of ingredients into the capsule to be assessed.

Capsule filling

The method by which the capsules are filled is also important. The available processes vary in complexity and hence the resultant powder-bed structures will be very different. For fine particles, decreasing the porosity resulted in a decrease in drug release (Newton and Rowley, 1970). This effect applied equally to capsules containing a drug with 10 or 50% lactose (Newton, 1972). For more complex formulations, packing seemed to be less important than other formulation factors (Newton *et al.*, 1971b). The loose and tight filling of capsules of cefalexin did not influence serum levels (O'Callaghan *et al.*, 1971), but it has been claimed that a high salicylate plasma level was obtained when the same quantity of aspirin and dibasic calcium phosphate was packed into a smaller size shell (McGee *et al.*, 1970). It would seem, however, desirable to ensure that the ingredients are filled into a capsule in such a manner as to allow rapid deaggregation of the particles. This means that care should be taken with the level of lubricant added and the way in which it is incorporated; they should be filled with a minimum of compression and/or incorporate a disintegrant that will ensure deaggregation.

Drug release from soft gelatin capsules

As with hard gelatin capsules, soft gelatin capsules are predominately administered by the oral route, but there are several papers that describe their use as a rectal dosage form, e.g. Vibelli *et al.* (1977), Carp *et al.* (1980), Moës, (1981) and Möller, (1984). In this section, however, the oral route of administration will be considered. There are only a few drugs which are of the correct consistency and dose level to form the total content of the capsule. Thus the vehicle used to present the drug within the shell is important. In general, the system must have the correct rheological characteristics to be handled by the filling process and must be compatible with the gelatin shell. Water-immiscible oils are the most important type of vehicle although water-miscible PEGs and non-ionic surfactants have also been used.

Capsules containing drug as the major component

Drugs that are oils or that are highly soluble in oils can be readily presented as a unit dose in a soft gelatin capsule, e.g. cod liver oil, clofibrate, ethchlorvynol and paramethadione. There appears to be little published work establishing the 'absolute bioavailability' of such drugs and indeed the low water solubility would present difficulties for formulation of an intravenous preparation.

Taylor and Chasseaud (1977) compared the bioavailability of clofibrate in a soft capsule with that of calcium clofibrate in a hard gelatin capsule formulation; the results indicated the bioequivalence of the two formulations. In a later paper, Taylor *et al.* (1978) compared the bioavailability of a soft gelatin clofibrate capsule with that of a film-coated tablet containing calcium clofibrate and calcium carbonate (1:1 mixture) under conditions approaching steady state. The results suggested that the two formulations did not differ significantly. Fischler *et al.* (1973) compared soft gelatin capsule formulations containing clomethiazole base with a tablet formulation containing clomethiazole edisylate. They

established that there was a more rapid and complete absorption from the capsules than from the tablets, owing to drug forms of different solubility. The addition of arachis oil to the capsule formulation was found to result in a further increase in the peak drug concentration. As the dissolution rate was not improved by the addition of arachis oil, the authors suggested that a change in the absorptive conditions of the GI tract might take place in the presence of the arachis oil. Angelucci *et al.* (1976) compared the bioavailability of flufenamic acid formulated as hard or soft gelatin capsules. The soft capsule contained vegetable oil, hydrogenated vegetable oils, beeswax and soya lecithin in addition to the drug. Results using both dogs and humans indicated that the soft gelatin formulation produced consistently higher plasma concentration–time curves. Acerbi *et al.* (1998) compared the bioavailability of ipriflavone at steady state, after administration as a 300 mg soft gelatin capsule twice a day and with a 200 mg tablet three times a day. As the drug is extensively metabolised, drug plus metabolite were measured. Their results indicated that in terms of bioavailability and reduced fluctuations in drug and metabolite levels, the capsule formulation offered clinical advantages.

Capsules containing the drug as a minor component

The solution of a drug in a solvent allows the use of accurate methods to subdivide the drug into unit-dose systems. If the solubility of the drug is high and the dose relatively low, it is possible to contain the dose within the volume of a soft gelatin capsule. Alternatively, the drug, especially if low-dose, may be dispersed in a fluid to provide an emulsion or a suspension. Such dispersion must have suitable rheological properties to ensure that they will provide uniform filling of the capsules.

Water-immiscible oils, which are compatible with the shell, provide only limited dissolving power. There is also the question of whether the drug is absorbed with the oily vehicle, an unlikely possibility, or has to transfer to the aqueous phase before absorption can take place. There are however, water-miscible solvents, which are compatible with the shell, e.g. PEGs or non-ionic surfactants, that have solvent properties for a large range of drugs.

Mallis *et al.* (1975) administered soft gelatin capsules containing 0.2 mg of digoxin dissolved in PEG and propylene glycol and compared them with a rapidly dissolving tablet containing 0.2 mg and two commercially available tablets containing 0.25 mg, in a single- and a multiple-dose study. Serum-digoxin concentrations achieved with the capsules were similar to those obtained with both the 0.25 mg tablets and better than those achieved with the 0.2 mg tablets. The time to achieve peak concentration was equivalent for all four preparations, but the area under the curve was significantly greater for the capsules than for the equivalent dose of a rapidly dissolving tablet and equivalent to that achieved with the 0.25 mg tablets. There were no significant differences between any formulations in the serum concentrations on the 2nd, 8th or 10th days, or in the steady-state urinary excretion. In a second study, the capsules were compared with a solution of the capsule contents and with a commercially available elixir; results indicated that the mean area under the serum concentration–time curve was greater for the capsule than for either of the two liquids. The mean time to reach peak concentration was similar for all three preparations. These results imply that presenting the drug in a capsule has a beneficial influence on bioavailability, possibly caused by changes in the rate of movement within the GI tract.

Marcus *et al.* (1976) compared the bioavailability of a soft gelatin capsule of digoxin (in ethanol, water, propylene glycol and PEG 400) with an oral alcoholic solution and a standard commercial tablet, relative to an intravenous solution (0.25 mg mL^{-1} in 40% propylene glycol, 10% ethanol, 0.3% sodium phosphate and 0.08% anhydrous citric acid). The intravenous infusion was administered over 1 hr and 3 hr. The serum concentration peaked at 5 ng mL^{-1} at the end of the 1-hour infusion and at 3.5 ng mL^{-1} after the second hour of the 3-hour infusion. The six-day urinary excretion after the 3-hour infusion was 21% more than for the same dose given over 1 hr. Hence, the assessment of the absolute bioavailability will be affected by the choice of infusion

rate. In general, the soft gelatin capsule had the highest level of bioavailability. Lindenbaum (1977) compared a soft gelatin capsule formulation of digoxin (in PEG 400, ethanol, water and propylene glycol) with an equivalent dose in 10% aqueous ethanol and with a tablet formulation. It was established that the area under the serum concentration–time curve and the six-day cumulative urinary excretion, were greater after administration of the capsule than for the other two formulations. Wagner *et al.* (1979) compared the *in vivo* bioavailability of four commercial soft capsule formulations of digoxin, but no details of formulation were given. There were no significant differences in the amount of digoxin absorbed from each formulation, as judged by urinary excretion or area under the plasma concentration–time curve. There were significant differences in the time to reach peak plasma concentration, in the *in vitro* 'burst time' and in the time required to release 50 or 85% of digoxin in a dissolution test. The possible involvement of the shell in the absorption process can present problems if the shell characteristics change on storage. Johnson *et al.* (1977) studied freshly prepared digoxin capsules and capsules which had been stored at 5 and 37°C for 10 months. There was some evidence that storage delayed the onset of peak plasma concentrations but there was no evidence that the extent of absorption was reduced. Further indications of the enhanced bioavailability of digoxin when presented as soft gelatin capsules come from papers reporting the administration of commercially available formulations, e.g. Binnion (1976), Longhini *et al.* (1977), O'Grady *et al.* (1978) and Alvisi *et al.* (1979). Padeletti and Brat (1978), however, compared equivalent doses of commercially available soft capsules and tablets and reported that they were bioequivalent. Heizer *et al.* (1989) compared the relative steady-state bioavailability of a soft gelatin capsule formulation of digoxin with a tablet formulation in 17 subjects with malabsorption syndromes. They concluded, 'in contrast to previous reports, we observed that digoxin from Lanoxin tablets appears to be well absorbed in subjects with malabsorption. Nevertheless, these subjects absorbed digoxin from capsules better than from tablets, with the greatest differences occurring in subjects without a colon and in those subjects with the lowest carotene concentrations'.

The use of polyhydric alcohols with non-ionic surfactants to provide solutions or dispersions in soft gelatin capsules was proposed as a method of improving the bioavailability of water-insoluble drugs by Hom and Miskel (1970). The suggestion was based on the *in vitro* dissolution performance of soft gelatin capsules of dicoumarol, diethylstilboestrol, digitoxin, ethinylestradiol, hydrocortisone, phenobarbital, phenylbutazone, propylthiouracil and sulfadiazine. In all cases, the drug appeared in solution at least as quickly and often at a faster rate from capsules than from commercial tablets. However, an improvement in bioavailability with this type of formulation does not always occur. Albert *et al.* (1974) found that soft capsule formulations of paracetamol and nitrofurantoin-containing surfactants were only equivalent in bioavailability and not superior to commercial tablets. Gundert-Remy *et al.* (1975a) found that soft capsule formulations of diphenhydramine hydrochloride based on PEG and surfactants were no more bioavailable than commercial sugar-coated tablets or hard capsule formulations. When the same drug was presented in soft capsule formulations in oil/wax surfactant mixtures, no reduction in bioavailability occurred. Similarly, oil/wax surfactant formulations of phenobarbital were found to be bioequivalent to tablet formulations (Gundert-Remy *et al.*, 1975b). In these latter cases, the drugs were relatively water soluble and the rate of drug release could be retarded by the presence of oil. In spite of a series of carefully designed experiments applying a pharmacological approach to identify the cause of the limited bioavailability (about 50% absorbed) of soft gelatin capsule formulations of etoposide, Joel *et al.* (1995) were unable to identify the source. They proposed therefore that the level of bioavailability was related to the drug and not to the formulation.

Further indications of the improved bioavailability of soft capsules come from the work of Fuccella *et al.* (1977) and Fuccella (1979), although no details of formulation are given. Comparison of commercial, hard and soft capsules of temazepam in a single-dose study established that the soft capsule produced faster absorption, with earlier and higher peak plasma

concentrations. However, there was no difference in the relative bioavailability as judged by the total area under the plasma concentration–time curve. In a comparison of 4×7.5 mg capsules and 1×30 mg capsules of temazepam, Abolin *et al.* (1993) found that there was no difference in the bioavailability of the two systems although the 7.5 mg capsules reached peak plasma concentrations slightly faster than the 30 mg dosage form. Soft gelatin capsule formulations of the highly water-insoluble dihydrotachysterol were compared with liquid formulations, one of which was a solution in middle-chain triglyceride, the other as with the capsule formulations, a solution in peanut oil or peanut oil and glycerin (Koytchev *et al.*, 1994). The preparation with the lowest bioavailability was the solution in the middle-chain triglyceride and all the other preparations were equivalent in their bioavailability. Hong *et al.* (1998) found that a soft gelatin capsule containing a suspension of gliclazide in PEG 400 and Tween 20 had a faster absorption (T_{max} 2.5 ± 1.3 hr for the capsule and 3.5 ± 0.9 hr for the tablet), yet equivalent AUC_{0-24}. Savio *et al.* (1998) found that the application of a soft gelatin capsule formulation was able to enhance the absorption of a natural product, silybin, complexed with phosphatidyl cholate. Schettler *et al.* (2001) compared the bioavailability of a new soft gelatin capsule formulation of ibuprofen with a standard sugar-coated tablet and a fast-dissolving ibuprofen lysinate formulation. The capsule and the fast-dissolving formulation had faster rates of absorption and higher C_{max} values than the standard tablets, but the extent of absorption was equivalent in the three preparations. In view of this rapid absorption, the authors claimed that the soft gelatin capsule provides a clear therapeutic advantage.

The fact that the formulation of the contents of soft gelatin capsules can influence their performance is demonstrated by the drug ciclosporin. Original formulations employed oil and alcohol solutions of the drug (Bleck *et al.*, 1990). When this was replaced with the drug dispersed in a microemulsion dose linearity was improved (Mueller *et al.*, 1994) and there was reduced inter- and intra-individual variability in the pharmacokinetics (Kovarik *et al.*, 1994). A bioequivalence study established that the microemulsion could be administered either as a capsule or in the liquid form (Kovarik *et al.*, 1993). A study in patients with rheumatoid arthritis by van den Borne *et al.* (1995), confirmed the improved bioavailability of the new formulation relative to the old. These findings were further confirmed in a bioavailability study in heart-lung transplant candidates with cystic fibrosis by Tan *et al.* (1995). Kaplan *et al.* (1999), however, found that recipients with small bowel transplants had very low bioavailability of the microemulsion formulation. This was despite the patients' normal absorption function for D-xylose and vitamin B12. It was suggested that an increase in either intestinal P450 3A4 cytochrome activity or P-glycoprotein counter-transport activity offered the best explanation for the decreased bioavailability. Lee *et al.* (1998) established, in an appropriately designed bioavailability study that, a formulation of ciclosporin from Korea, Neoplanta, met the bio-equivalence criteria of both the European Committee for Propriety Medicinal Products and the guidelines for the US Food and Drug Administration when tested against an original micro-emulsion formulation, Sandimmune Neoral. No formulation details are supplied, but presumably this new formulation was also based on a microemulsion approach.

Liquid-filled hard gelatin capsules

Walker *et al.* (1980) have described a method whereby molten and thixotropic formulations can be filled into hard capsules on a conventional filling machine. Such a process allows all the advantages previously claimed for soft gelatin preparations such as the use of an accurate liquid feed and the inclusion of PEGs and surfactants to improve the bioavailability of highly water-soluble drugs. Similarly, formulations containing waxes, which retard drug release, can be filled into hard capsules by this process, e.g. François (1983) and Brossard *et al.* (1991). Thus the formulation techniques for both hard and soft capsules are now similar and this widens the scope for controlling the release of drugs from capsule formulations and the range of dosage forms available.

Controlled-release capsule preparations

As well as providing a solid unit oral dosage form for immediate release formulations, capsules make an important contribution to oral controlled-release products. It is possible to fill hard gelatin capsules with pellets (see Chapter 6), which have been coated with a controlled-release coating, without the potential of damage to the coat which can occur if such pellets are compacted to form tablets. There are several major products, which fall into this category, e.g. Losec and Inderal LA, and there is a growing perception that multipar-ticulate controlled-release systems as opposed to monolithic devices are preferable for the administration of this type of dosage form. Graffner *et al.* (1986) and Josefsson *et al.* (1986) for example, demonstrated that enteric-coated pellets of ery-thromycin administered in a capsule provided better reproducibility of serum levels than the same drug administered as enteric-coated tablets. Reduced inter-subject variability was also reported for an enteric-coated pellet formulation of diclofenac when compared to an enteric-coated tablet formulation (Scheidel *et al.*, 1993) in healthy volunteers. Evidence that a capsule forms a readily available system for providing a unit dose from pellets is described in Chapter 6.

Treated capsules for controlled-release products

It has been known for some time that capsules can be coated on a manufacturing scale to provide an enteric-coated dosage form (Jones, 1970) and there are enteric-coated capsules containing peppermint oil on the market, e.g. Colpermin and Mintec. Murthy *et al.* (1986) have shown that it is possible to coat successfully capsules with aqueous enteric polymers. In a bioavailability study in young healthy volunteers, Roldan *et al.* (2002) compared an enteric-coated soft gelatin capsule (Aminomux, Gador SA Buenos Aires, Argentine), which contained a micronised pamidronate, a bisphosphonate, in oily phase mainly of lecithin, with an enteric-coated tablet developed by the University of

Leiden exclusively for research. They found that both products delayed drug release, with a slightly longer delay for the capsule, that the value of the C_{max} was slightly higher for the tablet, but the bioavailabilities were approximately equivalent.

An alternative to applying a coat onto the capsule is to treat the shell with an ethanolic solution of formaldehyde (Pina *et al.*, 1996, 1997). Such a process was found to be able to provide capsules whose disintegration and dissolution did not change on storage for a range of capsule formulations. There are no *in vivo* evaluations of such capsules.

However, there are no major products that control the release of drug over the entire GI tract produced by coating the capsule itself. In addition to the enteric coating systems, there have been several reports of special coating to ensure that a drug can be released in the colon (Watts and Illum, 1997). Such coats are not specifically designed for coating capsules nor have the majority of the systems been shown to be effective in humans. A coating system, which has been applied to hydroxypropyl methylcellulose capsules, has however been shown by Tuleu *et al.* (2002) to stay intact in healthy volunteers until it reached the colon, at which point it disintegrates, as measured by gamma scintigraphy. This coating system is based on a mixture of an aqueous ethanolic dispersion of amylose in a solution of ethyl cellulose in ethanol. Such systems had been shown by *in vitro* experiments to be capable of preventing dissolution in pancreatic media but are subject to digestion by bacterial amylases, which occur in the colon (Siew *et al.*, 2000). In a series of papers (Niwa *et al.*, 1995; Takaya *et al.*, 1995, 1997, 1998) Takada and co-workers have described a pressure-controlled device formed by coating the inner surface of a capsule with ethylcellulose, the thickness of which controls the time at which the capsule will disrupt and release its liquid content as a result of peristaltic pressure in the GI tract. It is claimed that in the fluid content of the upper GI tract, these pressures will not disrupt the balloon-like structure formed by the ethylcellulose membrane, while in the lower GI tract, the peristaltic pressure, caused by absorption of water, will be more effective in causing disruption of the system. While evidence of extended release times

of model drugs in dogs is provided, the actual mechanism by which the delay is achieved does not appear to be conclusively established and the release may be due to time effects rather than pressure. The complexity of the method of preparing the capsules does not appear to be commercially attractive.

In contrast to gastric emptying, there are a limited number of studies on the transit of capsules through the colon, as in most instances capsules do not survive to reach the colon intact. In a gamma scintigraphy study, Parker *et al.* (1988) enclosed capsules in a tight-fitting rubber sheath to ensure that the capsules remained intact during transit through the GI tract. They concluded that the density (0.7, 1.1 or 1.5 g cm^{-3}) and volume (0.3, 0.8 or 1.8 cm^3) of such devices had less effect on transit through the colon than the variation between subjects. In view of the system used to enclose the capsule, it is questionable whether it is strictly correct to call these dosage forms 'capsules'. An alternative approach to providing capsules for delivering drugs to the colon is to form the capsules themselves from material that is resistant to dissolution in the stomach and small intestine, but susceptible to the bacterial enzymes. Such systems are described by Brøndsted *et al.* (1998) using cross-linked dextran. While there appears to be evidence for success *in vitro*, there is as yet no evidence that such systems will indeed function in humans.

Excipient blends in extended-release formulations

Ojantakanen (1992) and Ojantakanen *et al.* (1993), using the findings of Baveja *et al.* (1985), established that it was possible to provide a delayed release form of ibuprofen by mixing approximately equal amounts of the drug with hydrophilic polymers, such as hydroxypropyl methylcellulose (HPMC, K100 and K15) and sodium carboxymethylcellulose (NaCMC, LV, MV and HV) and filling into hard gelatin capsules. The addition of the polymers led to sustained drug absorption with no loss of bioavailability when compared to a reference capsule of the drug when the formulations were administered to healthy volunteers. Such simple formulations could offer advantages to the preparation of sustained-release oral dosage forms.

Conclusions

Hard and soft gelatin capsules have a role to play in the oral administration of drugs in providing a unit solid dose system for immediate release and for release targeted at different regions of the GI tract. The two types of capsule are able to accommodate a variety for excipients, allowing many different possibilities to the formulator. The desired release performance of the dosage form can be achieved with capsule formulations with a knowledge of the drug molecules' physical chemical properties, permeability of the membrane of the GI tract, and metabolism and elimination.

References

Abolin, C., Hwang, D-S. and Mazza, F. (1993). Bioavailability of temazepam: Comparison of four 7.5 mg capsules with a single 30 mg capsule. *Ann. Pharmacother.*, 27, 695–699.

Acerbi, D., Poli, G. and Ventura, P. (1998). Comparative bioavailability of two oral formulations of ipriflavone in healthy volunteers at steady-state. Evaluation of two different dosage schemes. *Eur. J. Drug. Metab. Pharmacokinet.*, 23, 172–177.

Aguiar, A. J., Wheeler, L. M., Fusari, S. and Zelmer, J. E. (1968). Evaluation of physical and pharmaceutical factors involved in drug release and availability from chloramphenicol capsules. *J. Pharm. Sci.*, 56, 1844–1850.

Albert, K. S, Sedman, A. J., Wilkinson, P., Stoll, R. G., Murray, W. J. and Wagner, J. G. (1974). Bioavailability studies of acetaminophen and nitrofurantoin. *J. Clin. Pharmacol.*, 14, 264–270.

Al-Dujuaili, H., Florence, A. T. and Salole, E. G. (1986). The adhesiveness of proprietary tablets and capsules to porcine oesophageal tissue. *Int. J. Pharm.*, 34, 75–79.

Alpsten, M., Bogentoft, C., Ekenveld, G. and Solvell, L. (1979). On the disintegration of hard gelatin capsules in fasting volunteers using a profile scanning technique. *J. Pharm. Pharmacol.*, 31, 480–481.

Alvisi, V., Longhini, C., Bagni, B., Portaluppi, F., Ruina, M. and Fersini, C. (1979). An improvement in digoxin bioavailability. Studies with soft gelatin capsules containing a solution of digoxin *Arzneimittelforschung*, 29, 1047–1050.

Amidon, G. L., Lennernas, H., Shah, V. P. and Crison, J. R. (1995). A theoretical basis for a biopharmaceutic drug classification: the correlation of *in vitro* drug product dissolution and *in vivo* bioavailability. *Pharm. Res.*, 12, 413–420.

Angelucci, L., Petrangeli, B., Celletti, P. and Flavilli, S. (1976). Bioavailability of flufenamic acid in hard and soft gelatin capsules. *J. Pharm. Sci.*, 65, 455–456.

Aoyagi, N., Kaniwa, N. and Ogata, H. (1990). Effects of food on the bioavailability of two indomethacin capsules containing different sizes of particles. *Chem. Pharm. Bull.*, 38, 1338–1340.

Bailey, Jr. R. T., Bonavina, L., McChesney, L., Spires, K. J., Muilenburg, M. I., McGill, J. E. and Demesster, T. R. (1987). Factors influencing the transit of a gelatin capsule in the esophagus. *Drug Intell. Clin Pharm.*, 21, 282–285.

Baveja, S., Rao, K. V. R. and Devi, K. P. (1985). Sustained release tablet formulation of diethylcarbamazine. Part III. *Int. J. Pharm.*, 27, 157–162.

Binnion, P. F. (1976). A comparison of the bioavailability of digoxin in capsules, tablet and solution taken orally with intravenous digoxin. *J. Clin. Pharmacol.*, 16, 461–467.

Bleck, J. S., Nashan, B., Christians, U., Schottomann, R., Wonigeit, K. and Sewing, K-F. (1990). Single dose pharmacokinetics of ciclosporine and its main metabolites after oral ciclosporine as oily solution or capsule. *Arzneimittelforschung*, 40, 62–64.

Borne, van den B. E. E. M., Landeewe, R. B. M., Goeithe, H. S., Mattie, H., Breedveld, F. C. and Dijlmans, B. A. C. (1995). Relative bioavailability of a new oral form of cyclosporin A in patients with rheumatoid arthritis. *Br. J. Clin. Pharmacol.*, 39, 172–175.

Brice, G. W. and Hammer, H. F. (1969). Therapeutic non-equivalence of oxytetracycline capsules. *JAMA*, 208, 1189–1190.

Bron, J. (1988). Relative bioavailability of carbocysteine from three dosage forms, investigated in healthy volunteers. *Biopharm Drug. Dispos.*, 9, 97–111.

Brøndsted, H., Andersen, C. and Hovgaard, L. (1998).

Crosslinked dextran – a new capsule material for colon targeting of drugs. *J. Control. Release*, 53, 7–13.

Brossard, C., Ratsimbazafy, V. and des Ylouses, D. L. (1991). Modelling of the theophylline compound release from hard gelatin capsule containing Gelucire matrix granules. *Drug Dev. Ind. Pharm.*, 17, 1267–1277.

Brown, J., Madit, N., Cole, E. T., Wilding, I. R. and Cadé, D. (1998). The effect of cross-linking on the *in vivo* disintegration of hard gelatin capsules. *Pharm. Res.*, 15, 1026–1030.

Bullingham, R., Monroe, S., Nicholls, A. and Hale, M. (1996). Pharmacokinetics and bioavailability of mycophenolate mofetil in healthy subjects after single-dose oral and intravenous administration. *J. Clin. Pharmacol.*, 36, 315–324.

Carp, G. B., Chemtob, C. and Chaumeil, J. C. (1980). Availability from rectally administered solid dosage forms: use of soft gelatin capsules. *Second International Conference on Pharmaceutical Technology*, Paris, V. 68–80.

Casey, D. L., Beihn, R. M., Digenis, G. A. and Shambu, M. B. (1976). Method of monitoring hard gelatin capsule disintegration times in humans using external scintigraphy. *J. Pharm. Sci.*, 65, 1412–1413.

CDER (Center for drug evaluation and research) (1997). *Guidance for industry; Dissolution testing of immediate release solid oral dosage forms*. US Department of Health and Human Services (USDSHH), Food and Drug Administration (FDA), August, 1997, Rockville, MD, pp 11.

CDER (2000). *Guidance for industry; Bioavailability and bioequivalence studies for orally administered drug products – general considerations*. USDSHH, FDA, October, 2000, Rockville, MD, pp 25.

Chaulet, J. F., Nony, P., Bevalot, F., Girard, P., Chaubaud, S., Mounier, C., Clair, P., Boissel, J. P. and Grelaud, G. (2002). Bioequivalence evaluation of a fixed combination of chloroquine and proguanil in a capsule formulation versus a standard medication. *Arzneimittelforschung*, 52, 407–412.

Chouchane, N., Barre, J., Toumi, A., Tillement, J. P. and Benkins, A. (1995). Bioequivalence study of two pharmaceutical forms of rifampicin capsules in man. *Eur. J. Drug Metab. Pharmacokinet.*, 20, 351–360.

Digenis, G. A., Sandefer, E. P., Page, R. C., Doll, W. J., Gold, T. B. and Darwazeh, N. B. (2000). Bioequivalence study of stressed and nonstressed hard gelatin capsules using amoxycillin as a drug marker and gamma scintigraphy to confirm time and GI location of *in vivo* rupture. *Pharm. Res.*, 17, 572–582.

Eadie, M. J. (1968). 'Dilantin' overdosage. *Med. J. Aust.*, 2, 515.

Eckert, T. (1967). The pH-endoradio probe: a method for determining the *in vivo* disintegration of oral dosage forms. Part 1. The disintegration of hard gelatin capsules *in vivo* (in German). *Arzneimittelforschung*, 17, 645–646.

Eerikainen, S., Leino, J. Harjula, M., Klinge, E. and Marvola, M. (1996). Use of hard gelatin capsule as a rectal dosage form. *S.T.P. Pharma Sci.*, 6, 435–440.

Fell, J. T., Calvert, R. T. and Riley-Bentham, P. (1978). Bioavailability of griseofulvin from a novel capsule formulation. *J. Pharm. Pharmacol.*, 30, 479–482.

Fincher, I. H., Adams, J. G. and Beal, H. M. (1965). The effect of particle size on the gastrointestinal absorption of sulfisoxazole in dogs. *J. Pharm. Sci.*, 54, 704–708.

Finholt, P., Kristiansen, H, Schmidt, O. C. and Wold, K. (1968). Effect of different factors on the dissolution rate of drugs from powders, granules and tablets. I. *Medd. Nor. Farm. Selsk.*, 28, 17–30, 31–47.

Fischler, M., Frisch, E. P. and Ortengren, B. (1973). Plasma concentrations after oral administration of different pharmaceutical preparations of clomethiazole. *Acta Pharm. Suec.*, 10, 483–492.

François, D., 1983. Technology of pastes and oils in hard gelatin capsules. *Labo-Pharma Probl. Tech.*, 31, 944–949.

Fuccella, L. M. (1979). Bioavailability of temazepam in soft gelatin capsules. *Br. J. Clin. Pharmacol.*, 8(Suppl. 1), 31S–35S.

Fuccella, L. M., Bolcioni, G., Tamassia, V., Ferrarioo, L. and Tognoni, G. (1977). Human pharmacokinetics and bioavailability of temazepam administered in soft gelatin capsules. *Eur. J. Clin. Pharmacol.*, 12, 383–386.

Gillies, H. C., Herriott, D., Liang, R., Ohashi, K., Rogers, H. J. and Harper, P. G. (1987). Pharmacokinetics of idarubicin following intravenous and oral administration in patients with advanced cancer. *Br. J. Clin. Pharmacol.*, 23, 303–310.

Graffner, C., Josefsson, K. and Stockman, O. (1986). Intra- and inter-subject variation of erythromycin absorption from single-unit and multiple-unit enteric-coated products. *Biopharm. Drug Dispos.*, 7, 163–171.

Gundert-Remy, U., Bilzer, W. and Weber, E. (1975a). Absorption of diphenylhydramin-HCl from different dosage forms. *Drugs Made in Germany*, 18, 99–104.

Gundert-Remy, U., Bilzer, W. and Kilgenstein, R. (1975b). On the absorption of phenobarbitone from various dosage forms. *Pharm. Ind.*, 37, 905–909.

Hagenlocher, M., Hannula, A. M., Wittwer, F., Solvia, M. and Speiser, P. (1986). Hard gelatin capsules for rectal drug delivery. *Fourth International Conference on Pharmaceutical Technology*, Paris, I, 398–405.

Hamlin, W. E., Northram, J. I. and Wagner, J. G. (1965). Relationships between *in vitro* dissolution rates and solubilities of numerous compounds representative of various chemical species. *J. Pharm. Sci.*, 54, 1651–1653.

Hansford, D. T., Newton, J. M. and Wilson, C. G. (1980). The influence of formulation on the absorption of orally administered griseofulvin preparations in rabbits. *Pharm. Ind.*, 42, 646–650.

Heizer, W. D., Pittman, A. W., Hammond, J. E., Fitch, D. D., Bustrack, J. A. and Hull, J. H. (1989). Absorption of digoxin from tablets and capsules in subjects with malabsorption syndromes. *Ann. Pharmacother.*, 23, 764–769.

Hey, H., Jørgensen, F., Sørensen, K., Hasselbalch, H. and Wamberg, T. (1982). Oesophageal transit of six commonly used tablets and capsules. *Br. Med. J.*, 285, 1717–1719.

Hom, F. S. and Miskel, J. J. (1970). Oral dosage form design and its influence on dissolution rates for a series of drugs. *J. Pharm. Sci.*, 59, 827–830.

Hong, S. S., Lee, S. H., Lee, Y., Chung, S. J., Lee, M. H. and Shim, C. K. (1998). Accelerated oral absorption of gliclazide in human subjects from a soft gelatin capsule containing a PEG 400 suspension of gliclazide. *J. Control. Release*, 51, 185–192.

Honkanen, O., Pia, L., Janne, M., Sari, E., Raimo, T. and Martti, M. (2002). Bioavailability and *in vitro* oesophageal sticking tendency of hydroxypropyl methylcellulose capsule formulations and corresponding gelatin capsule formulations. *Eur. J Pharm. Sci.*, 15, 479–488.

Hunter, E. Fell, J. T., Calvert, R. T. and Sharma, H. (1980). '*In vivo*' disintegration of hard gelatin capsules in fasting and non-fasting subjects. *Int. J. Pharm.*, 4, 175–183.

Joel, S. P., Clark, P. I., Heap, L., Webster, L., Robbins, S., Craft, H. and Slevin, M. L. (1995). Pharmacological attempts to improve the oral bioavailability of oral etoposide. *Cancer Chemother. Pharmacol.*, 37, 125–133.

Johnson, B. F., Macauley, P. V., Smith, P. M. and French, J. A. G. (1977). The effect of storage upon the *in vitro* and *in vivo* characteristics of soft gelatin capsules containing digoxin. *J. Pharm. Pharmacol.*, 29, 576–578.

Jones, B. (1970). Production of enteric coated capsules. *Manuf.Chem.*, 41(5), 53–54, 57.

Jonkman-de Vries, J. D., Herben, V. M. M., van Tellingen, O., Dubbelman, A. C., ten Bokkel Huinink, W. W., Rodenhuis, S., Bult, A. and Beijnen, J. H. (1996). Oral bioavailability of low dose 'home made' etoposide capsules. *Clin. Drug Invest.*, 12, 298–307.

Josefsson, K., Levitt, M., Kann, J. and Bon, C. (1986). Erythromycin absorption from enteric-coated pellets given in multiple doses to volunteers, in comparison with enteric-coated tablets and film coated stearate tablets. *Curr. Ther. Res.*, 39, 131–142.

Kamimori, G. H., Karryekar, C. S., Otterstetterr, R., Cox, D. S., Balkin, T. J. Belenkey, G. L. and Eddington, N. D. (2002). The rate of absorption and relative bioavailability of caffeine administered in chewing gum versus capsules to normal healthy volunteers. *Int. J. Pharm.*, 234, 159–167.

Kaplan, B., Lown, K., Craig, R., Abecassis, M., Kaufman, D., Leventhal, J., Stuart, F., Meier-Kriesche, H-U. and Fyer, J. (1999). Low bioavailability of cyclosporine microemulsion and tacrolimus in a small bowel transplant recipient. *Transplant*, 67, 333–338.

Kim, Y. G., Lee, Y. J., Lee, E. D., Lee, S. D., Kwon, J. W., Kim, W. B., Shim, C-K. and Lee, M. G. (2000). Bioequivalence assessment of cycloserin capsule to Dura seromycin capsule of cycloserine after a single oral dose administration to healthy male volunteers. *Int. J Clin. Pharmacol. Ther.*, 38, 461–466.

Kovarik, J. M., Meuller, E. A., Johnston, A., Hitzenberger, G. and Kutz, K. (1993). Bioequivalence of soft gelatin capsules and oral solution of a new cyclosporine formulation. *Pharmacotherapy*, 13, 613–617.

Kovarik, J. M., Mueller, E .A., van Bree, J. B., Tetzloff, W. and Kutz, K. (1994). Reduced inter- and intra-individual variability in cyclosporine pharmacokinetics from a microemulsion formulation. *J. Pharm Sci.*, 83, 444–446.

Koytchev, R., Alken, R-G., Vagaday, M., Kunter, U. and Kirkov, V. (1994). Differences in the bioavailability of dihydrotachysterol preparations. *Eur. J. Clin. Pharmacol.*, 47, 81–84.

Kubo, H., Osawa, T., Takashima, K. and Mizobe, M. (1996). Enhancement of oral bioavailability and pharmacological effect of 1-(3,4-dimethoxyphenyl)-2,3-bis(methoxycarbonyl)-4-hydroxy-6,7,8-trimethoxynaphthalene (TA-7552), a new hypocholesterolic agent, by micronization in co-ground mixture with D-mannitol. *Biol. Pharm. Bull.*, 19, 741–747.

Lapham, G., Aranko, K., Hanhijarvi, H. and Humphreys, D. M. (1996). Bioavailability of two clodronate formulations. *Br. J. Hosp. Med.*, 56, 231–233.

Lee, Y-J., Chung, S-J. and Shim, C-K. (1998). Bioequivalence of Neoplanta capsule to Sandimmune Neoral, microemulsion formulations of cyclosporin A in humans subjects. *Int. J. Clin. Pharmacol.*, 36, 210–215.

Lerk, C. F., Lagas, M., Boelstra, J. P. and Broersma, P. (1977). Contact angles of pharmaceutical powders. *J. Pharm. Sci.*, 66, 1480–1481.

Lerk, C. F., Lagas, M. and Fell, J. T. (1978). Effect of hydrophilization of hydrophobic drugs on release rate from capsules. *J. Pharm. Sci.*, 67, 935–939.

Lerk, C. F., Lagas, M., Lie-A-Huen, L., Broersma, P. and Zuurman, K. (1979). *In vitro* and *in vivo* availability of hydrophilized phenytoin from capsules. *J. Pharm. Sci.*, 68, 634–638.

Levy, G. (1970). Pharmaceutical considerations in dosage form design. In: *Sprowls, Prescription Pharmacy*. 2nd Edition. Philadelphia, USA: J. B. Lippincott Company, 74.

Lin, J. H. (1996). Bisphosphonates: a review of their pharmacokinetic properties. *Bone*, 18, 75–85.

Lindenbaum, J. (1977). Greater bioavailability of digoxin solution capsules. Studies in the post-prandial state. *Clin. Pharmacol. Ther.*, 21, 278–282.

Ljungberg, S. and Otto, G. (1970). Particle size and intestinal absorption of acetylsalicylic acid in dogs. *Acta Pharm. Suec.*, 7, 449–456.

Longhini, C., Avisi, V., Bagni, B., Portluppi, F. and Fersini, C. (1977). Bioavailability of digoxin improved by pharmaceutical manufacturing process. *Curr. Ther. Res.*, 21, 909–912.

Lu, K., Savaraj, N., Cavanagh, J., Fuen, L. G., Burgess, M., Bodey, G. P. and Loo, T. L. (1986), Clinical pharmacology of 4-demethoxydaunorubicin (DMBR). *Cancer Chemother. Pharmacol.*, 17, 143–148.

Mallis, G. I., Schmidt, D. H. and Lindenbaum, J. (1975). Superior bioavailability of digoxin solution in capsules. *Clin. Pharmacol. Ther.*, 18, 761–768.

Marathe, P. H., Sandefer, E. P., Kollia, G. E., Greene, D. S. Barbhaiya, R. H., Lipper, R. A., Page, R. C., Doll, W. J., Ryo, U .Y. and Digenis, G. A. (1998). *In vivo* evaluation of the absorption and gastrointestinal transit of avitriptan in fed and fasted subjects using gamma scintigraphy. *J. Pharmacokinet. Biopharm.*, 26, 1–20.

Marcus, F. I., Dickenson, J., Pippin, S., Stafford, M. and Bressler, R. (1976). Digoxin bioavailability: Formulations and rates of infusion. *Clin. Pharmacol. Ther.*, 20, 253–259.

Martin, C. M., 1968. Brand, generic drugs differ in man. *JAMA*, 205(9), 23–24, 30.

Marvola, M., Vahervuo, K., Sothmann, A., Marttilia, E. and Rajaniemi, M. (1982). Development of a method for study of the tendency of drug product to adhere to the esophagus. *J. Pharm. Sci.*, 72, 975–977.

Marvola, M., Rajaniemi, M., Marttilia, E., Vahervuo, K. and Sothmann, A. (1983). Effect of the dosage form and formulation factors on the adherence of drugs to the esophagus. *J. Pharm. Sci.*, 72, 1034–1036.

Mason, W. D., Conklin, J. D. and Hailey, F. J. (1987a). Relative bioavailability of four macrocrystalline nitrofurantoin capsules. *Int. J. Pharm.*, 36, 105–111.

Mason, W. D., Conklin, J. D. and Hailey, F. J. (1987b). Between lot and within-lot comparisons of bioavailability of macrocrystalline nitrofurantoin capsules. *Pharm. Res.*, 6, 499–503.

McGee, B. J., Kennedy, D. R. and Walker, G. C. (1970). Some factors affecting the release and availability of drugs from hard gelatin capsules. *J. Pharm. Sci.*, 59, 1430–1433.

Meyer, M. C., Straugh, A. B., Mhatre, R. M., Hussain, A., Shah, V. P., Bottom, C. B., Cole, E. T. Lesko, L. L., Maliouski, H. and Williams, R. L. (2000). The effect of gelatin cross linking on the bioequivalence of hard and soft acetaminophen capsules. *Pharm. Res.*, 17, 962–966.

Moës, A. J. (1981). Formulation of highly available theophylline rectal suppositories. *Pharm Acta Helv.*, 56, 21–25.

Mojaverian, P., Chan, K., Desai, A. and John, V. (1989). Gastrointestinal transit of a solid indigestible capsule as measured by radiotelemetry and dual gamma scintigraphy. *Pharm. Res.*, 6, 719–724.

Molinaro, M., Corna, G., Fiorito, V., Spreafico, S., Bartoli, A. N. and Zoia, C. (1997). Bioavailability of two different oral formulations of amoxycillin in healthy subjects. *Arzneimittelforschung*, 47, 1406–1410.

Möller, H. (1984). *In vitro* and *in vivo* dissolution of rectal indomethacin dosage forms. *Pharm. Ind.*, 46, 514–520.

Mueller, E. A., Kovarik, J. M., van Bree, J. B., Tetzloff, W., Grevel, J. and Kutz, K. (1994). Improved dose linearity of cyclosporine pharmacokinetics from a microemulsion. *Pharm. Res.*, 11, 310–304.

Murthy, K. S., Enders, N. A., Mahjour, M. and Fawzi, M. B. (1986). A comparative evaluation of aqueous enteric polymers in capsule coating. *Pharm. Technol.*, 10, 36–46.

Nelson, E. (1960). Urinary excretion kinetics for evaluation of drug absorption II. Constant rate-limited release of tetracycline after ingestion by humans. *J. Am. Pharm. Assoc., Sci. Edn.*, 49, 54–56.

Neuvonen, P. J. (1979). Bioavailability of phenytoin: clinical pharmacokinetic and therapeutic implications. *Clin. Pharmacokinet.*, 4, 91–103.

Newton, J. M. (1972). The release of drug from hard gelatin capsules. *Pharm. Weekbl. Ned.*, 107, 485–498.

Newton, J. M. and Bader, F. (1980). The influence of drug and diluent particle size on the *in vitro* release of drug from hard gelatin capsules. *J. Pharm. Pharmacol.*, 32, 167–171.

Newton, J. M. and Razzo, F. N. (1974). The influence of additives on the *in vitro* release of drugs from hard gelatin capsules. *J. Pharm. Pharmacol.*, 26(Suppl.), 30P–36P.

Newton, J. M. and Razzo, F. N. (1977a). The *in vitro* bioavailability of drugs formulated as hard gelatin capsules. *J. Pharm. Pharmacol.*, 29, 205–208.

Newton, J. M. and Razzo, F. N. (1977b). The influence of additives on the presentation of drugs in hard gelatin capsules. *J. Pharm. Pharmacol.*, 29, 294–297.

Newton, J. M. and Rowley, G. (1970). On the release of drug from hard gelatin capsules. *J. Pharm. Pharmacol.*, 22(Suppl.), 163S–168S.

Newton, J. M. and Rowley, G. (1975). *Drug formulations*. US Patent 3 859 431.

Newton, J. M., Rowley, G. and Törnblom, J-F. V. (1971a). The effect of additives on the release of drug from hard gelatin capsules. *J. Pharm. Pharmacol.*, 23, 452–453.

Newton, J. M., Rowley, G. and Törnblom, J-F. V. (1971b). Further studies on the effect of additives on the release of drug from capsules. *J. Pharm. Pharmacol.*, 23(Suppl.), 156S–160S.

Niwa, K., Takaya, T., Morimoto, T. and Takada, K. (1995). Preparation and evaluation of a time-controlled release capsule made of ethylcellulose for colon delivery of drugs. *J. Drug Targeting*, 3, 83–89.

O'Callaghan, C. H, Tootill, J. P. R. and Robinson, W. D. (1971). A new approach to the study of serum concentrations of orally administered cephalexin. *J. Pharm. Pharmacol.*, 23, 50–57.

Ogata, H., Aoyagi, N., Kaniwa, N., Ejima, A., Sekine, N., Kitamura, M. and Inoue, Y. (1986). Gastric acidity dependent bioavailability of cinnarizine from two commercial capsules in healthy volunteers. *Int. J. Pharm.*, 29, 113–120.

O'Grady, J., Johnson, B. F., Bye, C. and French, J. (1978). The comparative bioavailability of Lanoxin tablets and Lanoxicaps with and without sorbitol. *Eur. J. Clin. Pharmacol.*, 14, 357–360.

Ojantakanen, S. (1992). Effect of viscosity grade of polymer additive and compression force on dissolution of ibuprofen from hard gelatin capsules. *Acta Pharm Fenn.*, 100, 211–218.

Ojantakanen, S., Marvola, M., Hannula, A.-M., Klinge, E. and Naukkarinen, T. (1993). Bioavailability of ibuprofen from hard gelatin capsules containing different viscosity of hydroxypropylmethylcellulose and sodium carboxymethylcellulose. *Eur. J. Pharm. Sci.*, 1, 109–114.

O'Reilly, I. and Nelson, E. (1961). Urinary excretion kinetics for evaluation of drug absorption. IV Studies with tetracycline absorption enhancement factors. *J. Pharm. Sci.*, 50, 413–416.

Parker, G., Wilson, C. G. and Hardy, J. G. (1988). The effect of capsule size and density on transit through the proximal colon. *J. Pharm. Pharmacol.*, 40, 376–377.

Padeletti, L. and Brat, A. (1978). Bioavailability of digoxin capsules. *Int. J. Clin. Pharmacol. Biopharm.*, 16, 320–322.

Paul, H. E., Hayes, K. K. J., Paul, M. F. and Borgmann, A. R. (1967). Laboratory studies with nitrofurantoin. Relationship between crystal size, urinary excretion in the rat and man and emesis in dogs. *J. Pharm. Sci.*, 56, 882–885.

Perkins, P. J., Wilson, C. G., Blackshaw, P. E., Vincent, R. M., Danserau, R. J., Juhlin, K. D., Bekker, P. J. and

Spiller, R. C. (1994). Impaired esophageal transit of capsule versus tablet formulations in the elderly. *Gut*, 35, 1363–1367.

Pina, M. E., Sousa, A. T. and Brojo, A. P. (1996). Enteric coating of hard gelatin capsules. Part 1. Application of hydroalcoholic solutions of formaldehyde in preparation of gastro-resistant capsules. *Int. J. Pharm.*, 133, 139–148.

Pina, M. E., Sousa, A. T. and Brojo, A. P. (1997). Enteric coating of hard gelatin capsules. Part 2. Bioavailability of formaldehyde treated capsules. *Int. J. Pharm.*, 148, 73–84.

Rail, L. (1968). 'Dilantin' overdosage. *Med. J. Aust.*, 2, 339.

Randolph, W. C., Beg, M. M. A. and Swagzdis, J. E. (1985). Bioavailability of a modified formulation capsule containing 25 mg hydrochlorthiazide and 50 mg triamterene. *Curr. Ther. Res.*, 38, 990–996.

Rawashdeh, N. R., Najib, N. M. and Jalal, I. M. (1997). Comparative bioavailability of two capsule formulations of mefenamic acid. *Int. J. Clin. Pharmacol. Ther.*, 35, 329–333.

Ribeiro, W., Zappi, E. A., Moraes, M. E. A., Bezerra, F. A. F., Lerner, F. E. and de Nucci, G. (2000). Comparative bioavailability of two fluconazole capsules formulations in healthy volunteers. *Arzneimittelforschung*, 50, 1028–1032.

Ridolfo, A. S., Thompkins, L., Bechholt, L. D. and Carmichael, R. H. (1979). Benoxaprofen, a new anti-inflammatory agent: particle size effect of dissolution rate and oral absorption in humans. *J. Pharm. Sci.*, 68, 850–852.

Roldan, E. J. A., Quattrocchi, O., Zanchetta, J., Plotkin, H., Arjaujo, G. and Piccinni, E. (2002). Comparable bioavailability of two esophago-gastric protective formulations containing pamidronate. *Medicina (Buenos Aires)*, 62, 317–322.

Rowley, G. and Newton, J. M. (1970). Limitations of liquid penetration in predicting the release of drugs from hard gelatin capsules. *J. Pharm. Pharmacol.*, 22, 966–967.

Rubino, T., Halterein, L. M. and Blanchard, J. (1985). The effects of aging on the dissolution of phenytoin sodium capsule formulations. *Int. J. Pharm.*, 26, 165–174.

Ryder, J. and Thomas, A. (1977). A comparison of the effectiveness of several disintegrants in capsules of 4-ethoxycarbonylphenoxy-2'-pyridyl methane (BRL 10614). *J. Pharm. Pharmacol.*, 29, Suppl. 63P.

Samyn, J. C. and Jung, W. Y. (1970). *In vitro* dissolution from several experimental capsule formulations. *J. Pharm. Sci.*, 59, 169–175.

Sasahara, K., Nitanai, T., Habara, T., Kojima, T., Kawahara, Y., Morioka, T. and Nakajima, E. (1980). Dosage form design for improvement of bioavailability of levodopa IV: Possible cause of low bioavailability of oral levodopa in dogs. *J. Pharm. Sci.*, 69, 261–265.

Savio, D., Harrasser, P. C. and Basso, G. (1998). Softgel capsule technology as an enhancer device for the absorption of natural principles in humans. A bioavailability cross-over randomised study on silybin. *Arzneimittelforschung*, 48, 1104–1106.

Sechaud, R., Graf, P., Bigler, H., Gruendl, E., Letzkus, M. and Merz, M. (2002). Bioequivalence study of valsartan tablet and capsule formulation after single dose in healthy volunteers using replicated crossover design. *Int. J. Clin. Pharmacol. Ther.*, 40, 35–40.

Scheidel, B., Blume, H., Walter, K., Nieciecki, A. and Babej-Dolle, R. M. (1993). Bioavailability study of enteric coated diclofenac formulations. 1st Communication: Bioavailability study following administration of a multiple unit formulation compared with a single-unit formulation. *Arzneimittelforschung*, 43, 1211–1215.

Schettler, T., Paris, S., Pellett, M., Kidner, S. and Wilkinson, D. (2001). Comparative pharmacokinetics of two fast-dissolving oral ibuprofen formulations and a regular-release ibuprofen tablet in healthy volunteers. *Clin. Drug Invest.*, 21, 73–78.

Siew, L. F., Basit, A. W. and Newton, J M. (2000). The properties of amylose-ethylcellulose films cast from organic-based solvents as potential coatings for colonic drug delivery. *Eur. J. Pharm Sci.*, 11, 133–139.

Smith, D. B., Margison, J. M., Lucas, S. B., Wilkinson, P. M. and Howell, A. (1987). Clinical pharmacology of oral and intravenous 4-demethoxydaunorubicin. *Cancer Chemother. Pharmacol.*, 19, 138–142.

Stewart, A. G., Grant, D. J. W. and Newton, J. M. (1979). The release of a model low- dose drug (riboflavine) from hard gelatin capsules. *J. Pharm. Pharmacol.*, 31, 1–6.

Stewart, D. J., Grewaal, D., Green, R. M. Verma, S., Maroun, J. A. Redmond, D., Roobillard, L. and Gupta, S. (1991). Bioavailability and pharmacology of oral idarubicin. *Cancer Chemother. Pharmacol.*, 27, 308–314.

Studebaker, M. L. and Snow, C. W. (1955). The influence

of ultimate composition upon the wettability of carbon blacks. *J. Phys. Chem.*, 59, 973–976.

Sun, J. X., Walter, B., Sanderfer, E. P., Page, R. C., Degenis, G. A., Ryo, U. Y., Cipriano, A., Maniara, W. M., Powell, M. L. and Chan, K. (1996). Explaining the variable absorption of a hypolipidemic agent (CGP 43371) in healthy subjects by gamma scintigraphy. *J. Clin. Pharmacol.*, 36, 230–237.

Takaya, T., Ikeda, C., Imagawa, N., Niwa, K. and Takada, K. (1995). Development of a colon delivery capsule and the pharmacological activity of recombinant human granulocyte colon-stimulating factor (fhG-CFS) in beagle dogs. *J. Pharm. Pharmacol.*, 47, 474–478.

Takaya, T., Sawada, K., Suzuki, H., Funaoka, A., Matsuda, K. and Takada, K. (1997). Application of colon delivery capsule to 5-aminosalicylic acid and the evaluation of the pharmacokinetic profile after oral administration to beagle dogs. *J. Drug Targeting*, 4, 271–276.

Takaya, T., Niwa, K., Muraoka, M., Ogita, I., Nagai, N., Yano, R., Kimura, G., Yoshikwa, H. and Takada, K. (1998). Importance of dissolution process on the systemic availability of drugs delivered by colon delivery system. *J. Control. Release*, 50, 111–122.

Tan, K. K. C., Trull, A. K. and Uttridge, J. A. (1995). Relative bioavailability of cyclosporin from conventional and microemulsion formulations in heart–lung transplant candidates with cystic fibrosis. *Eur. J. Clin. Pharmacol.*, 48, 285–289.

Tannenbaum, P. J. Rosen, E., Flanagan, T. and Crossley, A. P. (1968). The influence of dosage form on the activity of a diuretic agent. *Clin. Pharmacol. Ther.*, 9, 598–604.

Taylor, T. and Chasseaud, L. F. (1977). Plasma concentrations and bioavailability of clofibric acid from its calcium salts in humans. *J. Pharm. Sci.*, 66, 1638–1639.

Taylor, T., Chasseaud, L. F., Darragh, A. and O'Kelly, D. A. (1978). Bioavailability of *p*-chlorophenoxyisobutyric acid (clofibric acid) after repeated doses of its calcium salt to humans. *Eur. J. Clin. Pharmacol.*, 13, 49–53.

Terao, T., Oga, T., Ota, S., Yamamoto, S., Otsubo, Y., Zamami, M. and Okado, M. (1994). Comparison of single dose pharmacokinetics between lithium capsules and tablets (Limas): a Japanese study. *Lithium*, 5, 157–159.

Tuleu, C., Waddington, W., Lui, D. and Newton, J. M. (2000). Gamma camera resolution: potential application for the evaluation of coated capsule disintegration. *Proceedings 3rd World Meeting APV/AGPI*, Berlin 3/6 April. 877–878.

Tuleu, C., Basit, A. W., Waddington, W. A., Ell, P. J. and Newton J. M. (2002). Colonic delivery of 4-aminosalicylic acid using amylose-ethylcellulose-coated hydroxypropylmethylcellulose capsules. *Aliment. Pharmacol. Ther.*, 16, 1771–1779.

Tyrer, J. H., Eadie, M. J., Sutherland, J. M. and Hooper, W. D. (1970). Outbreak of anticonvulsant intoxication in an Australian city. *Br. Med. J.*, 4, 271–273.

Vibelli, C., Cherrichetti, S., Sala, P., Ferrari, P. and Pasotti, C. (1977). Feprazone. Bioavailability in a new suppository preparation. *Clin. Trials J.*, 14, 83–88.

Villikka, K., Perrttunen, K., Ronsell, J., Ikavalko, H., Vaho, H. and Pylkkanen, L. (2002). The absolute bioavailability of clodronate from two different oral doses. *Bone*, 31, 418–421.

Wagner, J. G., Gerard, E. S. and Kaiser, D. G. (1966). The effect of dosage form on serum levels of indoxole. *Clin. Pharmacol. Ther.*, 7, 610–619.

Wagner, J. G, Stoll, R. G., Weidler, D. J., Ayers, J. W., Hallmark, M. R., Sakmer, E. and Yacobi, A. (1979). Comparison of the *in vitro* and *in vivo* release of digoxin from four different soft gelatin capsule formulations. *J. Pharmacokinet. Biopharm.*, 7, 147–158.

Walker, S. E., Ganley, J. A., Bedford, K. and Eaves, T. (1980). The filling of molten and thixotropic formulations into hard gelatin capsules. *J. Pharm. Pharmacol.*, 32, 389–393.

Watts, P. and Illum, L. (1997). Colonic drug delivery. *Drug Dev. Ind. Pharm.*, 23, 893–913.

Wolen, R. L., Carmichael, R. H., Ridolfo, A. S., Thompkins, L. and Ziege, E. A. (1979). The effect of crystal size on the bioavailability of benoxaprofen: studies utilizing deuterium labeled drug. *Biomed. Mass Spectrum.*, 6, 173–178.

York, P. (2002). The design of dosage forms. In: Aulton, M. E. (Ed.), *Pharmaceutics: The science of Dosage Form Design*. 2nd edition., Edinburgh, UK: Churchill & Livingston.

13

Capsule standards

Brian E Jones

The standards that are applied to capsules can be divided into two categories. Pharmacopoeial standards control the quality of capsules in relation to their medicinal use, i.e. ensure that they contain the correct drug in the correct dosage and that it is available for absorption. Industrial standards control the quality of the capsule shell and its contents in order to ensure the efficiency of the manufacturing process and to produce a product that is acceptable to the consumer.

Pharmacopoeial standards

The official tests are designed to ensure that capsule products comply with a minimum acceptable standard. They fall into three groups: raw materials, product quality, which covers the content of active ingredient and uniformity of weight, and tests that measure the release of product from the capsules, e.g. disintegration and dissolution. The plethora of different tests in the past is being actively rationalised by the three main pharmacopoeias, European (PhEur), Japanese (JP) and the United States (USP), as they work towards harmonisation through the International Conference on Harmonisation (ICH) initiatives. The best example of this is the progress that has been made on the standardisation of dissolution test apparatuses. Other pharmacopoeias have also adopted the two main standard apparatuses, basket and paddle; Indian (1996) and the People's Republic of China (2000).

Types of capsules, definitions

The entries in the pharmacopoeias that are concerned with capsule products define more than one sort of capsule. For instance, in the PhEur five categories of capsules are distinguished: hard capsules, soft capsules, gastroresistant capsules, modified-release capsules and cachets. This list is slightly confusing because the third and fourth categories in this list can be either hard or soft capsules whose mode of release of their active contents has been modified by formulation and/or processing. The last category, the cachet, has historically been used in different ways from the hard capsules to which it is most akin. It was not included in the PhEur monograph until the 1998 edition. Its inclusion in a capsule monograph could be seen as more logical if its name in other languages is considered; in German it is *Stärkekapseln* and in Latin *capsulae amylacae*.

Raw materials

Pharmacopoeial monographs give only brief details, if any at all, of the substances from which capsules can be made. They make the implicit assumption that all the materials shall be of pharmacopoeial quality. Apart from the base polymer, gelatin or hydroxypropyl methylcellulose, other materials that may be used include colouring agents, plasticizers, preservatives and surfactants. Colouring agents and other additives are considered in Chapter 4.

Product quality

Empty hard capsules

Most pharmacopoeias have tests for filled capsule products but most have none for the empty capsules themselves. The exceptions are the Chinese and Japanese pharmacopoeias. The entry in the JP in its capsule monograph has a test called 'purity'. This test uses a sample of five capsules. The capsules are tested individually. Each one is taken apart, placed in a 100 mL conical flask, 50 mL of water is added and the flask is shaken repeatedly. During this test the water is maintained at 37 ± 2°C. All the capsules must completely dissolve within 10 min. The resulting solutions must be odourless and neutral or slightly acidic. The Chinese pharmacopoeia has a full monograph that also includes tests on filled capsules and this is discussed in a later section.

Disintegration

Standard oral products

This test was introduced originally as a means of measuring the disintegration of tablets as an indicator that *in vivo* they would break-up into their original granules. One of the first pharmacopoeias to introduce a test for tablets was the British Pharmacopoeia (BP) in 1948. The apparatus consisted of a test tube, 15 cm long and 2.5 cm wide, filled with sufficient water at 37°C, to almost completely fill it so that there was an airspace of about 1 cm when the tube was closed with a plug. The tube was then placed in a water bath and inverted repeatedly at such a speed that the tablet travelled through the water but without hitting the ends. The end-point was when the tablet had 'dissolved or completely disintegrated'. A sample of five tablets was tested and the limit was less than 15 min, with a retest allowed if one failed. There were obviously problems in determining this end-point because in the next edition of the British Pharmacopoeia (1953) the end-point was changed to read, 'to dissolve or disintegrate or to soften throughout so that it disintegrates at the slightest touch'. The first monograph on capsules in the BP appeared in 1958 and included the statement that 'The shell is soluble in water at 37°C'. In this volume

of the pharmacopoeia, the apparatus for testing tablet disintegration was changed to the form and dimensions of that used in all pharmacopoeias today: a set of tubes, with mesh bases, held in a pair of plastic plates and attached to a mechanism for raising and lowering it in a container of water. This was apparently developed from an apparatus used for shaking-out extractions that was first used to test tablets by the US army medical department in 1940 (Gershberg and Stoll, 1946).

The first disintegration test for capsules was introduced in the BP in 1968. This used the standard tablet disintegration apparatus, a sample of five capsules tested together in one tube and had a limit of not more than 15 min. The determination of the end-point for capsules is more problematic than for tablets because of the adhesive nature of the shell. When capsules disintegrate, they split and allow the contents to escape, leaving empty capsule shells. The shell pieces may then agglomerate to form a large mass of gelatin, which will take longer to dissolve because of its thickness and may adhere to the mesh partly blocking the holes. Thus the determination of a precise end-point is more problematic than for tablets because of lumps of gelatin adhering to the mesh. In addition it was known that the solubility of gelatin is influenced by temperature. This problem was recognised by the BP and they commissioned a study to try to improve the reproducibility of the test (Jones and Cole, 1971). These workers examined the effects of the nature of the test solution, its temperature and whether single capsules should be tested instead of groups of five. They included in their tests both hard and soft gelatin capsule products of the BP. To measure the effect of temperature on gelatin capsule disintegration they used the ball bearing test of Boymond *et al.* (1966, 1967). This test involves putting a metal ball bearing in the body of a hard capsule, suspending it in a stirred container of water and then measuring the time for the ball bearing to be released from the capsule shell. This method avoids the influence of the fill material inside capsules that is the rate-controlling factor in disintegration (Jones, 1972). They found that over the permitted temperature range of the BP test there was an approximately 30% change in solubility and that when the temperature fell to about 30°C the capsules became insoluble. The recommendation from this study was that single capsules instead of groups of five should be tested and that the temperature range should be reduced to 37 ± 1°C and this was adopted in the BP in 1973. Following the lead of the USP the major pharmacopoeias have standardised on an apparatus with six tubes, which is now the normal sample size.

Several test fluids are permitted to be used. Water is always the first choice. The PhEur, when justified and authorised, permits the use of 0.1 M hydrochloric acid or artificial gastric juice. The volume of test fluid is not precisely defined other than its depth must be such that the assembly at its highest point must be at least 15 mm below the surface. The temperature of the fluid must be maintained between 36 and 38°C. Adding a disc to the tube to help keep the capsules below the surface of the liquid is allowed by both the PhEur and the USP 26 if permitted by the individual monograph. The JP test uses a disc for all capsule products.

The PhEur (2002) and the USP 26 have adopted a special apparatus for large-size capsules and tablets that has only three tubes with larger diameters than the standard ones. The USP 26 defines large capsules and tablets as being longer than 18 mm. The PhEur has a wider limit on the temperature control for this apparatus than for the standard test of 37 ± 2°C.

The end-point of the disintegration test has over the years been modified to try and cover all eventualities in order to make it easier to judge whether on not the capsule has complied with the test. The PhEur considers it to be achieved when there is no residue on the screen, or if present, it is soft and has no palpably firm or dry core and that there are only fragments of gelatin shell remaining on the screen or attached to the lower surface of the disc. The JP defines the end-point as when no residue of the capsules or only the film of the capsules or only a small amount of soft residue or a muddy substance remains in the tube. The USP 26 states that within the time limit all of the capsules have disintegrated except for fragments of shell. Both the JP and the USP 26 permit a retest. The JP describes a failed capsule as one that remains intact or shows no evidence of leakage of its contents even though the film has dissolved, been ruptured or ripped off. A further six capsules are tested and all must pass the test for the sample to be accepted. The USP 26 describes the non-compliance of the sample if one or two capsules fail to disintegrate completely. An additional sample of 12 capsules is tested and not less than 16 of the total must have disintegrated for the sample to pass the test. The time limit for capsule product disintegration varies between the pharmacopoeias. For the standard test the JP uses a limit of 20 min and the PhEur, USP and the BP have a limit of 30 min. There is one product in the USP 26, Cyanocobalamin Co 58, where the sample size is one. The reason for this is that it is a radiopharmaceutical used for special diagnostic procedures and only small lots are prepared at any one time.

More recently the pharmacopoeias have moved away from disintegration testing to dissolution testing. The British Pharmacopoeia (2002) has a test using water for 30 min for all products except for three standard release hard capsule products, cefalexin, erythromycin estolate and mefanamic acid; two are carried out in hydrochloric acid solutions and two have a shorter 15 min time limit. The USP 26 includes a disintegration test for several soft capsule products but these are performed in dissolution apparatus no. 2, the rotating paddle. The sample size is six capsules and they are tested individually. The test medium is 500 mL of water with a paddle speed of 50 rpm. The test instructs that the capsule is dropped into the apparatus and the paddle only started after it has sunk to the bottom. The time taken for the capsule to rupture is noted and the limit for this is 15 min. If two of the six capsules fail this but release their contents between 15 and 30 min then a further sample of 12 capsules is tested. The sample passes if not more than two of the 18 rupture in more than 15 but less than 30 min. There is only one disintegration test for a hard gelatin capsule product, salsalate. This uses the standard disintegration apparatus with simulated gastric fluid with no added pepsin as the test medium and has a limit of not more than 15 min.

Gastroresistant products

Many pharmacopoeias have included tests for enteric capsules over the years, even though there are few products available and those are restricted to a few countries (Jones and Törnblom, 1975). Gastroresistant capsules, sometimes referred to as enteric capsules, are examples of a delayed-release product, which are intended to resist the action of the gastric fluids and release their active substances in the intestinal fluid. They can be manufactured either by coating or modifying the shell, or by filling the capsules with particles covered with a gastroresistant coating.

The European Pharmacopoeia (2002) has two tests that depend on how the product is formulated. For capsules that contain particles with gastroresistant coating a dissolution-type test is carried out. For capsule products with gastric-resistant shells a modified disintegration test is carried out. The standard apparatus is used. The test uses initially a 0.1 M hydrochloric acid solution and the apparatus is operated for 2 hr, after which time the capsules are examined and must show no signs of disintegration or rupture that permit the escape of the contents. The test medium is then replaced with a phosphate buffer solution pH 6.8, discs are added to the tubes and the apparatus operated for 1 hr. The sample passes the test if all the capsules have disintegrated. A retest is permitted if capsules have stuck to the discs during the test. For the retest no discs are used and all six capsules must comply. The PhEur has only one gastroresistant capsule product, peppermint oil.

The JP has a similar set of two tests to the PhEur. The tests are performed on separate samples of six capsules both using the standard disintegration apparatus. The first one uses test fluid no. 1 (sodium chloride/hydrochloric acid mixture, pH 1.2) and the apparatus is operated for 120 min. The capsules are then examined and pass if no more than one of the capsules shows signs of disintegration; the enteric coating film is ruptured, peeled off, or otherwise broken and active substance has leaked. If two of the capsules fail then a repeat test is permitted and all six of these must pass. The second one uses as test fluid no. 2. (phosphate buffer, pH 7.5) with discs and it is operated for 60 min. The capsules pass if they meet the standard J P end-point for disintegration. For capsules that contain enteric-coated granules, these are first removed from the

capsules and placed on a 500 µm sieve. A sample of 0.1 g of the residue on the sieve is transferred to each of the six tubes in the disintegration apparatus. The test medium is fluid no. 1 and the apparatus is operated for 60 min. The sample complies with the test if no more than 15 particles have fallen through the mesh. The test is repeated using test fluid no. 2 and a time of 30 min. The sample complies with the test if no residue remains in any of the tubes or only one sample remains intact, or if the residue is the film of the samples, or is only a small amount of soft residue or muddy substance.

The USP 26 does not have a disintegration test for enteric-coated products and they are tested by using a dissolution test. This is discussed in the appropriate section of this chapter.

Non-oral products

Capsules that are for rectal or vaginal administration may also have to comply with a disintegration test. The PhEur applies the test that is routinely used for suppositories and pessaries unless they are intended for prolonged local action. This consists of two perforated metal discs set 30 mm apart held in a glass or plastic cylinder. The capsule to be tested is placed in this device which in turn is placed in a vessel containing at least 4 L of water at 36–37°C. The water is agitated with a slow stirrer. The device is held vertically and must be positioned not less than 90 mm below the surface of the test fluid. It is inverted, without being removed from the water, every 10 min. Disintegration is said to be complete when the gelatin shell ruptures, allowing the release of the contents. The same limit of 30 min is applied to both rectal and vaginal capsules.

Dissolution testing

The dissolution test measures the rate at which a drug is released into solution from a dosage form and is used as an indication of the bioavailability of the product and of product quality. Over the years many different types of apparatus have been devised. However, in order to simplify testing procedures the major pharmacopoeias have chosen to standardise on three apparatuses: the revolving basket, the paddle and the flow through cell. The first two apparatuses are the ones of choice for the majority of tests and the latter one is used for products whose active component is poorly soluble. In addition the USP 26 has a fourth apparatus based on the standard disintegration apparatus and this is used for extended-release products that require the dissolution rate to be determined over a range of pHs. This apparatus allows for the easy change over of the test medium. The original concept adopted by the pharmacopoeias was to use water as the test medium in order to simplify the procedures. However, because of the wide range of solubilities and the need to provide sink conditions it has been found necessary to use special test media in order to obtain realistic values for this test.

There is a significant difference between the USP and the JP, and the PhEur in that all three pharmacopoeias describe the apparatuses and the test conditions that can be used but the PhEur does not have monograph details on the tests to be used for individual capsule products. The PhEur in the preamble to the 'dissolution test for solid dosage forms' states 'unless otherwise justified and authorised, either the paddle apparatus or the basket apparatus or in special cases the flow-through cell apparatus may be used'. The general standard test conditions to be used are listed and are referred to as 'prescribed for each preparation' and none is individually listed (European Pharmacopoeia, 2002). In Europe, individual countries in their national pharmacopoeias have listed specific test conditions in product monographs, e.g. the British Pharmacopoeia.

One of the problems that occurs during dissolution testing is that capsule products, especially those in hard capsules, float because their fill density is less than $1 \ g \ mL^{-1}$. This is particularly true for apparatus no. 2, the rotating paddle, and the USP states it is the preferred one for tablets. Apparatus 1 is preferred for capsules or products that tend to float, however, in practice the USP uses the rotating paddle for about one-third of its capsule products. The pharmacopoeias have attempted to overcome the problem of capsules floating by the use of a sinker and there are differences in those prescribed by the three major pharmacopoeias. The JP authorises the use of one for any product that floats and it has the most precise specification for one. It is constructed from an acid-resistant metal wire 1 mm in diameter and is described as being a metal cage, the body of which has an inside diameter of 12.0 ± 0.2 mm, a length of 25–26 mm and the sides are held together by double wires in the shape of a cross. It is constructed so that one of the ends can be opened to place the capsule inside and is closed with a clasp. The PhEur simply states that for products that float when using apparatus no. 2, a glass or metal helix should be used to keep the dosage form horizontal at the bottom of the vessel. This sinker must be made from substances that are chemically inert in the media for it to be used in a dissolution test. The USP 26 makes a similar statement to the PhEur for products that float in apparatus no. 2. It states that a small loose piece of non-reactive material such as not more than a few turns of wire helix can be attached to the dosage unit. However, it does permit in addition the use of any validated sinker device. The USP 26 in the general information section 1088 '*In vitro in vivo*' evaluation' describes the use of a sinker made from 'a few turns of platinum wire'. The USP has one test in apparatus no. 1 that uses a sinker, diazepam extended-release capsules. It instructs that the capsule should not float but be kept on the base of the basket by wrapping around it a piece of copper wire, which is of 18-gauge thickness, 10 cm long and is bent into four loops. It has been shown that the nature of the sinker can influence the dissolution results (Avgoustakis *et al.*, 1992).

Standard-release products

The USP 26 has set dissolution standards for the majority of its standard capsule products; of the 143 monographs, 128 have a dissolution test, 12 have a disintegration test and only three products have neither test, benzonatate, beta-carotene and isotretinoin. One product, a triamterene hydrochloride and hydrochlorthiazide capsule, has three different dissolution tests. Another two products, extended phenytoin sodium and nitrofurantoin macrocrystalline form capsules, are not described as extended-release products *per se* but the dissolution limits are in the style of such products. It is presumed that this is because the original products were launched during the 1960s before dissolution testing was made official and that these were the formulations and performance that they had when launched. The range of test conditions and limits applied for standard-release capsule products are shown in Table 13.1 and 13.2. The PhEur does not list individual test conditions. The BP has 71 monographs for products in capsules, of these 67 are for standard-release products and of these 26 have defined dissolution tests and limits; summaries of the test conditions are shown in Tables 13.3 and 13.4. The JP has only one capsule dissolution test, which is for indometacin. This uses apparatus no. 1, revolving at 100 rpm with 900 mL of pH 7.2 phosphate buffer and the sample passes if 75% is released in 20 min. The BP has special conditions for one product isotretinoin, which is an unstable compound. The dissolution test is carried out in the standard disintegration apparatus, which it requires to be flushed out with nitrogen before use and be protected from light during the test.

The USP 26 has a special procedure for 16 capsule products. The filtered solutions taken from the dissolution apparatus are pooled together to form the test solution that is used to determine the average amount in the sample by repeat testing in either automated apparatuses or in those which compare samples with a standard such as in gas/liquid chromatography. This can

Table 13.1 USP 26 summary of dissolution test conditions for standard-release capsule products in apparatus no. 1, rotating basket

Basket speed (rpm)	No. of products	Volume test medium (mL)	No. of products	Test media	No. of products
50	7	500	12	Water	43
100	72	750	2	HCl solutions	21
150	2	900	61	Phosphate buffer	9
		1000	6	Acetate buffer	4
				SGF*	3
				Mixed systems	2
				Tris buffer	1

SGF* is simulated gastric fluid with no enzymes.

Table 13.2 USP 26 summary of dissolution test conditions for standard-release capsule products in apparatus no. 2, rotating paddle

Paddle speed (rpm)	No. of products	Volume test medium (mL)	No. of products	Test media	No. of products
50	34	500	7	Water	24
75	11	900	34	HCl solutions	10
100	4	1000	7	SLS* solutions	4
		4000	1	Phosphate buffer	4
				Mixed solution	4
				Acetate buffer	1
				SGF**	1
				SIF***	1

SLS* is sodium lauryl sulfate; SGF** is simulated gastric fluid with no enzymes; SIF*** is simulated intestinal fluid with no enzymes.

Table 13.3 BP (2002) summary of dissolution test conditions for standard-release capsule products in apparatus no. 1, rotating basket

Speed (rpm)	No. of products	Volume test medium (mL)	No. of products	Test media	No. of products
100	11	900	11	0.1 M HCl	7
				Water	2
				0.12 M HCl	1
				pH 7.5, phosphate buffer	1

Table 13.4 BP (2002) summary of dissolution test conditions for standard-release capsule products in apparatus no. 2, rotating paddle

Speed (rpm)	No. of products	Volume test medium (mL)	No. of products	Test media	No. of products
50	12	500	1	0.1 M HCl	6
75	2	900	14	pH 7.5 phosphate buffer	4
100	1			Water	2
				pH 7.2 phosphate buffer	1
				pH 4.7 acetate buffer	1
				pH 4.5 phosphate	1

best be illustrated by using as an example ampicillin capsules 500 mg. First, a standard test solution is prepared by dissolving an accurately weighed amount of USP ampicillin RS in water to give a solution with a concentration of $500/900$ mg mL^{-1}; 500 being the label claim quantity in milligrams of ampicillin in the capsules. This solution is used in an automated analysis system as the reference to calculate the strength of samples withdrawn from the pooled samples. Sampling and testing is continued until a series of criteria are met that are based on the label claim amount of ampicillin and the amount (Q) that has to be released in the stated

Table 13.5 USP 26 acceptance tables for capsule dissolution tests

Type	Level	Number tested	Acceptance criteria
Single capsule	S_1	6	Each capsule is not less than $Q + 5\%$.
	S_2	6	Average of 12 capsules ($S_1 + S_2$) is equal to or greater than Q and no individual capsule is less than $Q - 15\%$.
	S_3	12	Average of 24 capsules ($B_1 + B_2 + B_3$) is equal to or greater than Q, not more than 2 capsules are less than $Q - 15\%$, and capsule is less than $Q - 25\%$.
Pooled sample	S_1	6	Average amount dissolved is not less than $Q + 10\%$.
	S_2	6	Average amount dissolved ($S_1 + S_2$) is equal to or greater than $Q + 5\%$.
	S_3	12	Average amount dissolved ($S_1 + S_2 + S_3$) is equal to or greater than Q.

S is sample; Q is target quantity to be dissolved (75% or as specified in the individual monograph); B is test in buffer solution as specified or pH 6.8.

time in the dissolution test. This system allows testing to be finished early if the sample meets a more stringent limit than the final test, see Table 13.5.

Capsule shells can sometimes interfere with the assay of actives. If this occurs the USP uses a correction factor. This is obtained by analysis of a solution prepared from at least six capsule shells that have been carefully emptied of product and dissolved in the dissolution medium. The correction factor cannot be used if it is greater than 25% of the label claim.

Modified-release products

The PhEur has a category of capsule, apart from standard hard and soft capsules, called 'modified-release'. Modified-release capsules are defined as 'hard or soft capsules in which the contents, or the shell, or both, contain special excipients or are prepared by a special process designed to modify the rate, place or time at which the active substance(s) are released'. The term modified-release can be applied to either prolonged-release or delayed-release capsule products. The latter description could be applied to gastroresistant capsules, which confusingly is another category of the PhEur. The USP 26 has two similar categories, extended-release articles and delayed-release (enteric-coated) articles. There are 12 monographs for extended-release capsule products and four for delayed-release (enteric-coated) products.

Extended-release products

There is currently a marked divergence in the pharmacopoeias in the test applied to these products. In Europe the BP has only one prolonged-release product, propranolol and to test this product it states that 'A suitable test is carried out to demonstrate the appropriate release of propranolol. The dissolution profile reflects the *in vivo* performance, which in turn is compatible with the dosage schedule recommended by the manufacturer'. In the USA on the other hand the USP 26 has 13 monographs for extended-release capsule products. The formulations of registered products can differ significantly for the same active ingredient because different manufacturers make them. This factor has been

allowed for by giving more than one test method and limits when required for such products. To do this, the test methods and limits included by the manufacturers of these products in their dossiers for regulatory approval have been used. The other significant difference from standard-release products is that the length of time over which the tests are carried out is much longer. These are normally given in hours and a tolerance of ± 2% on the stated time is permitted.

The USP 26 uses four apparatuses for these tests. The standard dissolution apparatus no. 1, the rotating basket, apparatus no. 2, the rotating paddle, apparatus no. 3, the reciprocating cylinder, which is based on the disintegration apparatus, and apparatus no. 4, the flow-through cell. There are monographs for 13 products and only apparatuses no. 1 and 2 are used. The limits for each product are given in the format of a range of the percentage amounts of labelled content released at each time point. The way in which this is interpreted to reduce the number of samples is shown in Table 13.6. The best way to describe the diversity of the test methods is to go through some product examples.

Diazepam extended-release capsules have a single test method and limit. Apparatus no. 1 is used with 900 mL of simulated gastric fluid TS, with no enzymes and is operated at a speed of 100 rpm for 12 hr. It instructs, rather unusually, that the capsule be wrapped in a small piece of copper wire to ensure that it remains at the base of the basket and does not float. The test fluid is sampled at 1, 4, 8 and 12 hr and the following limits are applied to each time point: 15–27%, 49–66%, 76–96% and 85–115%, respectively. Chlorphenamine (chlorpheniramine) maleate extended-release capsules have two test methods and limits. The first method uses apparatus no. 1 with 500 mL of water and is operated at a speed of 100 rpm for 10 hr. The test fluid is sampled at 1.5, 6.0 and 10.0 hr and the following limits are applied to each time point: 15–40%, 50–80% and not less than 70%, respectively. Products that pass this test are labelled to indicate that it meets USP drug release test 1. The second method uses apparatus no. 2 at 50 rpm and it is operated according to method B for delayed-release (enteric-coated) products, see next section. This is a two-stage test, first in an acid solution and

Table 13.6   USP 26 acceptance table for extended-release capsule products

Level	Number tested	Criteria
L_1	6	No individual capsule lies outside each of the stated ranges and no individual capsule is less than the stated amount at the final test time.
L_2	6	The average value of the 12 capsules ($L_1 + L_2$) lies within each of the stated ranges and is not less than the amount at the final test time; none is more than 10% of the labelled content outside each of the standard ranges; none is more than 10% of labelled content below the stated amount at the final test time.
L_3	12	The average value of the 24 capsules ($L_1 + L_2 + L_3$) lies within each of the stated ranges, and is not less than the stated amount at the final test time; not more than 2 of the 24 capsules are more than 10% of labelled content below the stated content at the final test time; none of the capsules is more than 20% of labelled content outside each of the stated ranges or more than 20% of the labelled content below the stated amount at the final test time.

then in a buffer. The capsules are first exposed to 900 mL of 0.1 M hydrochloric acid solution for 1 hr then transferred to 900 mL of simulated intestinal fluid TS with no enzymes for a further 6 hr. Samples are tested after 1.0, 3.0 and 7.0 hr and the following limits applied at each time point: 30–60%, 55–85% and not less than 75%, respectively. Products that pass this test are labelled to indicate that they meet USP drug release test 2. Indometacin extended-release capsules have three test methods all carried out using apparatus no. 1, with a speed of 100 rpm and using a pH 6.2 phosphate buffer as the test solution. The differences between the tests are: two use 750 mL of solution and the other uses 900 mL, two tests are run for 24 hr and the other one for 12 hr and the final limit values were not less than 80% released for two (12 and 24 hr) and 85% for the other. The two most complex monographs from the testing point of view are theophylline extended-release capsules, which have nine test methods and diltiazem hydrochloride extended-release capsules, which have 12 tests, this being a reflection of the number of manufacturers and hence formulations that are on the market in the USA. Both cases are made complex because for each product there are different dosage regimens indicated on the labels, either 12- or 24-hour dosing. The sampling times for the diltiazem product for the 12-hour dosing product range from 8 to 24 hr and for the 24-hour product from 15 to 30 hr.

Delayed-release (enteric-coated) products

The USP 26 has monographs for four delayed-release products that are enteric coated in some way: aspirin, doxycycline hyclate, lansoprazole and pancrealipase. Each one has its own specific test requirements but all are tested using dissolution apparatus no. 1, the rotating basket. There are two test methods for this category of product, A and B, both of which involve two stages; the first stage involves exposure to an acid medium and the second to a less acid buffer.

Aspirin delayed-release capsules are described as having either the shell or the contents enteric coated. They are tested using method A at a rotation speed of 100 rpm. For the first stage 750 mL of 0.1 M HCl is used and the capsules are exposed for 2 hr. At the end of this time a sample of test medium is taken, the capsule is removed from the acid solution. This is replaced with a pH 6.8 buffer solution. The apparatus is operated for 90 min and then a further sample of test medium taken. The sample is considered to have passed the test if it complies with the conditions shown in Table 13.7. The value Q in the table for aspirin capsules is 75%.

Doxycycline hyclate delayed-release capsules contain enteric-coated granules and use test method B. The test is carried out by emptying the contents of the capsules and putting them into the basket of the apparatus. The acid and buffer stages of the test are carried out on separate samples. In this case the acid test uses 1000 mL of 0.06 M hydrochloric acid, a basket speed of 50 rpm and a time of 20 min. The first sample of six pass if no individual result is greater than 50% dissolved. If this is not met then a further six capsules are tested and pass if not more than two of the results are greater than 50%. The buffer stage uses a pH 5.5 neutralised phthalate solution for 30 min. The sample passes if not less than 85% of the labelled amount is released.

Lansoprazole delayed-release capsules contain enteric-coated granules and are tested by method A, the same as for aspirin, except that it is carried out in apparatus B, with the paddle revolving at 50 rpm in 500 mL of test media. The product must release not more 10% in 60 min in the acid stage and release not less than 80% after 60 min in the buffer stage.

Pancrealipase capsules contain enteric-coated granules and are tested using a standard dissolution method. The first stage is carried out in apparatus 1 at a basket speed of 100 rpm, using 800 mL of simulated gastric fluid TS, without enzymes for 60 min. At the end of this time the content of each basket is transferred to the glass vessel of apparatus no. 2. The test solution is 800 mL of pH 6.0 phosphate buffer, the paddle speed is 100 rpm and the operation time is 30 min. At the end of this a sample is removed for analysis of lipase activity. The sample passes if not less than 75% of the labelled USP units of lipase activity per capsule are dissolved. If the results

with these last three products do not meet the first case then retests are allowed and limits are given in acceptance tables for the total sample.

The dissolution of aspirin from a gastroresistant capsule product was measured in both apparatus no 2, the rotating paddle and USP apparatus no. 4, the flow-through cell, as part of a collaborative study between pharmaceutical company laboratories in Denmark and Sweden (Gjellan *et al.*, 1997). They found comparable results were obtained in all laboratories with both apparatuses and they commented that it was easier to change the pH conditions from 1.2 to 6.8 with the flow-through cell.

Application of dissolution testing

This area has provoked a lot of thought and investigational work since it has been realised that there are more efficient ways in which to make progress. Two factors came together in the 1990s, the Biopharmaceutical Classification System and the American FDA's Scale-Up and Post Approval Changes (SUPAC) initiative. Both of these point to the fact that some changes are simple and can be made without risk whilst others are much more complex and require significantly more effort to make them work. One of the factors in this is that dissolution studies can have an important part to play in reducing unnecessary testing. The views of the

Table 13.7 USP 26 acceptance tables for delayed-release (enteric-coated) capsules

Type	Level	Number tested	Criteria
Acid	A_1	6	No individual value exceeds 10% dissolved.
	A_2	6	Average of 12 capsules ($A_1 + A_2$) is not more than 10% dissolved and no individual capsule is greater than 25% dissolved.
	A_3	12	Average of 24 capsules ($A_1 + A_2 + A_3$) is not more than 10% dissolved and no individual capsule is greater than 25% dissolved.
Buffer	B_1	6	Each capsule is not less than $Q + 10\%$.
	B_2	6	Average of 12 capsules ($B_1 + B_2$) is equal to or greater than Q and no individual capsule is less than $Q - 15\%$.
	B_3	12	Average of 24 capsules ($B_1 + B_2 + B_3$) is equal to or greater than Q, not more than 2 capsules are less than $Q - 15\%$, and capsule is less than $Q - 25\%$.

Q is target quantity to be dissolved; A is test under acid conditions; B is test in buffer solutions as specified or pH 6.8.

regulatory bodies on both sides of the Atlantic have been given in a series of guidance documents (CDER, 1997, 2000a, 2000b; CPMP, 2001).

Some of the official thinking behind dissolution testing can be found in the general information included in the American and British pharmacopoeias. The views of the BP are explained in a Supplementary Chapter in section E entitled 'Dissolution testing of solid oral dosage forms' (BP, 2002) and those of the USP in section 1088 '*In vitro* and *in vivo* evaluation of dosage forms' (USP 26). The BP confirms the efforts that have been made to standardise European methodology with the USA. The three apparatuses of the PhEur and the BP have been standardised with the USP. For the BP the paddle apparatus is stated to be the method of choice for many preparations but they do not contemplate changing over the existing tests that use the basket method to this one. The principle behind the tests is to work under 'sink conditions', which the BP defines as a volume of test fluid 'at least five to 10 times the saturation volume' and the USP as 'not less than three times that required to form a saturated solution of the drug substance'. The test conditions are chosen to provide a 'gentle hydrodynamic regimen'. Test fluids that are closer to 'physiological media' are preferred to those composed of mixtures of water and organic solvents or to solutions containing surfactants. The BP considers that the limits applied to this test 'offer an acceptable degree of assurance of "total dissolution"'. They drew attention to the fact that every dosage unit in a sample must release 70% of the label claim of the active and if a retest is carried out then all in the second sample must comply. They described the standard time limit of 45 min as being somewhat arbitrary but which should be satisfactory for the majority of standard-release products. Compliance with the test was considered to give an 'assurance that most of the active ingredient would be dissolved in a reasonable time when the dosage form was subjected to mild agitation in an aqueous medium'. They commented on the fact that whilst compliance does not assure bioavailability it does reduce the chance of 'unsatisfactory bioavailability caused by inadequate dissolution'. There is a significant difference between the BP and the USP. Not all

monographs for solid oral dosage forms in the BP have dissolution limits and it was explained that 'the ultimate objective of dissolution testing' was to ensure 'adequate and reproducible bioavailability without the recourse to routine *in vivo* testing'. After products have been approved for use it was stated that dissolution testing played a role in quality assurance to ensure batch-to-batch consistency. Overall it was explained that although the objective of the BP was to set a standard dissolution test for all relevant products, there was a problem in applying this to well-established products. This was particularly true for widely used products made by a wide range of companies, who typically have their own dissolution specifications. A pragmatic approach was being adopted to try and find compromise conditions that satisfied all cases and in some cases the opportunity was taken to harmonise conditions with the USP.

The BP considers that one outstanding problem that needs to be solved is the problem of dissolution testing for low-solubility preparations. The USP approach to this problem has been to use modified media and these data have been correlated with *in vivo* results. However, the use of conditions that move away from the 'gentle hydrodynamic regimen' in an aqueous medium raises the question about the relevance of specifications as a quality indicator or as an indicator of bioavailability. Two sorts of media modifiers have been used, surfactants and organic solvents, which raise the question about whether the excipients in a product function in the same way in such a milieu, e.g. disintegrants. They reported that a consensus had been reached that if a modifier had to be used then it should be low concentrations of sodium lauryl sulfate. The USP 26 has nine tests for standard-release capsule products with test media that contain sodium lauryl sulfate. The concentrations are in the range 0.5–5%, except for ursodiol, which uses a phosphate buffer with 0.01%. The use of the flow-cell apparatus is seen as a method that is most suitable for low-solubility preparations. The BP commented that there was lack of experience in its usage in the UK. They had set-up a working party, consisting of representative industry, control laboratories and licensing assessors. Its initial findings were that the method had merits and should be pursued further.

When modified-release preparations are approved by the licensing authorities, data must be provided showing that their dissolution profiles reflect the *in vivo* performances. However, the BP after much discussion and practical trials had come to the conclusion that it was not possible for the pharmacopoeia to control the products through their dissolution profiles. They quoted the example of theophylline products where the problem is to avoid dose dumping and at the same time ensure sufficient release of the active. The USP follows a completely different course over this matter and for modified-release products it has multiple methods and limits for those products that are produced by more than one manufacturer.

Product testing *in vivo* and *in vitro*

The USP has a section '1090 *in vivo* bioequivalence guidances'. This has two parts, one on the general conduct of bioequivalence tests and the other on protocols for specific drugs. The general guidance gives a background to the statistical basis for the organisation of such tests and how to interpret results. It concentrates on oral extended-release dosage forms. It gives a series of drug protocols that give the precise details and background for carrying out *in vivo* bioequivalence and *in vitro* dissolution testing. There are 24 protocols and five are for capsule products: cefaclor, gemfibrozil, ketoprofen, piroxicam and tolmetin. All of these actives are well absorbed after oral administration. The protocols are aimed at providing generic manufacturers with the information required to prepare a dossier for comparing their proposed product against the FDA designated reference product.

The individual protocols describe the clinical usage and pharmacology of the drug and its pharmacokinetics. They give a detailed methodology for carrying out the bioequivalence study, including types of study, such as single dose under fasting conditions, the effects of food, etc. *In vitro* testing is described together with the conditions under which the results from this could be used to obtain a waiver to obviate the need for an *in vivo* study.

Uniformity of mass and of content

The uniformity of dosage units can be demonstrated by either measuring their mass or their content of active ingredient. The pharmacopoeias divide up dosage forms into various categories and apply limits to them that are related to both their application and their method of manufacture.

Uniformity of mass

The test for uniformity of mass is the simplest indicator of the content of active ingredient, assuming that the contents of the capsule are homogeneous. The specified weight limits, within which the capsule contents must fall, are symmetrically placed about the mean weight of the test sample in all pharmacopoeias. The uniformity of mass limits for capsules are for the contents only. The sample of capsules is first weighed individually and then the contents are emptied out. This is a more difficult procedure for soft capsules than for hard capsules. The USP instructs that the soft capsules be cut open using a suitable clean dry instrument such as a scalpel. The contents are removed from the shells by washing them in a suitable solvent and leaving them at room temperature for 30 min for any solvent retained to evaporate. Care has to be taken in performing this operation to try and ensure that water from the shells is not lost. The PhEur suggests that ether is a suitable solvent for carrying this out. The capsule mass is the difference between the two readings.

The PhEur uses a sample of 20 capsules and applies a 'double limit' test. For capsules containing less than 300 mg, not more than two out of the sample may be outside ±10% of the average and all must be within ±20%. For capsules with a greater weight, 300 mg or more, the PhEur specifies a stricter set of limits. In this case, not more than two out of the sample may be outside ±7.5% and all must be within ±15%. In this case the limit is based on the average mass of the sample taken.

The USP has a test that combines the weight of capsule contents with the product assay. It instructs that a sample of not less than 30 capsules be taken. Ten of these capsules are then weighed individually, emptied out and the shells weighed. The weight of the capsule contents is then calculated from these two weights. The content weights are used to calculate the content of active in each capsule by using the result of the product assay as prescribed in the individual monograph. These results must comply with the prescribed limits in the individual monographs. Some USP monographs do not have a content uniformity test and in these cases a weight variation test is applied. They define certain categories of products to which this applies; this test applies to those that contain 50 mg or more of active providing that it does not comprise more than 50% by weight of the capsule contents. If the product contains actives that are present at lower concentrations then a content of uniformity test must be applied to them. Liquid-filled soft capsules are also exempt except for those that are filled with suspensions.

A simple test in which the decision is taken on a single sample of a specified size is usually prescribed. However, the JP has a sequential test that is applied to soft capsules. An initial sample of 20 capsules is weighed; if they fail the test, a further sample of 40 is weighed and the results of both tests are pooled to reach a final decision. The aim is not to reduce sampling but to make allowance for the greater difficulty of handling soft capsules.

Uniformity of content

Tests for uniformity of content of capsules are included in tests for single-dose preparations. The tests are all based on taking an initial sample and applying a limit, if the product does not comply then a further sample is tested and a different limit applied to the total number of results. This type of process has been designed to reduce analytical effort but at the same time assure product quality. There is a significant difference between the pharmacopoeias in how they treat the results, whether they apply the limits to the mean of the assay samples or whether they apply them to the label claim.

The PhEur requires that the product be assayed using a suitable analytical method. Multivitamin and trace-element preparations are not included in this test. Three different criteria are applied to the results obtained depending upon the type of product. Test B is applied to capsules. A sample of 10 capsules is taken at random and assayed individually and the average content calculated. The sample passes the test if not more than one result is outside 85–115% of the average value and none is outside 75–125%. The sample fails if more than three results are outside of the limits 85–115% or if one result is outside 75–125%. A retest is permitted for samples where two or three individual results are outside the 85–115% limits but within the 75–125% limits. If this is the case then a further 20 capsules taken at random are assayed. The product then complies if not more then three of the individual results are outside the 85–115% and none is outside of the limits 75–125%.

The USP has an analytical test that applies to all individual dosage units. A sample of at least 30 capsules is taken. Ten capsules are then assayed individually as per the monograph for the preparation. For those products were there is some analytical difficulty, the monograph allows a composite sample to be made from the number of individual dosage forms sufficient to make an analytical sample. Aliquots of this sample, which correspond to the contents of single capsules, are assayed. From these results the weight of active ingredient equivalent to one average capsule is calculated (P). A correction factor F is calculated, which is the weight of active ingredient equivalent to one capsule obtained from the standard assay (A) divided by P. This correction factor is validated by calculating the percentage difference between A and P and dividing it by A. If this value is greater than 10, F is not valid. The F factor is used to calculate the weight of active ingredient in each dosing unit by multiplying the individual weights found in P by F. Two types of limit are applied to the results obtained, one based on percentages and the other on the relative standard deviation (RSD) both calculated from the label claim for the product. The acceptance criteria are that the sample passes if either all capsules are within the range 85.0–115.0% or if the RSD is equal to or less than 6.0%. A retest is permitted if one capsule is outside the range 85.0–115.0% but not outside 75.0–125.0% or if the RSD is greater than 6.0%. A further sample of 20 capsules is tested. The sample complies if two or three capsules are outside 85.0–115.0% but not outside 75.0–125.0% or if the RSD does not exceed 7.8%.

The JP uses a sample of 30 capsules and from the results calculates an acceptance value. This value is calculated as follows:

$$\text{Acceptance value} = \mid M - \bar{x} \mid - ks$$

where, M is the label claim, $\bar{x}$ is the mean of individual assays, s is the standard deviation of the sample and k is the acceptance constant. $k = 2.2$ for a sample of 10 capsules and $k = 1.9$ for a sample of 30 capsules.

The contents of 10 capsules are assayed first and if the acceptance value is equal to or less than 15% then the sample passes. If it does not, a retest is performed on a further 20 capsules and this passes if the final acceptance value is equal to or less than 15% and no capsule exceeds 25% of label claim.

The Indian Pharmacopoeia (1996) has an interesting adaptation of this test which recognises that sometimes a sample of 20 capsules may not always be available. The smallest sample that is permitted is five and to make the test more statistically correct the limits are widened accordingly. For example for a product with limits of 90–110%, which contains more than 0.12 g and less than 0.3 g of active the following values 0.2, 0.5 and 1.2 are subtracted from the lower value and 0.3, 0.4 and 0.8 are added to the upper values for samples sizes of 15, 10 and 5, respectively. For capsule products with other ranges of active content the values are adjusted proportionately.

Powders for inhalation in hard capsules

The first capsule product that contained a powder for inhalation was Intal developed by Fisons in the UK and registered in the early 1970s (Martindale, 1972). This contained micronised sodium cromoglicate dispersed on to the surface of lactose particles. These were also the first capsules to be produced where the weight of the fill was less than that of the shell. This presented some practical problems because most production control systems measure the gross weight of the product and use this to adjust the fill weight of capsules. The fill weights of this type of capsule are less than 40 mg, with some containing less than 10 mg. Microdosing systems had to be developed for automatic filling machines for these products. The difficulty in controlling the uniformity of fill weight for such products has

been recognised by the pharmacopoeias. The first to do so was the British in 1973 which included a new type of capsule product first called a 'cartridge'. The British Pharmacopoeia (1980) called it an 'insufflation'. These had much wider uniformity of fill weight limits than standard capsules. The British Pharmacopoeia (2002) has a monograph for 'sodium cromoglicate powder for inhalation' and while the word capsule does not appear in the title it does so in the text. The hard gelatin capsules are described as containing from 20.0 to 24.4 mg of active per capsule. It lists two different formulations, one that contains lactose and one without lactose. For the former the limits for the uniformity of weight are that when the contents of a sample of 20 capsules are weighed, all the weights should deviate by no more than 15% of the average weight for the sample except that two can be within 15 and 25%. For the latter the limits are that no weight must deviate by more then 25% from the average weight of the contents. There is an additional limit for moisture content determined by loss on drying at 100°C at a pressure of less than 0.7 kPa: the limits are for formulations containing lactose, 5.5–10.0% and for those with only active 9.0–18.0%. The reason for this is that sodium cromoglicate has an anomalous behaviour with regard to its equilibrium moisture content when filled into gelatin capsules (Bell *et al.*, 1973).

The USP has a special test for dry powder inhalations that are filled into hard capsules. In this case it is the uniformity of the capsule contents as delivered through the inhaler mouthpiece that is measured. The capsule is placed inside its inhaler and device operated to pierce or open the capsule as instructed in the product information. The inhaler mouthpiece is connected to the specified apparatus, B. This consists of a collection tube at one end of which there is a mouthpiece adapter to hold the inhaler tightly and to give an airtight seal. At the other end there is a filter paper covering an outlet leading to two valves and a timer that are connected to a vacuum pump. These control the airflow so that it produces a specific pressure drop, 4 kPa, across the inhaler and for a time that allows 4 L of air to be drawn through the apparatus. After the inhaler has been used, the contents of the collection tube and the filter paper are rinsed with a suitable solvent and assayed. The capsule sample of 10 passes the test if not more than one amount is outside the range 75.0–125.0% of the label claim and no amount is outside the range 65.0–135.0%. A retest on a further 20 capsules is allowed if two or three capsules are outside 75.0–125.0% but not outside the limit 65.0–135.0%. The sample passes if not more than three amounts are outside the limits 75.0–125.0% and no amount is outside 65.0–135.0%. There is one product in this category 'cromolyn sodium inhalation powder'. This is described as being a mixture of equal parts of cromolyn sodium (sodium cromoglicate) and lactose and the content of active ingredient per capsule is between 95 and 125%.

Capsules for nutritional supplements

The USP/NF 24/19 (2001) included for the first time a special series of tests for nutritional supplements. These include disintegration and dissolution tests and a weight variation test.

Disintegration testing

This uses the standard apparatus for larger capsules and tablets. The initial sample size for capsules is six. The test medium for both types of gelatin capsules is 0.05 M acetate buffer, pH 4.5, kept at $37 \pm 2°C$. One capsule is added to each and the apparatus operated for 45 min. At the end of this time the baskets are removed from the fluid and the capsules are considered to have disintegrated if the standard end-point definition is met. If one or two capsules fail to disintegrate completely then a further 12 capsules are tested and the sample passes if no fewer than 16 of the capsules pass the test. There is a separate test for botanical dosage forms. For these capsule products the same test medium is used, but the apparatus is only operated for 20 min before the capsules are examined.

Dissolution testing

These tests are for standard-release products and not those intended for delayed or extended release and apply to all USP classes of nutritional supplements for vitamin–mineral dosage forms II

to VI that have been prepared as capsules. Class I products contain oil-soluble vitamins and are thus exempt as is the requirement for dissolution standards for oil-soluble vitamins contained in other classes of product. Soft gelatin capsules of nutritional substances meet the requirements for disintegration.

Many of these gelatin capsule products contain many components and to measure the dissolution of every one would make an almost impossible task. To simplify the testing the USP has introduced the concept of 'index vitamins' and 'index minerals'. The index vitamin is chosen from the water-soluble ones present and is preferably riboflavin. If the latter is not present in the formulation then pyridoxine, then niacin, then thiamine and if none of these is present in the formulation ascorbic acid is chosen. Similarly for products that are combinations of minerals, the order of preference is iron, calcium, zinc and magnesium. The dissolution test conditions are, the use of apparatus no. 1, the test medium is 900 mL of 0.1 M HCl with the basket revolving at 100 rpm and the duration of the test in 1 hr. For products for which riboflavin is the index vitamin the volume of the test medium is increased to 1800 mL. The limit for the index vitamin and mineral is that not less than 75% of the labelled content is released in 1 hr.

Weight variation

The weight uniformity tests are those that were applied in former editions of the USP to standard capsules products. For hard capsules the initial test is to weigh individually a sample of 20 intact capsules. The sample passes if all the weights of the capsules fall within ±10% of the average weight of the sample. If they do not then a sample of 20 is weighed individually so the identity of each capsule is maintained, the capsule caps are removed and the contents carefully removed and the insides cleaned out using a cotton wool plug. The emptied shells are then reweighed and the weight of the contents calculated. The difference between the individual net weights and the average net weight of the contents is calculated. The sample passes if not more than two capsules deviate from the average between 10 and 25% and none exceeds 25%. If between two and not more than six capsules are between 10 and 25% then another retest with a further sample of 40 capsules is permitted. In the case of these pooled results a new average weight is calculated and the test criteria are applied to the new value. The sample passes if not more than six exceed 10% of the new average and none exceed 25%. The same criteria are applied to soft capsules and the test procedure includes instructions in how to clean the emptied capsule shells of residual fill material.

Miscellaneous pharmacopoeial requirements

Most pharmacopoeias include other statements in the monographs that refer to matters such as formulation of the contents, labelling of products, microbiological quality, stability, storage conditions and packaging.

Microbiological quality

Materials that are to be filled into capsules must conform to certain general criteria. In particular, additives should be innocuous and should not reduce the stability of the active ingredient or interact with the capsule shell. The microbiological quality of non-sterile pharmaceutical products is dependent upon their formulation and their method of use. In the European Pharmacopoeia (2002) capsule products are listed in category 3B, which is for products for oral administration and made from natural animal or vegetable origin materials: the limits per gram are a total viable count, not more than 10^4 bacteria, 10^2 fungi and 10^2 enterobacteria or other Gram-negative bacteria and an absence in a 1 g sample of *Salmonella*, *Escherichia coli* and *Staphylococcus*. The USP 26 comments that few of the raw materials used to manufacture pharmaceutical products are received in a sterile state and encourages the use of procedures and practices to minimise contamination. Normal industrial procedures use monitoring of materials and the environment to act as an indicator of performance. It concludes by stating that 'it remains essential to apply strict good manufacturing practices to assure a lowest possible load of microorganisms'.

Packaging and storage

Storage and packaging are closely interrelated because without one the other cannot be controlled. The pharmacopoeias have considered this factor and apply varying degrees of complexity to the advice that they offer. This varies from the simple statement in the JP that they should be stored in 'well-closed containers', to the longer PhEur advice of 'storage in well-closed containers at a temperature not exceeding 30°C' and to the extensive list in the USP that defines the container for dispensing capsules for each product. The USP has three categories of container, tight, well-closed and light-resistant. All the pharmacopoeias in the individual monographs have, where necessary, specific temperature storage advice.

The USP 26 has included a new section of general information: (1146) packaging practice – repackaging a single solid oral drug product into a unit-dose container. This note reflects the increasing use of such repacking operations for both capsule and tablet products in pharmacy practice not only in the USA but also in many other countries throughout the world. It defines what is meant by pharmacy, repackaging and a repackager. It defines the operation as the removal of a product from its original bulk package and putting it into another container, normally smaller in size, for general distribution and not for use by a named patient. There are details given about the construction of blister packages and of the types of films that can be used. It describes the process and the types of unit-dose packages that can be used, a thermoformed blister, a pouch and a preformed one. The requirements for qualifying the installation, operation and performance of the systems are given together with how inspections should be carried out.

Hard capsule monographs

The tenth edition of the Pharmacopée Française (1986) was the first pharmacopoeia that included a monograph that dealt solely with hard capsules. It included descriptions and some tests that are only applied to hard gelatin capsules. Methods were given to identify the colourants used. Organic colourants were identified by thin layer chromatography and changing the composition of the mobile phases and the extraction processes was recommended if one colourant was in a much higher concentration than the others. Colourants are extracted from the capsules using a mixture of ammonia, water and methanol and the extract concentrated using evaporation at reduced pressure. Silica gel or cellulose-coated plates were used on which to develop the extracts in two different solvent systems. The identity of each colourant was confirmed by comparing it with a spot of standard solution of colourant. It gave a test for 'lubrifiants', which are the mould lubricants used in the manufacture of the capsule shells that enable the films to be removed from them when they are dry. Lubricants are typically edible mixtures of soaps and oils. The test involved shaking a 1 g sample in 30 mL of methylene chloride and then evaporating off the solvent. The limit was a residue not greater than 0.5% of the weight of the sample. There was a test for the amount of sulfites, expressed as SO_2, but the sulfite preservatives are no longer used in France.

The People's Republic of China Pharmacopoeia English version (2000) has a monograph entitled 'Vacant capsules', which is a literal translation equivalent to empty capsules. It has a detailed description of hard gelatin capsules, which it divides up into three categories, transparent, semi-transparent and opaque. There are three performance tests, compactness, brittleness and disintegration, which are carried out on capsules that have been filled with talc. The 'compactness' could best be described as a measure of how well-filled capsules retain their contents when handled. A sample of 10 capsules is filled and closed manually and then dropped one at a time from a height of 1 m onto a wooden disc 2 cm thick. The capsules pass if 'not more than two show a leakage of powder'. If more than two fail then a further 10 are tested and all must comply. The brittleness test involves first conditioning a sample of 50 capsules by placing them in a dessicator for 24 hr at 25°C over a saturated solution of magnesium nitrate, which produces a relative humidity (RH) of 53% (Nyqvist, 1983) and would ensure that the capsules are within their normal moisture specification. The capsules are then

transferred one at a time to a glass tube, with an internal diameter of 24 cm and 20 mm tall, standing on a wooden disc, 2 cm thick. A disc of PTFE, 22 cm in diameter and weighing 20 g, is dropped on to each capsule and the test is passed if not more than 15 break. The disintegration test is carried out in the standard apparatus using filled capsules that all must disintegrate within 10 min. If one fails a repeat test is permitted with a further sample of six and all of them must pass to comply.

Industrial standards

Industrial standards are intended for and are agreed between, the capsule producers and users. Usually, they do not have official legal status, but they are taken as standards on which to base good manufacturing practice. Ruthemann (1986) published a useful bilingual monograph, in English and German, on how to evaluate defects in empty hard gelatin capsules. This uses a German quality standard DIN 40 080 as the basis for defining faults in terms of effect on their usage, e.g. 'Major defect leads to inefficient function and so to a deficiency of the capsules, "leads to complaints from user", or leads to reduced efficiency in production and impairs reliability of production tools and packaging equipment'. It gives information on the size of sample to be taken according to lot size and the accept/reject numbers for defects found in terms of acceptable quality levels (AQL). It describes how a full inspection should be made and includes details of visual defects, dimensional defects and aspects of packaging to enable a full test certificate to be produced. There is less published information on soft capsules because they exist to all intents and purposes as filled capsules so there is not the need to examine the empty soft capsule shells because they are an inherent part of the manufacturing process.

One official standard for quality control that can be applied to both hard and soft capsules is the American Federal Standard for Capsules (for Medicinal Purposes) (American Federal Standard No. 285A, 1976), which is frequently used as the basis for industrial quality control. This gives the requirements for American governmental users such as the armed forces and it applies to filled capsules. However, since the defects may arise from the empty shells, the standards are applied in the industry to the quality control of empty capsules. Because defects may arise in the filling process, there needs to be a lower level of defects in empty shells if the standard is to be achieved on filled capsules. The Federal Standard includes references to pharmacopoeial standards but extends their scope to many other aspects of physical testing, packaging, etc. In the application of in-process quality control, standard statistical sampling and inspection methods are used. In Britain these are set out in Defence Guide DG-7-A (Sampling Inspection) and Defence Specification DEF-131-A and, in the USA, in the Military Standard MIL-STD-105.

Physical standards for capsules

Dimensions

Hard capsules are made in a range of standard sizes and even though many manufacturers make them, their dimensions are sufficiently standardised to fit all filling machines. However, there are differences in design features, such as self-looking mechanisms. Some minor adjustments, particularly to closed joined lengths, need to be made to filling machines when switching between manufacturers. The basic dimensions of hard capsules from two major manufacturers are given in Tables 4.1 and 4.2 (Chapter 4). The dimensions for soft capsules are dependent upon those of the moulds in which they are formed. The outside dimensions of soft capsules are also dependent upon the amount of material that is filled into each because the shells are flexible and will expand if more material is filled.

Capsule shapes

The Federal Standard defines seven different shapes of capsules, three each for hard and soft and a further miscellaneous one called 'special'. The hard capsules are described as 'conventional', 'bullet-like', which have a parabolic-shaped body and 'tapered ends', which have flat ends with angled faces connecting them to the cylinder of the cap and body. These last two shapes were specially shaped hard capsules produced by Eli Lilly & Co. Ltd and Smith Kline & French, respectively and were designed to make products of their own manufacture distinctive and to prevent counterfeiting. The soft capsule shapes are described as elliptical (oval), oblong and round. These shapes reflect the method of manufacture. The elliptical ones are made in the rotary die process, the round one by the Globex process and the oblong ones, which were typically filled with powders, by a plate process.

Solubility

Pharmacopoeial solubility tests are usually performed on filled capsules. The results are dependent upon both the nature of the capsule wall and the formulation of the contents. Industrial tests are particularly concerned with the solubility of the capsule wall. The American Federal Standard requires capsules to remain undissolved after immersion in water at $25 \pm 1°C$ for 15 min, but to 'completely fall apart, dissolve, or disintegrate' after immersion in 0.5% w/w hydrochloric acid at 36–38°C for 15 min. The first is called a 'water resistance test' and the second an 'acid solubility test'. Whilst the standard does not give any explanation for the reasoning behind these two tests it could be assumed that the first one was used as a simple way to distinguish between gelatin capsules and those made from a material of vegetable origin. This specification was first published in the 1950s and at that time capsules were available in the USA made from methylcellulose. In fact they were included in the capsule monographs of both the British and Italian Pharmacopoeias in the 1950s (Jones and Törnblom, 1975). Cellulose capsules dissolve in water at 25°C but gelatin capsules will swell and distort but not dissolve (Chiwele *et al.*, 2000).

Odour

Capsules made from gelatin will always have a distinctive odour. When a sealed container of them is opened this will be very noticeable particularly to those unfamiliar with the odour of gelatin. However the odour will soon dissipate. A test can be made to check that the capsules have not picked up a foreign odour during transport or that they have not been spoiled in some way. The Federal specification has a simple test that takes into account the quantity of capsules in the container being examined. The container is opened at room temperature in a place that is free from draughts. The container is tested after leaving it open for a specified time that is dependent upon the number of capsules it contains, e.g. for 101 to 500 capsules, 10 min and for containers containing more than 1000, 25 min. Capsules are considered to be 'odorless' capsules if none can be detected at this time. Capsules that are flavoured or contain ingredients that have a characteristic odour themselves are excluded from this test.

Defects in the hard capsule shells

Empty hard capsules can suffer from a number of defects that may be produced during manufacture. These defects are classified in three ways according to type and relative importance. Critical defects are those which would interfere with the filling process. This category includes capsules that are too short or too long, squashed capsules and capsules with holes, cracks and flat areas. Major defects are those which would cause a problem in use and reduce the effectiveness of the filled capsule. These include separated capsules, double caps, thin walls and splits. Minor defects are those which do not affect the performance of the capsule as a dosage form but which spoil the appearance of the product. These are surface blemishes such as pits, specks, bubbles or other marks. The precise definitions of these faults are made by the individual empty shell manufacturers who supply full documentation to their customers.

Ruthemann (1986) in his monograph on quality assurance gave a good example of the controls that can be applied to the visual defects that occur in hard capsules. In this he defines the faults and gives them acceptance levels depending on the seriousness of the problem that they may cause. The faults arise during the manufacturing process and come either from the gelatin/polymer solution, e.g. bubbles or specks of poorly dispersed pigment, or are caused by some mechanical action of the machine, e.g. poorly cut cap edges or improperly joined capsules.

In addition to the faults described above, other visual defects can be caused during the process of printing the capsule shells with some form of information to help identify the products that are to be filled into them. The print visual defects can either be those that prevent the capsule being identified, e.g. print missing or totally illegible, or those that are legible enough to enable the product to be identified but whose visual quality is poor, e.g., smudged print, ink specks, parts of logos. Each manufacturer supplies specifications for limits on all of these defects.

The visual quality of capsules is a compromise between what the process is capable of producing and what the customer is willing to accept. Over the years as the capsule manufacturers have improved the efficiency and reliability of their processes the visual quality of their products has also improved. This is seen in the reducing AQL values for fault levels. Another factor that drives the manufacturers to improve general quality is the increasing performance of the capsule-filling machines. These have progressed from semi-automatic manually operated machines to fully automatic machines that are sometimes run unattended over prolonged periods, e.g. lights-out manufacture. In the former machines it was comparatively easy for the operator to make an intervention and correct for problems caused by faulty capsules and faults at a level of one per thousand did not significantly reduce output. In the latter machines whilst there are automatic systems to remove blockages, etc., if the faults are at levels of one per thousand this would significantly affect outputs. The quality of empty capsules produced by several manufacturers is such that filling machines can be operated at speeds of up to 200 000 capsules per hour at full efficiency.

Standards for filled hard capsules

A widely available example of a quality system for filled hard gelatin capsules is given in the American Federal Standard No. 285A (1976). This lists capsule defects in three categories and can be used as a basis for setting up an inspection plan.

Major A defects[1]

- Capsule not type-specified (i.e. hard shell).
- Capsule not free of cracks, breaks, pinholes or splits when leakage of contents may occur.[4]
- Capsules not uniform in appearance.
- Base and/or cap of capsule not as specified.
- Capsule not uniform in colour(s).
- Capsule empty.[4]
- Capsule not free of embedded surface spots and contamination.[4]
- Capsule fill not free from foreign matter.[2]

Major B defects[1,3]

- Capsule does not maintain tight closure or seal in the immediate container, or during normal handling, or dispensing.[4]
- Immediate container not free from extraneous matter.[4]
- Capsule not intact (i.e. cap separated from body).[4]
- Capsule not free of foreign odour, other than characteristic odour.[4]
- Immediate container not internally or externally clean.[4]
- Void space of immediate container not filled, when required.[4]
- Immediate container not free of excess ingredient (capsule contents).[4]

Minor defects[1]

- Capsule not free of pits or dents.
- Capsule not free of thin areas.
- Capsule not free of specks, spots or blemishes.
- Capsule not free of cap and/or body cutting into one another.
- Capsule not smooth.[4]
- Capsule not free of adhering surface spots.[4]

The superscripts are explained as follows:

1 Inspection is not restricted to classified possible defects listed above.

2 For opaque capsules use a sample size of 20 capsules to examine contents. No capsule shall show evidence of foreign matter.

3 Capsules obtained from the bottles used for examination of Major B defects may be used for examination of Major A and Minor defects annotated '4'. Thus, no additional capsules need be selected.

4 Applies to the examination of capsules in filled, final (immediate) containers.

References

American Federal Standard (1976). No. 285A. *Capsules (for Medicinal Purposes)*. Washington DC 20407, General Services Administration, 1976.

Avgoustakis, K., Athanasiou, A. and Georgakopoulos, P. P. (1992). Effect of helix characteristics on the dissolution rate of hard gelatin capsules. *Int. J. Pharm.*, 79, 67–69.

Bell, J. H., Stevenson, N. A. and Taylor, J. E. (1973). A moisture transfer effect in hard gelatin capsules of sodium cromoglycate. *J. Pharm. Pharmacol.* 25, 96P–105P.

Boymond, P., Sfiris, J. and Amacker, P. (1966). Herstellung und Prüfung von darmlöslichen Gelatinekapseln. *Pharm. Ind., Berl.*, 28, 836–842.

Boymond, P., Sfiris, J. and Amacker, P. (1967). The manufacture and testing of entero-soluble gelatin capsules. *Drugs made in Germany*, 10, 7–19.

British Pharmacopoeia (1948). *Tabellæ, disintegration test*. General Medical Council, London: Constable & Co., 518–519.

British Pharmacopoeia (1953). *Tablets, disintegration test*. General Medical Council, London: Pharmaceutical Press, 553–557.

British Pharmacopoeia (1958). *Appendix XXIV, Disintegration test for tablets*. General Medical Council, London: Pharmaceutical Press, 964–965.

British Pharmacopoeia (1968). *Appendix XXI, C. Disintegration test for capsules*. General Medical Council, London: Pharmaceutical Press, 1368 and British Pharmacopoeia 1968 Addendum 1969. *Disintegration test for capsules*. p. xx.

British Pharmacopoeia (1973). General Medical Council, London: Pharmaceutical Press.

British Pharmacopoeia (1980). General Medical Council, London: Pharmaceutical Press, Volume 2.

British Pharmacopoeia (2002). Volume II, London: Stationery Office.

CDER (Center for drug evaluation and research) (1997). *Guidance for industry, Dissolution testing of immediate release solid oral dosage forms.* US Department of Health and Human Services (USDSHH), Food and Drug Administration (FDA), August 1997, Rockville, MD, pp 11.

CDER (2000a). *Guidance for industry, Waiver of in vivo bioavailability and bioequivalence studies for immediate-release solid oral dosage forms based on the biopharmaceutics classification system.* USDSHH, FDA, August, 2000, Rockville, MD, pp 13.

CDER (2000b). *Guidance for industry. Bioavailability and bioequivalence studies for orally administered drug products – general considerations.* USDSHH, FDA, October, 2000, Rockville, MD, pp 25.

Chiwele, I., Jones, B. E. and Podczeck, F. (2000). The shell dissolution of various empty hard capsules. *Chem. Pharm. Bull.*, 48, 951–956.

CPMP (Committee for proprietary medicinal products) (2001). *Note for guidance on the investigation of bioavailability and bioequivalence.* EMEA, CPMP/EWP/QWP/1401/98, 26 July, 2001, London.

European Pharmacopoeia (2002). 4th edition, Council of Europe, Strasbourg.

Gershberg, S. and Stoll, F. D. (1946). Apparatus, for tablet disintegration and for shaking out extractions. *J. Am. Pharm. Assoc., Sci. Edn*, 35, 284–287.

Gjellan, K., Magnusson, A.-B., Ahlgren, R., Callmer, K., Christensen, D. F., Espmarker, U., Jacobsen, L., Jarring, K., Lundin, G., Nilsson, G. and Waltersson, J.-O. (1997). A collaborative study of the *in vitro* dissolution of acetylsalicylic acid gastroresistant capsules comparing the flow-through cell and the USP paddle method. *Int. J. Pharm.*, 151, 81–90.

Indian Pharmacopoeia (1996). Government of India Ministry of Health & Family Welfare, Controller of Publications, Delhi.

Jones, B. E. (1972). Disintegration of hard gelatin capsules. *Acta Pharm. Suec.*, 9, 261–263.

Jones, B. E. and Cole, W. V. J. (1971). The influence of test conditions on the disintegration time of gelatin capsules. *J. Pharm. Pharmacol.*, 23, 438–443.

Jones, B. E. and Törnblom, J.-F. V. (1975). Gelatin capsules in the pharmacopoeias. *Pharm. Acta Helv.*, 50, 33–45.

Martindale, the extra pharmacopoeia (1972). In: Blacow, N. W. (Ed.), *Sodium cromoglycate*. London: Pharmaceutical Press, 1702–1703.

Nyqvist, H. (1983). Saturated salt solutions for maintaining specific relative humidities. *Int. J. Pharm. Tech. & Prod. Manuf.*, 4(2), 47–48.

Pharmacopée Française (1986). *Gélules (enveloppes de) en gélatine*, addition July 1986, pp 5. In Pharmacopée Française, 10th Edition, 1982, L'Adrapharm, Paris.

Pharmacopoeia of the People's Republic of China (2000). English edition. Beijing: The State Pharmacopoeia Commission of P. R. China, Chemical Industry Press.

Ruthemann, H. D. (1986). *The pharmaceutical plant, volume 33, Quality assurance of hard gelatin capsules, defect evaluation list.* Aulendorf, Germany: Editio Cantor, pp 31.

United States Pharmacopoeia/United States National Formulary (2001). 24th/19th edition. Washington: United States Pharmacopeia Commission.

United States Pharmacopoeia/United States National Formulary (2003). 26th/21st edition, and Supplement 1, January 2003. Washington: United States Pharmacopeia Commission.

Index

Page numbers in *italic* refer to figures or tables; those in **bold** indicate main discussions.